Science Education Research in Latin America

Cultural and Historical Perspectives on Science Education

HANDBOOKS

Series Editors

Kenneth Tobin (*The Graduate Center, City University of New York, USA*)
Wolff-Michael Roth (*University of Victoria, Canada*)

VOLUME 7

The titles published in this series are listed at *brill.com/pseh*

Science Education Research in Latin America

Edited by

Charbel N. El-Hani, Eduardo F. Mortimer,
Mauricio Pietrocola and María Rita Otero

BRILL
SENSE

LEIDEN | BOSTON

All chapters in this book have undergone peer review.

The Library of Congress Cataloging-in-Publication Data is available online at http://catalog.loc.gov

Typeface for the Latin, Greek, and Cyrillic scripts: "Brill". See and download: brill.com/brill-typeface.

ISSN 2589-6180
ISBN 978-90-04-40854-8 (paperback)
ISBN 978-90-04-40855-5 (hardback)
ISBN 978-90-04-40908-8 (e-book)

This book is printed on acid-free paper and produced in a sustainable manner.

Contents

PART 3
Science Teaching and Teacher Education

PART 4
Discourse Analysis and Argumentation in Science Education

PART 5
The History, Philosophy, and Sociology of Science in Science Teaching

PART 6
Science Education in Non-Formal Settings

Preface

The road to this book was long and winding, but, in the end, we feel satisfied with its broad coverage of research on science education in Latin America. It is certainly impossible to capture in any single volume all the breadth and diversity of science education research in this world region, and we regret that there were so many colleagues, lines of research, and topics we were unable to include. We tried, however, to distribute the chapters among as many different research topics, countries, and scientific communities as possible. We hope the result will benefit readers by providing them with a panoramic view of research in Latin America delving into both the challenges and vast potential of science education.

Figures and Tables

Figures

Tables

Notes on Contributors

Agustín Adúriz-Bravo
obtained his Ph.D. (2001) from Universitat Autònoma de Barcelona. He is Professor of Didactics of Physics at the Universidad de Buenos Aires. He is author of over 400 publications, including *Una introducción a la naturaleza de la ciencia* (FCE, 2005).

Maria José Pereira Monteiro de Almeida
is graduated in Physics and has a Ph.D. in Science. She is Professor at the Postgraduate Programs in Education and in Science and Mathematics Teaching at the State Universidade Estadual de Campinas (UNICAMP) and Researcher at the National Council for Scientific and Technological Development (CNPq), Brazil. She mainly works in Science/Physics Teaching and Cultural Practices.

José de Pinho Alves Filho
is Professor of Science Education at Universidade Federal de Santa Catarina. He graduated in Physics (1969) at Universidade Federal do Rio Grande do Sul, and obtained a Master's degree (1990) and a Ph.D. in Education (2000) from Universidade Federal de Santa Catarina. He does research on Science Teaching and is dedicated to Teaching Science and Scientific and Technological Literacy.

Adela Molina Andrade
graduated in B.A. in Biology (1977). She obtained a Master's in Education (1993) from the Pontificia Universidad Javeriana, Bogotá-Colombia, and has a Ph.D. (2000) in Education, Didactics area, from the Universidade de São Paulo, Brazil. She is Full Professor of Science Education at Universidad Distrital Francisco José de Caldas, Bogotá-Colombia. Her research interests include science teaching research, context, diversity, cultural difference, and classroom experiences developed by teachers who recognize the existence in their classrooms of students coming from culturally diverse backgrounds. She has published several research articles, chapters and books in her areas of interest.

Maria Maite Andrés
holds a Ph.D. (2005) in Science Education from Universidad de Burgos, Spain, a M.Sc. in Instructional Psychology (1998) from Universidad Central de Venezuela and a degree in Physics Teacher (1979) from Universidad Pedagógica experi-

mental Libertador (UPEL). She is Professor of Physics and Physics Education at IPC-UPEL, and coordinates the research group "Learning and Teaching in Physics Lab Work". Her publications are in Science Education and Science Teacher Education.

José André Peres Angotti
is Professor of Science and Technology Education at Universidade Federal de Santa Catarina, where he graduated in Physics (1972). He obtained a Master's (1981) and a Ph.D. in Science Education (1991) at Universidade de São Paulo, Brazil. He has supervised 4 postdoctoral researchers ad published 63 works, comprising articles, books and book chapters, including *Teaching Physics with TDIC* (free e-book in Portuguese, 2015).

Irene Arriassecq
is Professor in Physics and Mathematics, M.Sc. in Epistemology and Methodology of Science, Ph.D. in Science Education, and an independent researcher at CONICET (National Scientific and Technical Research Council). Her present research interests include Nature of Science in Physics Education and the design of Teaching-Learning Sequences for Physics Education with a focus on teaching and learning Relativity Theory in high school. In addition, during the last years she started to study the complex aspects related to the their use, which involves a new way of scientific production, and their relationships with science classes.

Romulo Gallego Badillo
is Bachelor in Chemistry from Universidad Pedagógica Nacional, Colombia. He is Master's in University Teaching and Emeritus Professor at Universidad Pedagógica Nacional. He authored a hundred articles on science teaching. He has been recognized nationally and internationally for the impact of his investigations.

Walter Antonio Bazzo
is a Mechanical Engineer (1977) and obtained a Ph.D. in Education (1998), focusing on Science Education. He develops studies in Technological Education with emphasis on the contemporary civilizatory process. He is also Professor of Mechanical Engineering and is affiliated with the Postgraduate Program in Scientific and Technological Education at Universidade Federal de Santa Catarina. He has published 10 books and more than 200 scientific articles. He currently coordinates the Nucleus of Studies and Research in Technological Education (NEPET).

Nelio Bizzo
graduated in Biology (1981) and has a Master's in Science (1984) from Universidade de São Paulo (USP), Brazil. He obtained his Ph.D. from the same institution (1991), after spending a school term as a graduate student in the UK. He is a Senior Full Professor of Science Education at the School of Education (USP), and a researcher at CNPq. He was elected Fellow of the Royal Society of Biology (London, UK). His publications include several books and articles on Science Education and History of Science.

Silvia del Valle Bravo
graduated in Mathematics (1985) and obtained a Master's in Science Education (2007) from Universidad Nacional de Tucumán (UNT), Argentina. She obtained her Ph.D. (2016) from Universidad Nacional del Centro de la Provincia de Buenos Aires (UNICEN), Argentina. She is Professor at the Physics Department at Universidad Nacional de Tucumán. Argentina. She has experience in Science and Technology Education.

Antonia Candela
graduated in Physics at Universidad Nacional Autónoma de México, with a Master's and Ph.D. in Science with speciality in Educational Research at the Centro de Investigación y de Estudios Avanzados (CINVESTAV). She has been working in this center since 1977 and since 1991 as a National Researcher. She is the co-author of five national programs for science education with 36 compulsory textbooks for primary education of México. Her research deals with ethnography and discourse analysis of science classroom interactions. She published *Ciencia en el aula: Los alumnus entre la argumentación y el consenso* (Paidós, 1999), and more than 70 chapters and articles in international journals of the field.

César Carrillo-Trueba
is the editor-in-chief of the journal *Ciencias* (UNAM) and works on educational projects in indigenous communities from an intercultural perspective. He has a Bachelor Degree in Biology (Universidad Nacional Autónoma de México) and a Master's in Anthropology and Ethnography (École des Hautes Études en Sciences Sociales de Paris). He is the author of *Pluriver, essai sur le statu des savoirs indigènes contemporains* (L'Harmatan, 2013), as well as other books and articles.

Anna Maria Pessoa de Carvalho
obtained her Ph.D. from Universidade de São Paulo, Brazil (1972), São Paulo University, Brasil. She is Professor of Science Education at that university and Researcher at CNPq, Brazil. She has published many indexed research articles, chapters and books in the area.

Charbel N. El-Hani
graduated in Biology (1992) and has a Master's in Education (1996) from Universidade Federal da Bahia, Salvador, Brazil. He obtained his Ph.D. (2000) from Universidade de São Paulo, Brazil. He is Professor of History and Philosophy of Science at Universidade Federal da Bahia and Researcher at CNPq, Brazil. He coordinates the National Institute of Science and Technology in Interdisciplinary and Transdisciplinary Studies in Ecology and Evolution (INCT IN-TREE). He has been the book review editor of *Science & Education* from 2013 to 2019. His research interests are in science education research, history, philosophy and science teaching, philosophy of biology, ecology and conservation.

Maria de los Angeles Fanaro
Graduated as Mathematics and Physics Teacher (1999) at Universidad Nacional del Centro de la Provincia de Buenos Aires (UNICEN) and obtained a Ph.D. in Science Education (2009) from Universidad de Burgos, Spain. She is Associate Professor at UNICEN and Independent Research Scientist of the CONICET (National Scientific and Technical Research Council), Argentina. She has experience in Physics and Mathematics Education, Didactic of Physics and Mathematics, Science Teacher Education, and Educational Research.

Thaís Cyrino de Mello Forato
graduated in Physics (1985), has a Master's degree in History of Science (2003) and a Ph.D. in Education (2009). She is Associate Professor of Science Education at Universidade Federal de São Paulo, Campus Diadema, and is affiliated with the Postgraduate in Science and Mathematics Teaching at this university, Campus Diadema. She is Adjunct Editor of *Prometeica – Revista de Filosofia y Ciencia*. Her main research interest is on history of sciences on science education. She has published many indexed research articles, book chapters and books in the area.

Mario Quintanilla Gatica
graduated in biology and chemistry (1985) and has a Ph.D. in Science Education (1997) from Universidad Autónoma de Barcelona, Spain, He is Associate Professor at the School of Education from Universidad Católica de Chile. He is Director of the Research Laboratory in Science Education (www.laboratoriogrecia.cl). He is researcher and evaluator of advanced human capital in CONICYT – Chile (National Scientific and Technical Research Council). He is director and co-researcher of several research projects in science education with Europe and Latin America. He is President of the Latin American Research Network in Science Didactics (REDLAD).

Ileana M. Greca
obtained a Ph.D. (2000) in Physics Education from Universidade Federal do Rio Grande do Sul, Brazil. She is Full Professor of Didactics of Experimental Sciences at Universidad de Burgos, Spain. Her research interests in science education include cognitive psychology, modern physics, applications of history and philosophy of science and professional development. She is coordinator of several national and European projects and has published many indexed research articles, chapters and books in the area. She is editor, with Flavia M. T. dos Santos of *Research in Science Education in Brazil and Its Methodologies* (Unijuí, 2012, 3rd edition).

Guaracira Gouvêa
graduated in Physics, and obtained a Master's in Education and a Ph.D. in Sciences. She is Full Professor at Universidade Federal do Estado do Rio de Janeiro, Brazil. She is coordinator of the research group Education, Discourse and Media and researcher at CNPq, Brazil. She has experience in the areas of science education and museum education. Her research includes subjects such as images, popularization of science, and museums.

Verónica Guridi
graduated in Mathematics and Physics (1993) from Universidad Nacional del Centro de la Provincia de Buenos Aires, Argentina, and has a Master's in Epistemology and Methodology of Science (1999) from Universidad Nacional de Mar del Plata, Argentina. She obtained her Ph.D. in Education with emphasis on Science and Mathematics Education (2007) from Universidade de São Paulo, Brazil. She is Professor at Escola de Artes, Ciências e Humanidades da Universidade de São Paulo. Her research interests are focused on science education, science teacher education, teacher's professional identities, and curriculum development in Science.

Deisy Baracaldo Guzmán
holds a Master's degree in Applied information Technologies to Education. She is Professor and Researcher at Universidad Libre and Universidad Pedagógica Nacional, Colombia. She has presented at national and international conferences in Europe, Asia, Central and South America. Her publications deal with Language Teaching and Science Teacher Education.

Stella M. Islas
has a Ph.D. in Education from the Universidad Nacional de Córdoba, Argentina and a Master's Degree in Science Research Methodology from the Universidad Nacional de Entre Ríos, Argentina. She has experience in Science Teacher Education and Epistemology.

Martha Marandino
is graduated in Biological Sciences, and has a Master's in Education and a Ph.D. in Education. She is Associate Professor in the School of Education at Universidade de São Paulo (USP) and Coordinator of the Research Group on Informal Education and Science Communication/GEENF. She is Director of the Center of Cultural Preservation from USP and Vice-coordinator of the Museum of Education and Toy at School of Education. She has experience in teaching and research in the areas of science education and museum education. Her research interests include science and particularly biology dissemination in out-of-school contexts, such as museums, science centers, zoos, aquaria, and botanical gardens.

André Ferrer P. Martins
graduated in Physics (1990), received his Master's (1998) and his Ph.D. (2004) in Education from Universidade de São Paulo. He is Professor at Universidade Federal do Rio Grande do Norte (UFRN) in Brazil. He has experience in Education, focusing on science education; physics teaching; history, philosophy and sociology of science in science education.

Isabel Martins
has worked as a secondary school Physics teacher in Brazil, as research officer at the University College London-Institute of Education, and as Associate Professor at Universidade Federal de Minas Gerais. She is Full Professor of Science and Health Education at Universidade Federal of Rio de Janeiro, where she has also served as Coordinator of Graduate Studies and Institute Director. She has led and collaborated in research projects funded by both Brazilian and international funding agencies. Her research interests concern the recontextualizations of sciencerelated discourses in curriculum policy documents, textbooks and popular science texts as well as in classroom contexts, and their role in the promotion of education for citizenship, scientific literacy and social justice. She published extensively in refereed journals, co-edited academic books and served as keynote speaker in international conferences.

Celia Medina
worked at Universidad Nacional de Jujuy, Universidad Nacional de Santiago del Estero, and now Works at the Faculty of Philosophy in Universidad Nacional de Tucumán, Tucumán, Argentina. Her specialization is in Philosophy of Science and Technology, but she also worked for many years in Philosophy of History. She has trained researchers in these areas for years, which she very much enjoys. She has been Visiting professor at the University of Saskatchewan, in Saskatoon, to give lectures on the formation of big bang theory. She is a

member of the National Association of Philosophy from Argentina and a member of Amnesty International.

Francisco Julián Betancourt Mellizo

graduated in Physics (1972), and obtained a Master's in Physics (1981), from Universidad Nacional de Colombia. He is Associate Professor at the Physics Department of Universidad Nacional de Colombia. He is creator (1984) and director of Museo de la Ciencia y el Juego (2019) of Universidad Nacional de Colombia.

Royman Pérez Miranda

is Bachelor in Biology and Chemistry by Universidad Pedagógica y Tecnológica de Colombia, and Master's in Chemistry Teaching from Universidad Pedagógica Nacional de Colombia. He is a professor and researcher in the field of science education at Universidad Pedagógica Nacional.

Eduardo F. Mortimer

graduated in Chemistry (1980) and has a Master's in Education (1988) from Universidade Federal de Minas Gerais, Belo Horizonte, Brazil. He received his Ph.D. (1994) from Universidade de São Paulo, Brazil. He is Professor of Science Education at Universidade Federal de Minas Gerais and Researcher at CNPq, Brazil. His publications include *Meaning Making in Secondary Science Classroom* (Open University Press, 2003), with Phil Scott, and *Conceptual Profiles: A Theory of Teaching and Learning Scientific Concepts* (Springer, 2014), with Charbel N. El-Hani.

Roberto Nardi

is Associate Professor at the School of Sciences, State University of São Paulo (UNESP). He graduated in Physics from UNESP (1972), and obtained a M.Sc. (Temple University, 1978) and a Ph.D. in Education (São Paulo University, 1990). He is researcher from the Brazilian National Research Council (CNPq), and has been a Coordinator of the Science and Mathematics Education Division in the Postgraduate Evaluation System (CAPES, Brazil, 2008–2011). He has been chair of the Physics Education Research Commission from the Brazilian Society of Physics – SBF (2011–2013), chair of the International Commission on Physics Education from the International Union of Pure and Applied Physics (ICPE-IUPAP) (2018–2020), and president of the Brazilian Association for Research in Science Education (ABRAPEC, 2003–2005). He is the editor of the journal *Ciência & Educação*.

María Rita Otero

graduated in Mathematics and Physics (1983), and obtained a Master's in Education (1998) from Universidad Nacional del Centro de la Provincia de Buenos

Aires (UNICEN) and a Ph.D. in Science Education (2003) at Universidade de Burgos, Spain. She is Professor at UNICEN and Principal Research Scientist of the CONICET(National Scientific and Technical Research Council), Argentina. She has experience in science (physics and mathematics) education, learning theories, science teacher education, and educational research.

Marta A. Pesa
graduated in Physics and has a Ph.D. in Physics (1997) from Universidad Nacional de Tucumán, Argentina. She is Professor at the Physics Department at Universidad Nacional de Tucumán and Director of Postgraduate School at the Facultad Regional Tucumàn of Universidad Tecnológica Nacional in Argentina. She has experience in Science and Technology Education

Mauricio Pietrocola
is a science educator and currently Professor at the School of Education, University of São Paulo, Brazil. He received his doctoral degree from The University of Paris 7 (Denis Diderot) in 1992 and has since been the author of numerous publications. His areas of work include curriculum development, pedagogical knowledge and innovative strategies of teaching and learning. His current focus is on connections between innovative education, decision-making and science education in a risk society. In 2013–2014 Mauricio was a Fellow at the Advanced Research Collaborative at the Graduate Center of The City University of New York (CUNY).

Elio Carlos Ricardo
graduated in Physics (1997). He obtained a Master's in Education (2001), a Master's in Scientific Culture and Innovation (2015), a Ph.D. in Scientific and Technological Education (2005), and worked in the Postdoctoral Program in History of Physics (2006), and the Postdoctoral Program in Philosophy of Technology (2016). He defended his Habilitation Thesis (Livre-Docência) in 2012. He is Associate Professor at Universidade de São Paulo, Brazil.

Flávia M. T. dos Santos
graduated in Chemistry (1990), has a Master's in Education (1996) from Universidade Federal de Santa Catarina and a Ph.D. (2001) in Education from Universidade Federal de Minas Gerais, Brazil. She is Associate Professor of Science Education at the Universidade Federal do Rio Grande do Sul, Brazil. Her research interests in science education include professional development, classroom discursive interactions, and teaching and learning scientific concepts. She has published many indexed research articles, chapters and books in the area.

Lúcia Helena Sasseron
graduated in Physics with emphasis in Science Education. She has a Master's in Science Education and a Ph.D. in Education. Currently, she is Professor of Science Education at Universidade de São Paulo. She develops research about scientific literacy, inquiry-based teaching and learning, and argumentation in science education.

Claudia Sepulveda
graduated in Biology (1994) and has a Master's (2003) and Ph.D. (2010) in History, Philosophy, and Science Teaching from Universidade Federal da Bahia and Universidade Estadual de Feira de Santana, Brazil. She is Professor of Science Education at Universidade Estadual de Feira de Santana. She coordinates the Collaborative Research Group in Science Education, where she develops studies on evolution teaching, classroom discursive interactions, educational design research, and education of ethnic-racial relations in science teaching.

Adriana Patricia Gallego Torres
is licensed in physics by the Universidad Pedagógica Nacional, de Colombia. She obtained a Ph.D. in Didactics of Experimental Sciences and is diretor of the line of research in Science, Technology and Society in the Inter-Institutional Doctorate in Education of Universidad Distrital Francisco José de Caldas, Colombia, where she is currently teaching.

Deise Miranda Vianna
graduated and obtained a Master's in Physics at Universidade Federal do Rio de Janeiro (UFRJ) and a Ph.D. in Education (Physics Teaching) from Universidade de São Paulo (USP). She is Associate Professor at UFRJ, teaching at the undergraduate and postgraduate programs in Physics Teaching at the Physics Institute, and the Oswaldo Cruz Foundation's Graduate Program in Biosciences and Health Teaching. She works in the area of research in physics education, science teaching, teacher training, inquiry teaching, and Science, Technology and Society.

PART 1

An Introduction to Science Education Research in South and Latin America

CHAPTER 1

Science Education Research in Brazil: Historical Aspects, Researchers' Representations, and the State of the Art

Maria José P. M. de Almeida and Roberto Nardi

Abstract

In this chapter, we describe some aspects of science education research in Brazil, highlighting historical aspects, researchers' representations and the state of art of this area of study. We adopted the position of Science Education researchers who, for many years, have witnessed the constitution, growth and consolidation of this area as an investigation field. From this viewpoint, it is clear to us that this constitution did not occur in the same way, or attained the same degree of growth in different Latin American countries, and we cannot even say that it is equally disseminated in different regions of a country. In turn, it is not our intention to speak in the name of a block of countries, the total area of which exceeds 20 million square kilometers, and in which more than 600 million inhabitants live. We would like to stress, also, the relevance that a detailed survey of events, institutions, research groups and journals in each of the countries of the block would have for the Science Education field. A complete record of this memory would not only contribute to acknowledging the relevance of all that has been done, but could also increment the large number of interactions already existing in the region. A study like that is a desirable and huge task. However, the intention here is merely to highlight some factors that may somehow justify such a study. For this, we concentrate on Brazilian institutions, events and journals, Regarding which we count on a larger amount of information. We thus intend to exemplify some of the factors associated to the origins, growth and consolidation of the Science Education area in Brazil.

Keywords

science education in Brazil – science education research memories – science education research in Brazil

 | DOI: 10.1163/9789004409088_001

1 Introduction

We wrote this chapter from the perspective of science education researchers who, for many years, have witnessed the constitution, growth, and consolidation of this area as a field of investigation. From this viewpoint, it is clear to us that this constitution did not occur in the same way or attain the same degree of growth in different Latin American countries, and we cannot even say that it is equally disseminated in different regions of a given country.

On the other hand, it is not our intention to speak unilaterally in the name of Latin America, a block of countries the total area of which exceeds 20 million square kilometers, and a region with more than 500 million inhabitants. We would like, however, to highlight a relevant point in this set of countries: the origin of the language. This immense population generally speaks Portuguese or Spanish, even if a large percentage thereof – officially or not – also speaks another language. An example of the latter may be found in Paraguay, where, besides Spanish, Guarani is also an official language. It should be highlighted here that, as the Spanish and the Portuguese languages are very close, linguistic proximity is undoubtedly an aggregating factor for joint projects, including many related to investigation in science education.

We would like to stress, also, the relevance that a detailed survey of events, institutions, research groups, and journals concerned with the research in each of the countries of the Latin American block would have for the science education field. Such a complete record would not only contribute to acknowledging the relevance of all that has been done, but could also augment the large number of interactions already existing in the region. Such a study would be a huge yet worthwhile task. However, the intention here is merely to draw attention to some factors that might justify such a study. To that end, we will basically concentrate on Brazilian institutions, events, and journals. We intend to thereby exemplify some of the factors associated with the origins, growth, and consolidation of the field of science education in Brazil.

In territorial terms, Brazil is the largest country in Latin America, comprising over eight million square meters, i.e. over 40% of the area. With a population of over 190 million inhabitants, it holds approximately 34% of the Latin American population. Moreover, it should be noted that, in the same way as we may observe very distinct cultural characteristics in the diverse countries of this block, we found different interests, cultural aspects, and economic characteristics from one region of Brazil to the next – and even found such divergences within the same municipality. The fact is, regionalization notwithstanding, science concepts and theories have a very universal character. For education, however, including science education, regional aspects cannot be ignored. Therefore, we believe that it is necessary to stress, from the very beginning, the necessity of being cautious about possible generalizations regarding historical aspects, researchers' representations, and the state of the art of investigation in science education.

2 Some Historical Aspects

In this section, we explore some aspects of the history of science education in Brazil which we understand as formative to this field of research, while observing that the establishment of a discipline as an area of knowledge/research does not occur at random, but due to socially and historically explainable occurrences.

At the beginning of the 17th century, several schools had been founded in colonial Brazil, the students of which belonged to the richer strata of the population (Miranda, 1998). Nevertheless, seeking to understand the historical roots of science education in Brazil, the author comments on the "extreme income concentration, the exclusion and the marginalization of the largest section of the population, and unavoidably, suffocating the technical and scientific development" (p. 206), and concludes by stating that the sequels left in scientific teaching for the republican period will have to be compensated at that period "to overcome the hindrances to the industrialization process increasingly demanded by both domestic and foreign conditions" (p. 207).

Note that the proclamation of the Republic occurred in Brazil in 1889, only 11 years before the beginning of the 20th century. In a periodization of the Brazilian education, Ribeiro (1982) points out that, in the period from 1870 to 1894, the belief in education is strengthened as a key to solution of the fundamental problems of the country. Also about this period, the author attributes to the republican phase the emergence of the trend of replacing the predominance of a literary-type teaching with that of the scientific type. The author, however, comments that, in fact, a reform favored either one or the other trend. From 1984 to 1920 the school organization reveals an oscillation between humanistic and scientific influence. The teaching had an encyclopedic character. The scientific knowledge taught in schools was based just in the scientific activity results (p. 85), whereas in the next period, from 1920 to 1937, "for secondary education, the objective propagated was that of developing the scientific spirit, the organization involving multiple types of courses and integrated primary and higher education (…)" (p. 95).

Evidently, all these trends were always associated to socio-political conditions, both of domestic and foreign influence. As from these brief comments related to science education in the Brazilian education, let us now verify some occurrences which we believe to have contributed to starting the researches in this field of teaching in Brazil.

The *Cultus* journal, edited by the Brazilian Institute of Education, Science and Culture (IBECC), as from 1949, was one of the sources that helped us understand a little about the ideas related to natural science teaching disseminated in Brazil. The IBECC had been established by federal decree at the Ministry of Foreign Affairs in 1946. According to the booklet of the Brazilian Foundation for the Development of Science Education, FUNBEC, it derived from the commitment assumed by the Brazilian.[1] At that meeting, with participants indicated by 42 countries, the United Nations Educational, Scientific and Cultural Organization (UNESCO), assumed the purpose of implementing the development of Education, Science and Culture in the different countries participating in the meeting. In the booklet there is also a comment about the event as the way found by the participants to implement the purpose of "(...) studying the possibility of destroying, once and for all, the war spirit existing in each man" (p. 1).

This booklet described the IBECC establishment and activities as well as teacher training centers established by the Federal Government in the 1960s in some Brazilian state capitals: Belo Horizonte, Porto Alegre, Recife, Rio de Janeiro, Salvador and São Paulo. FUNBEC[2] was established in the 1960s, apparently having as a main goal the production, dissemination and commercialization of pedagogical resources in the sciences areas. Bibliographic productions or translations made at the foundation were essentially edited by EDART – São Paulo Editora Ltda., which also edited the *Aula Maior* periodical. In the early 1980s, FUNBEC started to issue the *Revista de Ensino de Ciências*, the last issue of which was number 24, published in 1993. As to the teacher training centers, as stated by Krasilchik (1978), its primary purposes were to "(...) increase and decentralize the teaching renovation sources by means of regional solutions" (p. 48).

Cultus had its publishing started in Brazil by the IBECC back in the 1940s, and the examination of some issues of the following decade evidenced the plurality of themes in the sciences areas. In these issues, in the first back cover, one reads "Journal for Scientific Dissemination and Development of Scientific Teaching in Secondary Courses". The reading of several of its issues made us to note[1],[2] proposals and recommendations, which were not, however, followed by some report on what could have originated the action that were being proposed. Occasional tests or assessments there published are also not consistent with what we now consider to be a scientific research in science education.

If we shift now to science education materials and journals of the 1960s, we observed how common at that period was the questioning to the sciences disciplines teaching then practiced, considered traditional and verbalist by the critics. According to Almeida (2003), a solution to the problem considered at the time was to import and to translate educational projects elaborated in other countries, mainly in the United States of America. If the case is particularized to physics teaching, the project there organized by the Physical Science Study Committee – the called PSSC Project, with the first Brazilian translation issue dating back to 1963, was one of the most disseminated. Aspects of physics teaching present in the project would even influence the further development of Brazilian projects. Among the criticisms to the project, pros and cons can be found.

A questioning we understand to be relevant reporting here does not exactly refer to the project, but to the policy that allowed its dissemination in some countries. In the preliminary issue of the Brazilian translation, the sponsorship of the United States Mission of Economic and Technical Cooperation with Brazil[3] – USAID can be verified, through the Center of Technical Journals of the Alliance.[4] This agreement, named *Alliance for Progress*, was extensively criticized. Among the criticisms, we highlight those by Arapiraca (1982), an author who, in a study on the relations between USAID and the Brazilian Ministry of Education, published in the early 1980s, is stunning when admitting that, aiming at the internalization of the pedagogical philosophy developed in the United States of America, its implementation was conducted with: "(...) the intention of legitimating a whole *modernizing* transformation imposed on the Brazilian nationality, in the sense of guiding its rationality by capitalist production way" (p. 110).

Due to questionings of this nature, the 1st National Symposium on Physics Teaching in Brazil was held in 1970, promoted by the Brazilian Physics Society, which already counted on a teaching office. In the symposium proceedings, the influence of the PSSC can be noted in the ideas of those who thought about Physics teaching at secondary level in Brazil at the time. In the symposium, two sessions concerning secondary teaching were held, each one with ten relators. In the first, named "Physics Teaching at Secondary Level", three relators refer to the PSSC. In the second, named "New Curricula", six relators in their thematic include this project. In their speeches, they discuss the difficulties that make the use of PSSC unfeasible, describe how they used it and even attribute to it the power of influencing the teacher's attitude. It is also verified that some accept and others criticize its proposed adoption as a solution to teaching.

From the same proceedings, we highlight a proposal unanimously approved to be recommended to the Brazilian Physics Society, which promoted the symposium:

> That the Brazilian Physics Society, in cooperation with Physics Societies in other countries, elaborates a Latin-American cooperation project to coordinate the efforts made in physics teaching research. (p. 334)

Even without considering later occurrences, we consider the ideas contained in a section of the proposal: "(...) elaborate a Latin American cooperation project to coordinate the efforts made in Physics Teaching research (...)", a turning point in science education as a research area.

Nonetheless, soon after the symposium, some teams were dedicated to the elaboration of Brazilian teaching projects. It is also a fact that there are statements attributing the beginning of research in Latin America to an occurrence before the symposium. Cláudio Dib (1991), referring to the UNESCO Pilot Project for physics teaching, developed in Brazil in the 1960s, states:

> (...) There is no exaggeration in stating that the development of systematic researches in physics teaching/learning started at the Physics Department at the University of São Paulo, and very probably in Brazil and in Latin America, with the conduction of the Pilot Project for Physics Teaching which, sponsored by UNESCO and with the cooperation of different educational and scientific institutions, such as the Latin American Physics Center and the Institute of Physics of the University of São Paulo, occurred at this Institute from August 1963 to July 1964. (p. 22)

Nevertheless, we want to highlight the difference between a proposal assumed by the participants of a symposium in 1970 and an individual statement in 1991, when the research was already consolidated, referring to something that had occurred nearly thirty years before. This statement is not the only representation on what could have originated the area as a research field. Other statements are referred to in the next item. Before that, however, we are going to point out other occurrences we understand to have been important factors for the development of research into science education in Latin America.

Concerning to science education journals, in January 1979, the number zero of the *Revista de Ensino de Física* (Physics Teaching Journal) was launched in Brazil, a periodical of the Brazilian Physics Society, which would later be called *Revista Brasileira de Ensino de Física* (Brazilian Physics Teaching Journal). Although the journal's contents table included only a session entitled *Research in Physics Teaching* in 1985, the reading of previous issues with articles of epistemological, historical and educational policy nature, among others, indicates the path of research as it is seen nowadays. The following section of the number zero editorial is also a good indicator of that: "There is a relatively large number of people doing researches related to physics teaching, and conversely, rarely are these works turned into papers that could reach, through their publication" (p. 1).

However, dissertations and theses presented in Brazil in the 1970s, two of them in 1972, are the greatest evidences that research in science education was already being conducted during that period (Carvalho, 1972; Dib, 1972; Moreira, 1972; Nassif, 1976; Pacca, 1976; Teixeira Jr., 1976; Violin, 1976; Bittencourt, 1977; Saad, 1977). Even not intending to exhaust the studies presented at that time, we point out that we could find some of them concluded in four different institutions and we also note that some presentations occurred early in that decade. Another observation is that a large amount of these studies referred to the analysis of aspects associated to teaching projects.

As to graduate programs, we highlight here that still in the 1960s the Physics graduate program, in the physics teaching modality, is established at the Physics Institute of the Federal University of Rio Grande do Sul, in 1967.

3 Some Researchers' Representations

As part of the research "Formação da Área de Ensino de Ciências: memórias de pesquisadores no Brasil" (Formation of the Science education Area: researchers' memories in Brazil), a considerable number of representations on the science education area was analyzed, including their characteristics and originating factors, part of them already focused by Nardi (2005), Nardi and Almeida (2007) and Nardi and Almeida (2008). Information for building this study data was obtained through semi-structured interviews with former science education Brazilian researchers.

However, before narrating some of this research outcomes let us explicit the notion of *representation* in the theoretical trend on which the studies were based. In them, a discursive perspective was adopted, coherent with that started by Michel Pêcheux in France in the 1960s. The references consulted on this trend were mainly the ones developed by Eni Orlandi in Brazil, and the basic supporting notions were: *discourse, production and representation conditions*, as they are understood in this discourse analysis perspective.

Under this perspective, non-transparency of the language, discourse as an effect of meanings among interlocutors and the role of the production conditions, immediate and socio-historical, in the formulation of discourses, detaching *social representations,* are presupposed. In Orlandi (1983), language and discourse conceptions coherent with this discoursive perspective are found. The author considers *language* as a work, in the sense of not having a natural or arbitrary character. Thus thought it is a result of the interaction between man and the natural and social realities, that is, it is constituted by social production, and should be thought as a necessary mediation: "(…) mediation as a constitutive relation, a modifying and transforming action" (p. 18), being the processes, in language constitution, socio-historical.

It is also from the same text that we conceptualize *discourse* as an effect of meanings among interlocutors. The author evidences that the meaning is not only constituted by interlocutors, but is associated to the situation and to the historical-social context. She also comments on the relevance of social representation, associating the meaning of the discursive process to the *positions* in which interlocutors are situated, taking into account the notions of *production and representation conditions*:

> (…) the interlocutors, the situation, the historical-social context (i.e. production conditions) constitute the meaning of the verbal sequence produced. When something is said, someone says to somewhere in society to another someone also somewhere in society and this is part of meaning. As exposed by Pêcheux, there are projection rules in the mechanisms of all the social formation which establish the relation between concrete situations and the representations of these situations within the discourse. The place, thus understood as a space of social representations, constitutes the discursive meaning. It is necessary to say that every discourse derives from another discourse and resends it to another one, hence one cannot speak of discourse, but in a state of discursive process, and this state should be understood as a result of institutionalized sedimented discursive processes. Finally, it is part of the discursive strategy to foresee, put oneself in the place of the listener (anticipation of representations), as from the very place of speaker, which regulates the possibility of answers, the scope of discourse. (p. 19)

In the research mentioned, the interviewees were selected from indications made by peers, in an initial consultation to researchers in the area. These researchers were contacted via electronic mail previously obtained from the associations congregating them. Out of 973 e-mails sent to these researchers, 202 were answered. These answers met the request for indicating five names of colleagues they understood to be acting in the area since its beginning and who, in their opinion, had contributed to building the area. 501 researchers were mentioned and 24 were interviewed. This number followed two criteria: the largest number of citations by peers and the consideration that the person were indicated by at least one researcher of each of the science education sub-areas: biology, physics, geosciences and chemistry education.

E-mails were sent to addresses received from: Team of the Referees' Board from the Ciência & Educação (Science and Education Journal), Brazilian Physics Society (SBF), Brazilian Chemistry Association (ABQ), Brazilian Association of Researchers in Science Education (ABRAPEC), Brazilian Society of Biology Education (SBenBio) and Brazilian Chemistry Society (SBQ). The e-mail sent to researchers said:

> We are starting a research aiming at contributing to the memory of Science Education in Brazil. In it, the intention is to interview some colleagues who have been acting in the area since its beginning, contributing to its constitution. Could you please answer this e-mail now, mentioning the names of five people who should be interviewed?

All the interviews were conducted by the same interviewer, who is one of the authors of this chapter; the basic questions were established, even admitting the possibility of variations that could occur as a consequence of the very answers of the interviewees. Moreover, the interviews questions were open and the interviewer only intervened in the interviewees' statements aiming at causing the continuity of their statements. The question receiving the most relevant answers to the theme discussed in this chapter was: *"In your opinion, what factors were decisive to this area formation? Why?"*

As stated by Almeida (2004), the theoretical grounds determine the possibilities and limits of what can be concluded from the analysis of the information obtained with one or the other technique used. Taking this fact under consideration, we have to admit that the considerations synthesized from the interviews, imply the relevance of the interviewees' and the interviewer's representations in obtaining the information constituting the discourses analyzed.

The analysis of the answers evidenced common representations, yet also great differences as to what the researchers considered that could have originated the field of science education, the number of factors mentioned being considerable. For part of the researchers interviewed, the factors that originated the field of science education field in Brazil are associated to events occurred, such as the translation of teaching projects elaborated abroad, the conduction of specific meetings and the establishment of graduate courses in the country. The analysis of pieces of other researchers' interviews evidenced their representation of how the graduation studies taken abroad would have contributed to the research in Brazil, after the researchers returned to the country.

It is also interesting to note that some researchers, despite questioned about the factors originating the area, manifested concerns related to the delimitation of certain characteristics they considered associated to it in their answers, such as the articulation in researches into theoretical referential, methodological procedures, for data construction and analysis, results and conclusions. As to characteristics, according to Nardi (2005):

> The characteristics mostly mentioned are, in order, the following: (I) the area inter or multidisciplinary character; (II) the role of specific knowledge in the research and teaching activities; (III) the research character applied or that of research & development; and, due to that, its classification as Applied Human Sciences or Applied Social Sciences. In the case of the researchers who started later in the area or who had a close relationship with foreign researchers, one can observe (IV) a conception and assessment of research closer to models originated abroad in function of their academic formation. Also noted in the interviewees speech are references to other themes, such as: (V) the presence of History and Philosophy of Science in the research in Science Education; (VI) The diversity of perspectives in terms of research objects, theoretical referential, methodological referential and other relevant aspects besides (VII) the (old) tension to define the locus of the research in this area: closer to the "contents" area and of their epistemological referential or historical-philosophical or closer to the contributions that marked the "education" area (psychology, sociology etc.). (p. 137)

Therefore, from what was observed in several researchers' discourses, there was no unanimity, as to their representations, on what would have originated the area or even about its characteristics. Conversely, in these interviews, when discussing the area origins, the reference was not only to research, as it is considered nowadays, but to occurrences that might have contributed to its reaching the point it did at the moment of the interviews, conducted in 2003 and 2004. Thus, coherently with the theoretical support used, the position from which the interviewed spoke determined what they said. That is, it is not inappropriate for us to admit a significant influence of the production conditions of the investigations of the researchers interviewed on the positions they assumed in relation to what they considered research and to the factors that originated the science education field of research.

4 State of the Art

After the first one, in 1970, the National Physics Teaching Symposiums (SNEF) continued to occur in Brazil, in different regions of the country. In January 2009, the XVIII was held. Even largely constituted by the presentation of researches, these symposiums started to have a more encompassing character and in 1986 the first Physics Teaching Research Meeting (EPEF) was organized in Brazil, which is having its XII edition in 2010. In the VIII EPEF Scientific Report conducted in 2000, there is a history of the EPEF in which the first of them is considered an "acknowledgement of the research installed capacity in the Physics Teaching field". These meetings are part of the Brazilian Society of Physics calendar and the society journal, launched in 1979, issued its 32rd volume in 2010, with four issues per volume.

In 1982, the first National Chemistry Education Meeting (I ENEQ) occurred in Brazil and, since then, meetings have occurred every two year; the XVth was held in 2010. Meetings are also promoted by the Brazilian Society of Biology Education (SBEnBio), and meetings organized by geologists have also occurred. Specific journals are also issued in the chemistry, biology and geology teaching areas.

In 1997, the first National Science Education Research Meeting (ENPEC) occurred, congregating researches on the teaching of the different disciplines of the so-called Nature Sciences, the VII being held in 2009. In discussions in the first of these meetings, the Brazilian Association of Research in Science Education (ABRAPEC) was established and consolidated in the second ENPEC. As from 2001, the issuing of the ABRAPEC journal was started, the *Revista Brasileira de Pesquisa em Educação em Ciências* (RBPEC – *Brazilian Science Education Research Journal*), launched in 2001, with three yearly issues.

Another aspect of the science education field functioning to be registered is that, for a long time in Brazil, the research groups that were constituted did so in Education Faculties or in related Science Institutes, such as the Institute of Physics, Institute of Chemistry, etc. In turn, as to the classification in a knowledge area in regulating organisms such as the Coordination for the Improvement of Higher Education Personnel (CAPES), Science education was initially part of the Education field and, in 2000, a partnership was created with researchers from Mathematical Education: the Science and Mathematics Education field, with a specific assessment committee. An immediate consequence was the growth in the number of graduate programs with specific characteristics of this area. According to a document by CAPES, Brazil (2009), the Area programs, which started with seven courses, now (2010) reach the expressive number of 61 programs, in the different regions in Brazil, totaling 78 courses, among master's (29), academic doctorate (19) and professional master's programs (30). According to the same document, this list includes "programs having as the focus of their academic work the areas of health education and their relations with environmental themes, and of Earth Sciences Education". The programs involve 885 teachers, 683 of whom permanent and 190 cooperators; 6039 students, with 2260 academic Master's, 220 Doctorate and 735 Professional Master's programs concluded. Figure 1.1 shows the evolution in the number of programs along the years.

These programs, as an example of what occurs to most programs in other graduate areas, are unevenly distributed in the different regions in Brazil, as shown in Figure 1.2

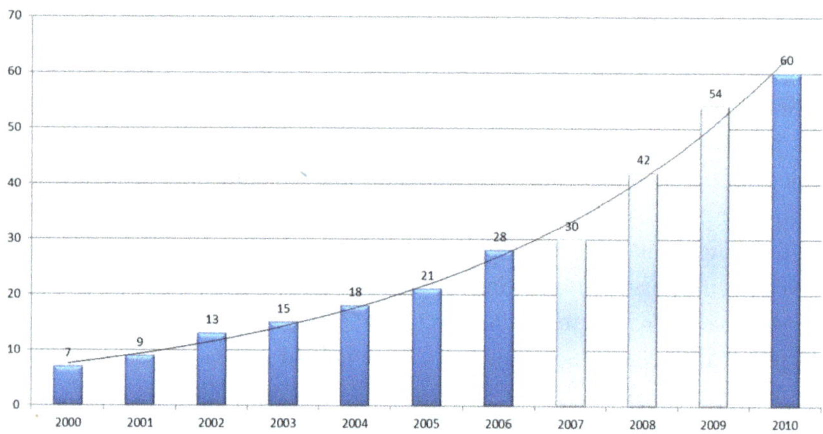

FIGURE 1.1 Evolution of the number of graduate programs in Science and Mathematics Education (SME) in Brazil in the period 2000–2010 (Brazil, CAPES, 2007–2010-Trienium Report)

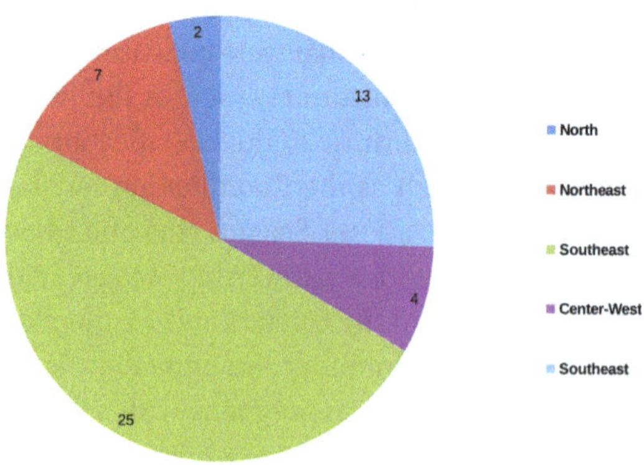

FIGURE 1.2 Distribution of Science and Mathematics Education Graduation Programs in Brazil throughout country's macro regions (Brazil, CAPES, 2007–2010-Trienium Report)

Another relevant aspect of Science education research field in Brazil is the large number of journals in the area or subareas published in the country, some strictly investigative and others including reports of experiences and other sessions. Besides those already commented herein, some more are now listed in the chronological order of their launching year: *Caderno Catarinense de Ensino de Física* (1984), today *Caderno Brasileiro de Ensino de Física*; *Ciência & Educação* (1995); *Química Nova na Escola* (1995); *Investigações em Ensino de Ciências* (1996); *Ciência & Ensino* (1996); *Ensaio Pesquisa em Educação em Ciências* (1999); *Física na Escola* (2000); *Revista Latino-Americana de Educação em Astronomia* (2004), the publication management of which is in Brazil, but with an editorial body from five countries in Latin America; *Experiências em Ensino de Ciências* (2006); *Revista de Ensino de Biologia* (2007); *Revista Brasileira de Ensino de Ciência e Tecnologia* (2008) *Alexandria* (2008) *Ciência em Tela* (2008). It should also be considered that a number of researches and even of dossiers or specific issues are published in Education journals in general or in areas such as Physics, Chemistry, etc. That is the case of the thematic issue Education in Physics from 1996, in the *Pro-Positions journal*, of the Science education dossier: history and language, published in the same journal in 2006.

It is important to note here that, in Latin American terms, even without a broader survey, which would certainly very much increase this number, some journals related to Science Education published in different countries are listed as follows: *Revista de Educación Química* (México, 1989); *Revista de Enseñanza de la Física* (Argentina, 1995); *Tecne Episteme y Didaxis* (Colômbia, 1996); *Física y Cultura: Cuadernos sobre Historia y Enseñanza de las Ciencias* (Colômbia, 1996); *Latin-American Journal of Physics Education* (México, 2007); *Revista Electrónica de Investigación en Educación en Ciencias* (Argentina, 2008).

The elaboration of catalogs of dissertations and thesis is another indicator of the research development stage in a country.[5]

Also to be highlighted is the great exchange among Brazilian research teams and those of other countries in Latin America, including the participation of a large number of researchers from different countries in meetings of the area in Brazil. As to Brazilian researches, going to meetings outside Brazil, in Latin America or in countries out of this block, this is an indicator of how many Brazilian researches present their work abroad. As an example, in the last Enseñanza de las Ciencias meeting held in Spain in 2009, the largest percentage of communications from other countries was undoubtedly the Brazilian ones.

Concerning research lines, apparently these have multiplied as the area matures. If in the 1980s, despite coexisting with other types of researches, researches in Alternative Conceptions and Conceptual Change could be said to be dominant, for an idea of the present comprehensiveness, 10 thematic lines of the Physics Teaching Research Meeting to be held in October 2010 are listed as follows: Teaching/Learning/Assessment in Physics; Physics Teachers' Formation and Professional Practice; Philosophy, History and Sociology of Science and Physics Teaching; Physics and Communication in Formal, Informal and Non-Formal Practices; Information Technology and Communication, and Physics Teaching; Didactics, Syllabus and Educational Innovation in Physics Teaching; Language and Cognition in Physics Teaching; Science, Technology, Society and the Environment and Physics Teaching; Public Policies in Education and Physics Teaching; Theoretical-Methodological Issues and New Demands in Physics Teaching Research.

With regard to support for scientific research, in Brazil there are several foundations in different states, as well as some federal agencies, such as the *National Council for Scientific and Technological Development* (*CNPq*) and the *Coordination for the Improvement of Higher Education Personnel* (*CAPES*). In relation to the latter one, responsible for legislation, accreditation and evaluation of postgraduate programs in the country, we still remember here that the recognition of research on teaching of science and mathematics in Brazil, can be noticed by the fact that in the period 2000–2010 these areas were established in one of 46 areas of assessment of graduate programs in the country. We recall also that, in 2011, this area, called 'Teaching of Sciences and Mathematics' became transformed in 'Teaching', aimed at accommodating academic and professional masters and PhD programs in other areas not just Science and Mathematics.

5 Final Considerations

The aim was to situate some historical aspects, researchers' representations and the state of the art of Science education research. Despite admitting the relevance of approaching the issue in different Latin American countries, the complexity of this approach is considered from the start and the study is essentially restricted to occurrences in Brazil.

As a perspective, we advocate the need for an international project of the countries in the block, with which histories and state of the art can be established, encompassing all the countries. It is a fact that research journals circulate internationally, yet the scope of our research cannot certainly be limited to the papers they issue.

However, under the light of international reviews, such as those by Welch (1985) and Gil-Pérez (1995), comparatively, we understand to have attested the representativity of the research reported in this chapter, even if limited to occurrences almost restricted to a single country.

We would also like to stress that the significant production of knowledge in Science education in Brazil, even if considering only the Master's thesis and Doctoral dissertation, already in accordance with the Megid catalog (1995), has comprehended very diversified thematic focuses and has approached different teaching levels and content subareas. Nonetheless, some questionings remain, such as the study by Delizoicov (2004), which proposes a greater approximation of the problems investigated with those faced by the Brazilian educational system.

This is a very polemic issue, once the functioning of an institution is not certainly limited to the knowledge accumulated about what occurs or may occur within it. However, it seems pertinent to admit, given the way teaching functions, that the research in Science education allows verifying, in specific situations, the researchers' community should also share some responsibility in the sense of contributing to making public policies directed towards obtaining the access to research feasible. And not merely concerning to its outcomes, but mainly to the procedures involved in its realization.

Notes

1 Brazilian Foundation for the Development of Science education. *Um esforço para o desenvolvimento*. São Paulo: FUNBEC, May 1968, 10 pp.
2 Another source of information on the IBECC activities which supported us was an IBECC booklet called: *An effort to improve Science Education in Brazil 1950–1965* (São Paulo: USP, 45 pp.).
3 United States Agency for International Development – USAID.
4 The agreement then signed by the IBECC and the University of Brasília, with USAID, is reported in the paper by Barry and Lorenz (1986).
5 In this sense, we highlight the catalog by the University of São Paulo, resulting from the USP/BID/CECAE Science Teachers Formation Project. Sub-Project: Advisory to graduate teaching programs in Physics: *Ensino de Física no Brasil – Catálogo de Dissertações e Teses* (1972–1992). São Paulo: s.n., 1992; more recently published as Kawamura and Salem (2008). As well as the catalog by Megid (1995).

References

Almeida, M. J. P. M. (2003). *Meio Século de Educação em Ciências: uma leitura de recomendações ao professor de física* (PhD Thesis). Universidade Estadual de Campinas, Campinas.

Almeida, M. J. P. M. (2004). *Discursos da Ciência e da Escola: ideologia e leituras possíveis*. Campinas, SP: Mercado de Letras.

Arapiraca, J. O. (1982). *A USAID e a educação brasileira.* São Paulo: Cortez – Autores Associados.

Barra, V. M., & Lorenz, K. M. (1986). Produção de materiais didáticos de ciências no Brasil, período: 1950 a 1980. *Ciência e Cultura, 38*, 1970–1983.

Bittencourt, D. da R. S. (1977). *Uma análise do Projeto de Ensino de Física-Mecânica* (Master's thesis). University of São Paulo, São Paulo.

Brasil. (2009). Coordenação de Aperfeiçoamento do Pessoal de Nível Superior – CAPES. Diretoria de Avaliação – DAV (Documento da Área de Ensino de Ciências e Matemática). Brasília. Retrieved February 22, 2010, from http://www.capes.gov.br/images/stories/download/avaliacao/ENSINO_CM_21dez09.pdf

Carvalho, A. M. P. (1972). *O ensino de física na grande* São Paulo: *estudo sobre um processo de transformação* (Doctoral dissertation). Faculty of Education, University of São Paulo.

Delizoicov, D. (2004). Pesquisa em Ensino de Ciências como Ciências Humanas Aplicadas. *Caderno Brasileiro de Ensino de Física, 21*(2), 145–175.

Dib, C. Z. (1972). *Tecnologia da educação e a aprendizagem de física* (Doctoral dissertation). University of São Paulo, São Paulo.

Dib, C. Z. (1991). Schenberg e a pesquisa educacional em física. *Boletim Informativo da Sociedade Brasileira de Física,* 21–29.

Gil-Pérez, D. (1995). New trends in science education. *International Journal of Science Education, 18*(8), 889–901.

Kawamura, M. R. D., & Salem, S. (2008). *Ensino de fisica no Brasil: dissertações e teses – catálogo analítico* (1996–2005). São Paulo: PROFIS:IFUSP.

Krasilchik, M. (1978). Uma experiência na renovação do ensino de ciências. In *Anais do Simpósio sobre Ensino de Biologia, Física e Química (1º e 2º graus) no Estado de São Paulo* (pp. 48–55). São Paulo: Academia de Ciências.

Megid, J. (Ed.). (1995). *O Ensino de Ciências no Brasil: catálogo analítico de teses e dissertações 1972–1995*. Campinas: Unicamp.

Miranda, A. C. (1998). *As Raízes do Ensino da Ciência no Brasil* (PhD Thesis in Education). Universidade Estadual de Campinas, Campinas.

Moreira, M. A. (1972). *A organização do ensino de Física na Universidade* (Master's thesis). Federal University of Rio Grande do Sul, Porto Alegre.

Nardi, R. (2005). *A área de ensino de Ciências no Brasil: fatores que determinaram sua constituição e características segundo pesquisadores brasileiros* (PhD Thesis). Faculdade de Ciências, Universidade Estadual Paulista, Bauru, São Paulo.

Nardi, R., & Almeida, M. J. P. M. (2007). Investigação em Ensino de Ciências no Brasil segundo pesquisadores da área: alguns fatores que lhe deram origem. *Pro-Posições, 18*(1), 213–226.

Nardi, R., & Almeida, M. J. P. M. (2008). Educación en Ciencias: lo que caracteriza el área de enseñanza de las Ciencias en Brasil según investigadores brasileños. *Revista Electrónica de Investigación en Educación en Ciências, 3*(1), 27–34.

Nassif, L. A. de L. (1976). *O conceito de ciência veiculado por materiais didáticos: uma análise do curso de Física do PSSC* (Master's thesis). Pontificial Catholic University of São Paulo, São Paulo.

Orlandi, E. P. (1983). *A linguagem e seu funcionamento: as formas do discurso*. São Paulo: Brasiliense.

Pacca, J. L. de A. (1976). *Análise do desempenho de alunos frente a objetivos do PEF* (Master's thesis). University of São Paulo.

Ribeiro, M. L. S. (1982). *História da Educação Brasileira: A organização escolar* (4th ed.). São Paulo: Editora Moraes.

Saad, F. D. (1977). *Análise do projeto FAI: Uma proposta de um curso auto-instrutivo para o 20 Grau* (Master's thesis). University of São Paulo, São Paulo.

Teixeira Jr., & de Souza, A. (1976). *Um projeto de ensino de ciências para o Brasil* (PhD thesis). University of Taubaté, Taubaté.

Violin, A. G. (1976). *O projeto de Ensino de Física (PEF): Mecânica I em um curso individualizado, 2v* (Master's thesis). University of São Paulo, São Paulo.

Welch, W. W. (1985). Research in science education: Review and recommendations. *Science Education, 69*(3), 421–447.

CHAPTER 2

Science Education Research Methods in Latin America over the Last Decade

Ileana María Greca and Flávia Maria Teixeira dos Santos

Abstract

This chapter presents the results of an inquiry about the research methods used in science education in Latin America during 2000–2009, as shown by the articles published in renowned Latin-American journals. Much of what we have been doing in Latin America is, certainly, influenced by what has been produced in English-speaking countries, and is mostly affected by what is presented in the most relevant international journals. Nevertheless, it seems that our hindrances, together with financial, social, and educational needs, might lead our research projects to look into the heart of our schools and classrooms, which in Latin America still display many shortcomings. The elected research methodology to approach these topics, so as to examine what was going on in schools and classrooms, was mostly qualitative, since this methodology seemed to offer more adequate solutions to the social problems involved in these inquiries. Nevertheless, the standard of the qualitative studies analyzed is heterogeneous, since there are some excellent articles mixed with others lacking that level of excellence. From the results it seems necessary to broaden the scope of methods, procedures and instruments for collecting and analyzing data. Also, it might be recommended to invest more time in the methodological formation of researchers and in the discussion of these issues in the area of research in science education in Latin America, so as to improve the general quality of research studies.

Keywords

research methods – science teaching – Latin America – review

 | DOI: 10.1163/9789004409088_002

subculture of the European-American culture (Baker & Taylor, 1995; Cobern, 1991; Jegede, 1994, cited by Medina-Jerez 2008). In this sense, science learning involves border crossing from non-scientific subcultures to the subculture of science. Based on the work of Costa (1995) and Aikenhead (2001), this research, developed in Colombia, focused on the description of crossing boundaries between the world of science and the learners' personal worlds. These boundaries became evident when learners expressed their attitude towards science and, also, towards how the class was conducted.

The study shows how differences in terms of border crossing between the different "subcultures" is determined by the cultural contexts of both rural and urban educational institutions. These institutions, as many others in Colombia, belong to communities affected by outlawed armed groups. In some of these institutions, the influence of the church and the military was observed. The boundaries encountered relate to sociocultural, socio-economic, psychosocial, linguistic, and structural aspects. In general, the speech of urban students is more influenced by the discourse of modern Western science. In contrast, students in rural areas are more concerned with environmental problems and their personal experiences with nature.

With the purpose of documenting some exchanges in culturally diverse societies like the Colombian, several investigations focused on the individuals' worldviews, such as the following two studies addressed here. Molina (2007) reports a study that looked for a cultural interpretation of children's ideas about cactus spines. A cultural perspective to approach the genesis of such ideas was adopted, claiming that explanations can be interpreted from multiculturalist or interculturalist frames of reference (Bruner & Haste, 1990). Eighteen children, between 8 and 12 years of age, were interviewed. It was established that some of their views derive from Western perspectives and others from non-Western perspectives on nature.

The second study, by Molina, Mojica, and Lopez (2005), compared children's ideas about nature in four culturally different communities: an urban school in Bogota; an immigrant peasant school in Bogota, composed of immigrant farmers from northwestern Boyacá; a U'wa school, in an indigenous community living in the northwestern department of Boyacá, Colombia; and a farmer school, influenced by the U'wa. The purpose of the study was to establish the shared criteria, values and descriptions of nature. An analysis of semantic content and of local history was made in order to establish the conglomerates of relevance that steer those ideas.

This notion, as already noted, is based on the intercultural nature of the Latin America countries (García Canclini, 2004). Accordingly, it is assumed that the subjects' conceptualizations of the natural world also reveal their origins in the heterogeneous and diverse societies to which they belong. Thus, the modeling of the subjects' ideas is inspired by the metaphor of collage proposed by Garcia Canclini. Research on conglomerates of relevancies emphasize the explanation of specific natural phenomena and the values on which such explanations are based. In this case the values refer to intentions, purposes, beliefs and worldviews that make it possible to establish the cultural bases of the perspectives that guide the subjects' conceptualizations (Molina, 2002, 2007, 2012a; Venegas, 2012; Pedreros, 2013). Finally, this approach to research also refers to the cultural and local historical interpretation of the ideas, for which some contexts are set to address their varied cultural origins and each subject's own perspectives about the natural world.

The results indicate that the ideas and values about nature overlap among the groups. The U'wa boys and girls and the immigrants share some naturalistic criteria on nature, that is, they express their interest in adjusting their descriptions to observable aspects of nature. In turn, city dwellers, immigrants and farmers share other naturalistic criteria. They share ethical, aesthetic and emotional values. City dwellers, immigrants and the U'wa share views about space and about the city. Immigrants and farmers share other criteria about space. The U'wa do not adhere to any utilitarian criterion about nature. City dwellers, immigrants and U'wa believe that nature is a national emblem, but differ in the criteria, the first two basing this conception on aspects of biodiversity and the latter on political aspects. In turn, an analysis of national historical texts shows that the presence of some contents in the children's narratives can be explained in historical terms.

6 As a Conclusion

As noted at the beginning of this chapter, it is true that the past tells us about the present, but it is also true that the future informs us about the present prospectively. Thus, the projects and researches analyzed here constitute a challenge to current educational and teaching concepts. However, these projects must also be understood as opportunities for the emergence of new concepts. This chapter has attempted to present some of the advances, yet the task is still far from complete.

The recognition of cultural diversity has already been established in many national Constitutions of the South American countries, but it is necessary to establish this recognition in all institutions and in the citizenry of our nations. As Marques (2000) argues, if school education is subject to the circumstances of its social context, it has some degree of autonomy and specificity in relation to other social practices. It also has concrete and objective conditions to influence, in its own way, the transformation of society. Thus, one can say that a pedagogical response may have a chance, as long as it takes into account the social dimensions of the educational phenomenon, i.e., it constitutes a response to social and cultural aspects. Science education research does not escape from this. Moreover, the recognition of cultural diversity and social cohesion also depends on the construction of new knowledge. Therefore, a central concern for research deals with the elaboration of methodological and theoretical categories to understand the building of school education as well as science teaching and learning in cultural contexts characterized by diverse and clearly differentiated cultures.

This situation defines particular scenarios in relation to the dialogue between different forms of knowledge and, thus, demands other forms of typifying science education. The works examined here, in fact, have been overcoming the distance among scientific knowledge, traditional (or ancestral) knowledge, and school knowledge. This is important, because the polarization among these forms of knowledge has resulted in inequalities, lack of recognitions and exclusion of communities clearly differentiated in cultural terms.

The relationship between disciplinary knowledge and pedagogical knowledge should include the definition of curriculum content and the sense that school communities assign to education, as well as the processes of social relations, in harmony with the cultural diversity the communities possess. Observations show that the construction of educational practices that can support the respect for, and appreciation of cultural differences depends crucially on the preparation of students and subject matter teachers to become individuals that are sensitive and respectful of cultural diversity. Another important factor in the success of culturally sensitive science education lies in teacher education, so as to allow the development of competencies that can help prospective teachers manage potential tensions and conflicts in the classroom and in the milieu of formal education institutions. It can be seen that there are several works advocating for a context-sensitive education which implies understanding of tensions involving the recognition of cultural diversity, like El-Hani and Mortimer (2007).

This perspective is expanded in Molina (2012b) through different references. Mosquera and Molina (2011) and Molina and Utges (2011) integrated the analysis of the science teachers' conceptions on cultural diversity (in Colombia) to Yuen's (2009) Developmental Model of Intercultural Sensitivity (DMIS). The results show the need to develop and strengthen skills such that, in teaching and learning processes, teachers are able to identify relationships between the 'I' and the 'you' (the students' own knowledge, visions, beliefs, and those of others). This is complemented by the work carried out by Gonzales and Contreras (2009), Baptista and El-Hani (2009) and Luigi, Aranguren, and Moncada (2008), which show different relationships between traditional ecological knowledge (TEK), school science knowledge and scientific knowledge, seeking the recognition of all of them. Castaño (2009) and Delizoicov (2008) coincide with Smolen and colleagues (2006) in recognizing the necessity that teachers discuss the adaptations of the curriculum, not only in terms of the students' individual characteristics, but mainly in terms of the diversity of cultural values. Weinstein, Curran, and Tomlinson-Clarke (2003) consider that the management of a culturally-sensitive class (MCSC) is part of a long-term training process, in which cultural diversity becomes a lens through which teachers see the classroom management tasks. The MCSC begins with the understanding of the 'I', the 'other' and the context.

Other important issues to address are the resulting ethical, psychological, and educational implications as well as the educational policies that may derive from research. Concerning educational proposals, they should not be oriented towards the construction of knowledge only, but also built from an epistemology that reflects the multicultural and intercultural relations. For the enrichment of educational proposals with a cultural perspective, aspects such as research on otherness need to be considered.

Notes

1 Research Project: Science education, context and diversity cultural: Perspectives on conceptual field. Funded by the Centre for Research and Scientific Development (CRSD), Distrital University Francisco José de Caldas.

2 For further information, see http://en.wikipedia.org/wiki/Mar%C3%ADa_Lionza

References

Aikenhead, G. (1996). Science education: Border crossing into the subculture of science. *Studies in Science Education, 27*, 1–52.

Aikenhead, G. (2001). Students' ease in crossing cultural borders into school science. *Science Education, 85*, 180–188.

Ávila, L. A. (2006). Crítica al análisis cultural predominante en la investigación educativa en México [Criticism of the dominant cultural analysis in education research in Mexico]. *Revista Mexicana de Investigación Educativa, RMIE, 11*(30), 1019–1036.

Baker, D., & Taylor, P. (1995). The effect of culture on the learning of science in non-Western countries: The results of an integrated research review. *International Journal of Science Education, 17*, 695–704.

Bolivar, A. (2004). Ciudadanía y escuela pública en el contexto de la diversidad cultural [Citizenship and public School in the context of cultural diversity]. *Revista Mexicana de Investigación Educativa, 9*(20), 15–38.

Bruner, J., & Haste, H. (Eds.). (1990). *La elaboración del sentido: Construcción del mundo por el niño* [The elaboration of meaning: The child's construction of the world]. Buenos Aires: Paidos.

Castaño, N. C. (2009). Construcción social de universidad para la inclusión. La formación de maestros con pertinencia y en contexto, desde una perspectiva intercultural [Social construction of the university for inclusion. The education of teachers with sense of belonging and of context from an intercultural perspective]. In D. Mato (Ed.), *Educación Superior, Colaboración Intercultural y Desarrollo Sostenible/ Buen Vivir. Experiencias en América Latina* [Higher Education, Intercultural Cooperation and Sustainable Development/Good Living. Experiences in Latin America] (pp. 183–206) Caracas: Instituto Internacional de la UNESCO para la Educación.

Cobern, W. (1991). *World view theory and science education research* (Monograph 3). Manhattan, KS: National Association for Research in Science Teaching (NARST).

Cobern, W., & Aikenhead, G. S. (1998). Cultural aspects of learning science. In B. J. Fraser & K. Tobin (Eds.), *International handbook of science education* (pp. 39–52). Dordrecht: Kluwer.

Cole, M. (1997). Poniendo la cultura en el centro [Placing culture at the center]. In *Psicología cultural* [Cultural Psychology]. España: Morata.

Costa, V. (1995). When science is "another world": Relationships between worlds of family, friends, school, and science. *Science Education, 79*, 313–333.

Delizoicov, D. (2008). La educación en ciencias y la perspectiva de Paulo Freire [Science education and Paulo Freire's perspective]. *Alexandria: Revista de Educação em Ciência e Tecnologia, 1*(2), 37–62.

Eagleton, T. (2005). *A idéia de cultura* [The idea of culture]. São Paulo: Editora UNESP.

Edward, D., & Mercer, N. (1988). *El conocimiento compartido. El desarrollo de la comprensión en el aula* [Shared knowledge. The development of understanding in the classroom]. Barcelona: Ed. Paidós.

El-Hani, C., & Mortimer, E. (2007). Multicultural education, pragmatism, and the goals of science teaching. *Cultural Studies of Science Education, 2*, 657–702.

Elkana, J. (1983). La ciencia como sistema cultural: Una visión antropológica [Science as a cultural system: An anthropological view]. *Boletín de la Sociedad Colombiana de Epistemología, 3*, XXX.

Freire, P. (1972). *Pedagogía del oprimido* [Pedagogy of the oppressed]. Buenos Aires: Siglo XXI Argentina Editores.

García Canclini, N. (2004). *Diferentes, desiguales y desconectados. Mapas de la Interculturalidad* [Different, disparate and disconnected. Maps of Interculturalism]. Buenos Aires: Gedisa.

Geertz, C. (1989). *La interpretación de las culturas* [The Interpretation of Cultures]. Barcelona: Gedisa.

George, J. (1999). World view analysis of the knowledge in a rural village: Implications for science education. *Culture and Comparative Studies, 83*, 77–95.

Giroux, H. (1997). *Cruzando límites* [Crossing boundaries]. Barcelona: Gedisa.

González, F., & Contreras, D. (2009). El concepto de diversidad vegetal desde la etnia Mapuche a la enseñanza formal en Chile [The concept of plant diversity from the Mapuche ethnicity to formal education in Chile]. *Revista Enseñanza de las Ciencias, Número Extra*, 389–393.

Jegede, O. (1994). *Traditional cosmology and collateral learning in non-Western science classrooms*. Toowoomba: Research and Evaluation Unit Distance Education Center, University of Southern Queensland.

Lopes, A. R. C. (1999). Pluralismo cultural em políticas de currículo nacional. In A. F. B. Moreira (Ed.), *Currículo: Políticas e práticas* [Curriculum: Policy and practices] (pp. 59–79). Campinas, SP: Papirus.

López, P. (2000). *Una oportunidad para aprender. La formación intercultural en la formación del profesorado* [An opportunity to learn. Intercultural training in teacher Education]. Cabildo de Tenerife: Área de desarrollo económico.

Luigui, M., Aranguren, J., & Moncada, J., A. (2008). El origen y el culto a María Lionza [The origin and cult to Maria Lionza]. *Revista de investigación, 63*, 19–42.

Machado, N. (1999). *Educação: projetos e valores* [Education: Projects and values]. São Paulo: Faculdade de Educação/USP.

Marques, F. A. (2000). A educação escolar e o resgate da identidade cultural das classes populares [School education and the rescue of popular classes cultural identity]. *Ciência & Educação, 6*(1), 65–73.

Molina, A. (2002). Conglomerado de relevancias y formación científica de niños, niñas y jóvenes [Conglomerate of relevance and scientific training of children and the young]. *Revista Científica, 4*, 187–199.

Molina, A. (2007). Relaciones entre contexto cultural y explicaciones infantiles acerca del fenómeno de las adaptaciones vegetales [Relationships between cultural contexts and children's explanations of plant adaptations]. *Nodos Y Nudos, 23*(3), 76–87.

Molina, A. (2010). Una relación urgente: enseñanza de las ciencias y contexto cultural [An urgent relationship: science education and cultural context]. In A. Claret (Ed.), *Memorias del II Segundo Congreso Nacional en Educación en Ciencias y Tecnología* [Proceedings of the II National Congress on Research in Science Education and Technology]. Santiago de Cali: EDUCyT.

Molina, A. (2012a). Contribuciones metodológicas para el estudio de las relaciones entre contexto cultural e ideas sobre la naturaleza de niños y niñas [Methodological contributions to the study of the relations between cultural context and ideas about the nature of children]. In A. Molina et al. (Eds.), *Algunas aproximaciones a la investigación en educación en enseñanza de las Ciencias Naturales en América Latina* [Some approaches to research in natural sciences education in Latin America] (pp. 63–88). Bogotá: Fondo Editorial Universidad Distrital.

Molina, A. (2012b). Desafíos para la formación de profesores de ciencias: aprender de la diversidad cultural [Challenges to science teacher education: learning from cultural diversity]. *Revista Internacional del Magisterio, 57*(6), 78–82.

Molina, A., Martínez, C. A., Mosquera, C. J., & Mojica, L. (2009). Diversidad cultural e implicaciones en la enseñanza de las ciencias: reflexiones y avances [Cultural diversity and implications for science education: reflections and progresses]. *Revista Colombiana de Educación, 56*, 103–128.

Molina, A, Mojica, L., & López, D. (2005). Ideas de los niños y niñas sobre la naturaleza de: un estudio comparado [Children's ideas on nature: a comparative study]. *Revista Científica, 7*(1), 41–62.

Molina, A., & Utges, G. (2011). Diversidad cultural, concepciones de los profesores y los ámbitos de sus prácticas. Dos estudios de caso [Cultural diversity, teachers' conceptions and the scope of their practices. Two case studies]. *Revista de Enseñanza de la Física, 24*(2), 7–22.

Moreira, F. B. (2002). Currículo, diferencia cultural e diálogo [Curriculum, cultural difference and dialogue]. *Educação e Sociedade, 23*(79), 15–30.

Mortimer, E. F. (1995). Conceptual change or conceptual profile change? *Science & Education, 4*, 265–287.

Mortimer, E. F., Scott, P., & El-Hani, C. N. (2011). Bases teóricas e epistemológicas da abordagem dos perfis conceituais [Theoretical and epistemological grounds of the conceptual profile approach]. *Tecné, Episteme y Didaxis, 30*, 111–125.

Mosquera, C. J., & Molina, A. (2011). Tendencias actuales en la formación de profesores de ciencias, diversidad cultural y perspectivas contextualistas [Current trends in science teacher education, cultural diversity and contextualist perspectives]. *Tecné, Episteme y Didaxis, 30*(2), 9–29.

Nieto, M. (2006). *Remedios para el Imperio: historia natural y la apropiación del nuevo mundo* [Remedies for the Empire: natural history and appropriation of the New World]. Bogotá: Universidad de los Andes.

Obregón, D. (Ed.). (2000). *Culturas científicas y saberes locales: asimilación, hibridación, resistencia* [Scientific cultures and local knowledge: assimilation, hybridization and resistance]. Bogotá: Universidad Nacional de Colombia.

Pedreros, R. I. (2013). Diálogo de las perspectivas de perfil conceptual y conglomerados de relevancia [A dialogue of the perspectives of conceptual profiles and conglomerates of relevance]. *Magis, Revista Internacional de Investigación en Educación, 6*(12), 119–131.

Quintrique, S., & McGinity, M. (2009). implicancias de un modelo curricular monocultural en la construcción de la identidad sociocultural de alumnos/as mapuches de la IX región de la Araucania, Chile [Implications of a monocultural curricular model to the construction of the sociocultural identities of the mapuche students from the IX region of Araucania, Chile]. *Estudios Pedagógicos Valdivia, 35*(2), 173–188.

Retamal, O. (1998). *Una educación para reconciliar al hombre con la tierra. Solo la educación holística hará posible la continuidad de la vida* [An education to reconcile man with Earth. Only a holistic education will make the continuity of life possible]. *Estudios Pedagógicos Valdivia, 24*, 107–121.

Santos, B. S. (1989). *Introdução a uma ciência pós-moderna* [Introduction to a postmodern science]. Rio de Janeiro: Graal.

Santos, B. S. (2001). As tensões da Modernidade [Tensions of modernity]. Biblioteca das Alternativas – Fórum Social Mundial. Retrieved November 18, 2014, from http://www.susepe.rs.gov.br/upload/1325792284_As%20tens%C3%B5es%20da%20Modernidade%20-%20Boaventura%20de%20Sousa%20Santos.pdf

Shweder, R. (1991). La rebelión romántica de la antropología contra el iluminismo, o el pensamiento es más que razón y evidencia [The romantic rebellion of anthropology against enlightenment, or the thought is more than reason and evidence]. In C. Geertz, J. Clifford, et al. (Eds.), *El surgimiento de la antropología posmoderna* [The emergence of postmodern anthropology] (pp. 78–113). México: Gedisa.

Smith, M. U., & Siegel, H. (2004). Knowing, believing, and understanding: What goals for science education? *Science & Education, 13*, 553–582.

Smolen, L., Colville-Hall, S., Liang, X., & Mac Donald, S. (2006). An empirical study of college of education faculty's perceptions, beliefs, and commitment to the teaching of diversity in teacher education programs at four urban universities. *Urban Review: Issues and Ideas in Public Education, 38*(1), 45–61.

Subercaseux, B. (2005). Diversidad cultural: el otro y el nosotros [Cultural diversity: The other and we]. In M. Net, I. Walker, et al. (Eds.), *Diversidad cultural: el valor de la diferencia* [Cultural diversity: the value of the difference] (pp. 45–50) Chile: LOM Ediciones.

Tyler, C. (1993). *El multiculturalismo y la política del reconocimiento* [Multiculturalism and the politics of recognition]. México: Fondo de Cultura Económica.

Velho, G. (1994). *Individualismo e Cultura. Notas para uma Antropologia da Sociedade Contemporânea* [Individualism and Culture. Notes for an Anthropology of Contemporary Society]. Rio de Janeiro: Jorge Zahar Editor.

Venegas, A. A. (2012). Criterios de análisis en la interpretación de las ideas de naturaleza con los conglomerados de relevancias [Analyitical criteria in the interpretation of ideas of nature with conglomerates of relevance]. *Revista Científica, 16*, 130–140.

Weinstein, C., Curran, M., & Tomlinson-Clarke, S. (2003). Culturally responsive classroom management: Awareness into action. *Theory into Practice, 42*(4), 269–276.
Wilson, B. (1981). The cultural contexts of science and mathematics education: Preparation of a bibliographic guide. *Studies in Science Education, 8*, 27–44.
Wolf, E. (1987). *Europa y la gente sin historia* [Europe and the people without history]. México: FCE.

PART 2

Teaching and Learning Science

CHAPTER 5

Emotions, Feelings, and Conceptualizations in the Didactics of Physics

María Rita Otero and Maria de los Angeles Fanaro

Abstract

In this chapter we propose the bases for a Didactics of Sciences that takes into account the emotions, feelings and reasoning in the reconstruction of a domain of scientific knowledge. The ideas of Humberto Maturana and Antonio Damasio are used as they offer neurobiological support to the relationship between emotions, actions, feelings and reasoning. Gérard Vergnaud's theory of conceptual fields is adopted as a theory of cognitive development to study the process of conceptualization in a certain domain of knowledge. Finally we present the example of a didactic sequence to teach the foundations of Quantum Mechanics for high school students, which has been designed and implemented by using the didactic principles proposed in this chapter.

Keywords

emotions – feelings – didactics of physics

1 Introduction

The Didactic of Physics[1] describes the process of physical knowledge reconstruction in any institution involving one or more teachers (Otero, 2006, 2007). The specific phenomena related to Physics teaching are analyzed, explained and described by the Didactic of Physics. The process of knowledge construction begins outside the scientific community and is influenced by a lot of constraints, which must be analyzed adopting a didactic viewpoint. In turn, these constraints greatly affect the knowledge taught.

The teacher and students integrate the class group (CG). The CG's work is conditioned by external standards – among others, pedagogical, epistemological, institutional, and political – which establish, in fact, a set of constraints for

 | DOI: 10.1163/9789004409088_005

its operation. Our didactic analysis focuses on the study of the CG members' actions related to the physical knowledge construction. The constructivist framework sets a biological continuity in the process of knowledge construction (Inhelder & Piaget, 1955/1976; Garcia, 2000; Maturana, 1995). The emotions involving coexistence and cooperation (Damasio, 1994, 2005; Maturana, 1995) are a necessary but not sufficient condition to enable the construction of knowledge in the CG. Being a cognitive and constructivist framework with profound didactic implications, Vergnaud's (1990, 1994, 2000, 2013) Theory of Conceptual Fields (TCF) helps describing and analyzing conceptualization in physics.

2 Constructivism and Knowledge

We take a position that coincides with Piaget's Genetic Epistemology and Humberto Maturana's epistemological point of view. Constructivism avoids aprioristic and empiricist conceptions and denies the existence of an absolute beginning to explain the origin and the essence of knowledge (Garcia, 2000). Considering that the cognitive activity is governed by a general mechanism, the constructivist theory establishes the principle of functional continuity that begins in the purely biological processes of a living being and its organization. Constructivism does not intend to give an "intrinsic definition" of knowledge because it is a process that makes sense in a social and historical context, where "levels" or "degrees" acquire significance in that context (Garcia, 2000). Constructivism rejects the subject-object duality, the Cartesian body-mind dualism and the emotion-reason dualism.

3 Maturana's Epistemology

Human beings assess cognition in any domain by specifying the domain with a question and demanding adequate action in that domain. If the answer satisfies us, as an adequate action in the domain specified by the question, we accept it as an expression of cognition in that domain and claim that those who answer our query know. Knowledge begins with the action of every organism interacting with its environment (Maturana, 1984, 1995).

In experience itself, however, we cannot distinguish between what we call an illusion and a perception: illusion and perception are experientially indistinguishable. It is only through the use of a different experience as a metaexperiential authoritative criterion of distinction, either of the same observer or of somebody else subject to similar restrictions, that such a distinction is socially made. Our incapacity to experientially distinguish between what we socially call illusion and perception is constitutive in us as living systems and is not a limitation of our present state of knowledge. The recognition of these circumstances should lead us to put a question mark on any perceptual certainty. If the properties of the observer are not questioned, if cognition and language are taken for granted, then the independent existence of what is known must be assumed. If the properties of the observer are questioned, if how cognition and language arise is questioned, then the experiential indistinguishability between illusion and perception must be accepted, taking as constitutive that existence is dependent on the biology of the observer (Maturana, 1984, 1995).

The objectivity without parentheses entails the assumption that existence is independent of the observer, that there is an independent domain of existence, and that the world is the ultimate reference for the validation of any explanation. Things and entities exist with independence of the observer that distinguishes them, and it is the independent existence of entities that specifies the truth. Objectivity without parentheses entails unity, and, in the long run, reductionism, because it entails reality as a single ultimate domain defined by independent existence. Those who have access to reality are necessarily right in any dispute, and those who do not have such access are necessarily wrong. Coexistence demands obedience to knowledge.

In contrast, objectivity with parentheses entails accepting that existence is brought forth by the distinctions of the observer, that there are as many domains of existence as kinds of distinctions the observer performs: objectivity in parentheses entails the multiverse, it entails that existence is constitutively dependent on the observer, that there are as many domains of truths as domains of existence she or he brings forth in her or his distinctions. Finally, within objectivity in parentheses each domain of knowledge is equally valid if not equally pleasant to be part of, and disagreements between observers will have to be solved not by claiming a privileged access to an independent reality, but through the generation of a common domain through coexistence in mutual acceptance. In the multiverse, coexistence demands consensus, that is, common knowledge.

3.1 *Emotions and rationality*

Emotions are kinds of relational behavior. As such our emotions guide moment after moment our doings by specifying the relational domain in which we operate at any instant, and give to our doings their character as actions. It is the configuration of emotioning that specifies our human identity, not our rational behavior. Rational behavior begun as a feature of the living of our ancestors with language in the use that they made of the abstractions of the coherences of their daily living as they operated as languaging beings (Maturana, 1995). But it was then as it is now that emotions specified the domain of rational behavior in which they operated at any instant. They were not aware of this then, but now we know that every rational domain is founded on basic premises accepted *a priori*, that is, on emotional grounds, and that our emotions determine the rational domain in which we operate as rational beings at any instant (Maturana, 1995).

Usually, the human beings are not fully aware of the emotions under which they choose their different rational arguments. They are rarely aware of the fact that what guides their living are the emotions even when they claim they are being rational.

3.2 *Emotions and Feelings*

Damasio (2005) stresses the role of emotions in human thought. The states of the body are modified by the emotions in a way that may or may not be evident. They are automatic, and sometimes modular. We are not always aware of their consequences once they are arranged. From a biological standpoint, emotions serve the well-being and survival of our body. Emotions precede feelings, both at the time to experience an emotion as in the historical evolution (Damasio, 2005, p. 34). Emotions are functional to a complex system of vital regulation; they are designed to avoid dangers, assist the organism to take advantage of a chance, or indirectly facilitate social relations. Emotions enable the organism to respond effectively but not creatively to favorable or threatening circumstances to survival. Feelings introduce a mental alertness and reinforce the impact of emotions affecting permanently attention and memory. Thus, together with memories, imagination and reasoning, feelings make possible the production of new, not stereotyped responses. "My hypothesis is that a feeling is the perception of a state of the body at once with the perception of a certain way of thinking and thinking with certain themes" (Damasio, 2005, pp. 85–86). Feelings and emotions are basic to social relations and decision-making like reasoning required in social life. The feelings that derive from positive and negative emotions are directly involved in our social experiences. It is important to emphasize that, against certain dualistic traditions, the emotions are inherently rational because they lead to the best solution in terms of survival.

4 Vergnaud's Ideas about Action, Conceptualization and Activity

The TCF is based on the idea of pragmatic knowledge construction. It cannot theorize about learning neither from symbolic representations nor from situations. Hence, situations and the sense of symbols must be considered, bearing in mind the students' actions in situations and their behavior organization. Then, the concept of scheme becomes extremely relevant. Vergnaud (1990) defines "scheme" as a functional dynamic *totality*, and as the invariant organization of activity and behavior for a certain *class of situation*. Schemes are made up of four kinds of components: *goals*, subgoals, and expectations; *rules* of action, information search, selection and control; *operational invariants* (concepts and theorems-in-action) and *inference* possibilities.

Conceptualization is part of activity organization and the scheme associated to a kind of situation necessarily involves conceptualizations. The operational invariants (OI) are the epistemological part of schemes whose function is to recognize the objects, their properties, relationships, and transformations. As regards OI, their main functions are to take and select the relevant information and infer the useful consequences for the action, the control and the subsequent information taking. The concepts-in-action, defined as objects or predicates, are neither true nor false, they are only pertinent or not. A theorem-in-action is a proposition considered true into the activity. In sciences, since a lot of theorems-in-action could be associated to a concept, then, it makes no sense to say that some people have understood certain concept. However, it does make sense to establish which theorem-in-action people are able to use into a given situation.

Without these four scheme components the structure of the activity could not be understood. It has a double characteristic: it is systematic and contingent. First, the activity is systematic for lots of situations because it is governed by strict rules. Second, it is contingent because the rules generate different activities and behaviors taking into account the parameters of situations. In the new situations, where the student has not a scheme, this characteristic of the activity is more evident. Physics situations could be regular or aleatory, especially in Quantum Mechanics (QM), increasing the uncertainty of conceptualization and, consequently, making both teaching and learning difficult.

The concept of scheme provides a theoretical reply to the knowledge construction problem. Adapting to new situations, the OI cover an essential function, when they are available in the cognitive structure to be combined and recombined, or when they appear in the situation mixing with the invariants formed before. The conceptualization function is assured by the OI; the schemes are the most important psychological tool for adapting to the new and diverse.

The students' performance in a situation is based on their implicit or explicit knowledge. As a result, it is necessary to pay attention to the cognitive development, its continuities and breaks, its obligatory ways, and the kind of problems, procedures and complex representations, analyzing the main mistakes and discoveries. A concept does not convey meaning in terms of only one kind of situation and a situation cannot be analyzed by means of only one concept. It is necessary to research into a large set of situations and concepts, classifying the kinds of relationships, the types of problems, the scheme treatments, the linguistic and symbolic representations, and the concepts organizing this set. The schemes organize the students' behavior for a given class of situations, and the action and the symbolic representation activity, especially the linguistic activity accompanying the action. Sometimes this is an internalized activity that becomes more and more important while the situations are newer, and the problems resolution is impossible without language, in particular when new conceptualizations and concepts are required. Language has communicational and representational functions. Moreover, language contributes to thought and action organization. Language and symbolic representations play an important role in conceptualization and action.

Activity is more than behavior: behavior is only the visible part of the activity. Therefore when analyzing physical and mathematical behavior, one must look into the representational activity underlying it. The concept of scheme is essential to cover this problem. The most important part of our knowledge consists of operating actions, and they cannot be put into words easily. This is true for every domain of knowledge, and it is even truer for any person and children, as they are unable to express the knowledge they use in action. A lot of properties in problems cannot be reduced to numerical structures, nor can they be considered as linguistic or symbolic entities only. They are concepts and theorems-in-action. The implicit character of a large part of our knowledge does not mean that explicit knowledge is not operational. But we cannot be satisfied with a theory that would consider physics only as an explicit body of knowledge.

Even when one is interested in the function of language and symbols in the development of the mind, it is necessary to identify safely which properties of the signifier represent which properties of the signified. We are aware today that words mean different things for different individuals, especially for the teacher and each student individually. Vygotsky explained 70 years ago that the "sense" given to words is different from their conventional "meaning".

Therefore, there is a theoretical need to analyze activity and representation as composed also of OI that may be different from the meaning of words. This problem can be solved only if we accept the idea that schemes involve concepts and theorems-in-action. It is our job to identify them, together with the other components of schemes, and representation.

5 Emotions, Feelings, and Conceptualizations in the Didactics of Physics

5.1 *Educational Institutions as Systems of Vital Self-Regulation and a Propitious Environment for Well-Being and Survival*

The educational institutions are human creations that should extend, at least, as a desire or intention, the mechanisms of vital self-regulation. These non-automatic mechanisms would have to contribute to our well-being and survival. The school would offer the possibility of coexisting in mutual acceptance of others to generate common domains of knowledge. This is essential for our well-being and it is also a necessary condition to cognitive development and science learning.

5.2 *The Class Group (CG) as Space of Coexistence*

Students and teachers integrate groups that would have to work as a space of coexistence. Coexistence in mutual acceptance is a necessary condition to knowledge reconstruction due to the fact that it allows the students to be committed to their own learning. Learning is neither obedience nor repetition. As Piaget said, it is necessary a team to think.

Coexistence requires that each member of the CG accept the legitimacy of the other ones (Maturana, 1995). Acceptance, the opposite of negation, is a basic emotion to learn in a consensual domain. Furthermore, acceptance calls for the state of being alert so as not to fall into the temptation of certitude. In the classroom, a lot of actions are usually taken from the habit of certitude. Acceptance is different from tolerance. The other one is tolerated when it is assumed that he or she is absolutely wrong while we are not. I am right, he is wrong; then I admit him to be wrong. Assuming objectivity with parentheses implies being aware that we do not have a transcendental access to the truth. The other one is as legitimate as we are and its reality is as legitimate as ours even if he or she does not like us or if we suppose he or she is dangerous. If we decide to deny the other, we shall have to assume our responsibility, but it is not because we can establish that he is wrong. Living in acceptance promotes self-awareness and self-identity.

5.3 *Teachers' Activity in Coexistence*

Even though the teacher has specific responsibilities, she or he cannot assume the students' responsibility for learning. Through the proposition of teaching situations and questions, teachers invite students to join a new world of shared meanings. Situations are complex tasks, in Vergnaud's words "*all complex situations could be analyzed as a combination of tasks*" (Vergnaud, 1990, p. 151). The cognitive processes are functions of the situations the students face and the answers they elaborate to respond to. There are a great number and variety of situations and classes of situations into a conceptual field. A set of classes is generated by analyzing the variables of situations. This task could be more or less difficult in certain domains. in QM and generally in Physics the task requires classifying the great complexity of mathematical and physical models. The students' knowledge is performed by the situations they face and manage progressively. In spite of the fact that learning is personal there are strong regularities among the students. They could be identified by analyzing the students' actions: How do they handle the same situation? Which previous ideas about objects, concepts and relationships do they have? Which are the steps of their knowledge construction processes?

A very important point in conceiving a teaching situation is to identify the questions and actions necessary to respond to it. Each situation could be thought as a combination of basic relationships between known and unknown data which lead to a set of possible questions. In Vergnaud's words (1990, p. 157) "*the didactic situations are an interesting and rich staging*" because they take nourishment from psychology, epistemology and physics. A good didactic staging is based on knowing the difficulties related to the cognitive tasks involved in the situations, the faced obstacles, the set of available knowledge and its possible representations. Cognitive Psychology is essential (Vergnaud, 1990, p. 157).

It is essential for the teacher to decide, select and design the situations that he or she will present to the students; this is his or her prime act of mediation (Vergnaud, 2000). Learning depends on the students accepting the teacher's invitation. Coexistence in acceptance entails teaching by taking into account the students' knowledge before questioning it instead of imposing the teacher's knowledge. When the teacher makes room for the students' activity he or she is teaching in coexistence, allowing them to assume their responsibility for learning. We use Vergnaud's idea of activity as a set of actions and perceptions so that we capture relevant information, OI and control mechanisms of actions. In addition, Vergnaud's idea of activity entails sharing meanings, views, and knowledge in verbal communication. How can the teacher teach something to someone without understanding the activity in the situation?

How can the teacher conceive the situations to teach without understanding the specific conceptual field where he or she is and the characteristic schemes related to these situations?

5.4 *Students as Self-Constructors of Knowledge*

The students have the responsibility to accept or reject the teacher's invitation to study physics. The students' learning consists in answering, evaluating, and deciding the course of action with the help of the teacher and the CG. The students share with the teacher the decision as to which ways to explore and with which instruments. They evaluate the quality of the answer too. Learning results are uncertain from a cognitive and epistemological point of view. Learning is a complex, not a linear process.

Not only is it necessary a coexistence in acceptance but also consensus to construct and validate the common knowledge in the CG. This constructed knowledge is not the same knowledge constructed in the scientific community of reference. Nevertheless, it is possible to exercise certain epistemological vigilance narrowing the gap with scientific explanations. The students' activity is a process that includes perceiving, selecting and gathering relevant information, carrying out actions, making gestures and inferences, conjecturing and answering, discussing the validity of the answers with others, voicing doubts and re-using the mistakes, sharing their knowledge and meanings with others.

5.5 *Emotion as Basis of Reason*

Rationality is based on emotion. An argument, an explanation will be considered rational as long as it satisfies the criterion of rationality of those who accept it. The emotions are the condition of possibility of reason. The scientific rationality goes further into satisfying certain logical principles which are also accepted by reaching an agreement by the scientific community. Science teaching at school cannot ignore that rationality is a historical construction. If teaching assumes the emotional bases of coexistence as necessary for knowledge reconstruction, it will contribute to develop institutions more compatible with a sense of well-being and cooperation.

5.6 *Mistakes Being "a Posteriori"*

Accepting the fact that mistakes are inevitable and that they are always "a posteriori" (Maturana, 1995, 2001) modifies our feelings about what we call mistaking. When do we realize we have made a mistake? We become aware of mistakes when thinking about the consequences of our actions. The traditional scholar culture does not acknowledge mistakes, understanding them in the objectivity without parentheses. This illusion of certainty causes negative feelings, devaluation and discouragement in the students. Assuming mistakes as being “a posteriori” contributes to the coexistence and mutual acceptance in the CG.

5.7 *Evaluation*

Evaluation is a process that allows the CG to analyze what is known and who knows it. Knowledge is reconstructed and analyzed by the CG taking into account the starting point. The CG will agree that someone knows when his or her actions satisfy the criteria of validity accepted by the group. Such criteria are a consensual product, being part of the public knowledge formulated and written in documents and in the joint effort activities. In these activities, teachers and students alike analyze which questions have been answered, which have not and which might be the new goals. Evaluation must be limited neither to tests nor to a moment.

5.8 *Well-Being, Creativity and Reasoning*

Well-being is functional to our survival mechanism. Social and educational institutions should be extensions of the non-automatic regulation of survival mechanisms (Damasio, 2005). The well-being of the CG is based on coexisting in acceptance and learning to avoid the emotions and actions that deny the other one. Coexistence collaborates in the process of reasoning, deciding the best way of acting in a given situation at any one time. In a state of emotional well-being the cerebral activity increases in the areas linked to reasoning and creativity while the opposite happens in a state of sadness. In other words, the students' knowledge and feelings have an impact on their emotional well-being. Nevertheless, in the traditional science classroom it is natural for the students to think that “not understanding” is normal due to the fact that they can consider this fact to be their own disability. The consequences are repetition as learning simulation, and, sometimes, discomfort and sadness, increasing incomprehension.

6 An Example about QM at Secondary School

Physics teaching researches (Cuppari et al., 1997; Fischler & Lichtfeldt, 1992; González, Fernández, & Solbes, 2000; Greca, Moreira, & Herscovitz, 2001; Hanc & Tuleja, 2005; Cabral de Paulo & Moreira, 2004; Montenegro & Pessoa, 2002; Moreira & Greca, 2000; Müller & Wiesner, 2002; Niedderer, 1997; Olsen, 2002; Osterman & Moreira, 2000; Ostermann & Ricci, 2004; Osterman, Prado &

Ricci, 2006; Pessoa, 1997; Pinto & Zanetic, 1999; Taylor et al., 1998; Taylor, 2003; Zollman, 1999) and the curriculum of many countries propose the study of the basic concepts of QM in secondary school (Lobato & Greca, 2005).

In Argentina, although the Physics syllabus for secondary school covers the basic concepts of modern physics, in practice these concepts are not studied. Both secondary and university teaching have forgotten that knowledge begins by questions; as a result, only answers are taught. Therefore, it is essential to focus on teaching questions and situations as complex tasks in order to teach a science alive. To that end, some conditions must be accomplished to study meaningful questions at school. These questions should have

- *cultural and social legitimacy*: the questions must be related to issues considered relevant by society.
- *physical legitimacy*: the questions must be related to basic situations in Physics.
- *functional legitimacy*: the questions must be related to other issues studied at school, in physics or in other science courses.

QM is transformed when it is taught at a given institution; this is the well-known phenomenon of didactic transposition (Chevallard, 1992, 1997, 1999). In Physics, there are a lot of conceptual fields (Verganud, 1990) in which at least one Conceptual Structure of Reference (CSR) can be distinguished and recognized (Otero, 2006). When a Physics teacher invites his or her students to study a specific conceptual field, he or she adopts more or less explicitly a particular CSR. A CSR is a set of concepts, the relationship between them, the principles, the knowledge claims, and the explanations relative to a conceptual field accepted by the scientific community of reference. Our investigation rebuilds a CSR based on Feynman's Paths Integral method (Feynman & Hibbs, 1965). A detailed analysis of this CSR can be consulted in Arlego (2008). The full proposal adapting a conceptual organization for high school students can be found in Fanaro and Otero (2008), and Fanaro, Arlego and Otero (2007). The CSR adopted will be partially or fully reconstructed by a CG or by someone who tries to study it in high school, or in basic and advanced courses at university. Moreover, the science teaching researcher needs to establish and rebuild a CRS. On the one hand, he needs to analyze the knowledge living in the scientific community and, on the other hand, the characteristics, constraints, and possibilities offered by the institution where this knowledge will be reconstructed.

Any attempt to reconstruct knowledge creates a different conceptual structure for the components and the relationship between them. In a more or less explicit way, each teacher of a certain group will reconstruct or select – based on an existing structure – one conceptual structure to be taught, and, in the best of the cases, he or she will invite his or her class to study it. We coined the term Proposed Conceptual Structure for Teaching (PCST) (Otero, 2006) to describe a set of concepts, the relationship between them, the knowledge claims, principles and situations related to a certain conceptual field that the teacher proposes to reconstruct based on a CSR.

There are characteristic structures related to diverse conceptual fields that are alive, adapted and accepted into certain institutions. They survive for all the time because they are viable. The design, analysis and rebuilding of a PCST related to QM, viable at high school, is a specifically didactic objective. We are also interested in replicability and adaptability in similar institutions. The structures are systems (components + organization) that include key concepts, like the relationships and fundamental principles that tie them together.

When we adopted Vergnaud's ideas about concepts and conceptualization, we included both the operating and the predicative form of conceptualization. The implicit aspects of knowledge are considered by the operating invariants involved in the conservation of the forms to organize the action. This idea of concepts related to action in all their variations makes it possible to build a bridge to the underlying emotions and feelings, also included in the conceptual structures. These structures are inseparable from the set of problems and situations that give sense to them. The PCST has the following components:

Teaching Situations: The situations are formulated around strong, personally, socially, scientifically, and institutionally relevant questions. The answers are provisional, not immediate, they require a lot of time, and, above all, they do not finish in formal schooling. The situations must be developed considering the scientific knowledge, the students' knowledge and the expected learning outcomes. The students' knowledge cannot be ignored by the designed situations. They are the result of a research activity which anticipates and controls their functioning, adaptability and viability. Teaching situations have an explicit didactic intention: they carry out activities concerning physical knowledge construction held by the students and the teacher in the class. The design, implementation and validation of teaching situations are complex processes, characteristic of the research activity in the didactic of Physics. In these processes students' activity and teachers' activity are analyzed according to a didactic framework, a cognitive framework, or both. In spite of this, it is necessary not to confuse the two ways to evaluate the obtained results.

Key Concepts: These are the main concepts that must be built. They are produced in the proposed situation and without them the posed problem cannot be resolved. We assume Vergnaud's ideas of concepts. Concepts are a short list of situations, OI, and referents (symbolic representations).

Key Questions: The situations proposed by the teacher are complex tasks. These situations and their derived questions will be discussed by the CG. The situations resolution calls for specific concepts that will be constructed by answering the questions.

Emotions: These are dynamic body dispositions determining our action domain (Maturana, 1995). Our conversations affect our emotions and our emotions affect our conversations. The PCST invites the students to enter upon a knowledge domain, where the denial of the other is avoided, and an appropriate emotional dynamics to knowledge construction is built. One of the main teacher's actions comes from acceptance regarding the students' knowledge, ideas, conceptions, and the need of room for students' learning activities.

Actions: They comprise three dimensions: the biological, mental and acting dimension. In the PCST we stress the last dimension. We are interested in the teacher and students' actions related to knowledge. We need to anticipate which actions are suitable for the knowledge domain that has been built. The different meanings of these concepts flow from the system of actions related to them in every domain and situation.

Symbolic representations: They refer to the external representation, verbal or not, used in language and the systems of symbols used to talk and write about scientific concepts of every knowledge domain.

The teacher and his or her CG will indeed reconstruct the PCST in a certain and specific institution generating the Conceptual Structure Effectively Reconstructed (CSER). The CSER is a set of concepts, relationships between them, principles and knowledge claims related to a certain conceptual field reconstructed by the CG. The teacher and the students interact in conversations characterized by an adapted emotional dynamic. Every member of the CG will relate to a personal conceptual structure and a unique network of personal and private meaning. Simultaneously, the conversation in the CG will result in the students creating a network of public and shared meaning. This consensual product is also known as "the process of meaning negotiation". This negotiation process can be more or less explicit and conscientious, depending on the professionalism of the teacher, and the distance between the CSR, the PCST and the CSER.

7 The Didactic Sequence

Putting into effect the previous ideas, we have designed, developed and implemented a sequence based on an alternative method for teaching the fundamentals of QM for high school students, focusing on Feynman's path integrals and highlighting the emergence of quantum behavior in the double-slit experiment (DSE). First, the didactic proposal was carried out in the last year of a high school Physics course. The group had thirty (30) students aged between 17 and 18. We analyzed in depth all the protocols of the CGs, synthesizing activities where the teacher and the students are interacting. Apart from that, we have also analyzed the students'activity, the students' replies in a final test (Fanaro, Otero, & Arlego, 2007), and the results of a test related to affective aspects after the last class (Fanaro & Otero, 2008). The sequence has been repeated three times since 2006. The steps of the sequence can be summarized as follows.

7.1 DSE *with Small Balls and Electrons*

First, the students imagined and predicted the results of the DSE where small balls were used. Afterwards, this experiment was simulated using the software "Doppelspalt".[2] This software allowed the students to appreciate the impact on the screen to generate the histogram of frequencies and visualize the theoretical curve of frequencies distribution, called I(x), generated by the software. Next, the students compared their predictions about the results of the experiment with the simulation results. Then, they solved a set of tasks to measure the effect on the form of the curve when the distance between the slits and the slit widths were changed. This led the group to accept the following conclusion:

> When both slits are open, the resulting curve is the sum of the individual curves, i.e., one slit open and the other one closed and vice versa.

After that, the students analyzed the DSE with electrons instead of small balls. The simulation allowed the students to assess the shape of I(x), which turned out to be very different from the curve obtained with small balls. The result was inexplicable from the classical theory and the naïve idea that electrons would behave like small balls.

Even though some students were unable to identify the distribution of the interference pattern observed in experiments with mechanical waves, in general, they were disturbed by the results of the simulation. This created the need to seek an explanation of the unexpected behavior of electrons. The group accepted another key principle in the sequence:

> Although the electrons arrive in discrete units when both slits are open, the resulting curve cannot be explained as if the electrons were small balls.

The distribution of electrons on the screen did not follow a pattern that could be produced by the separate contribution of particles emerging from each slit. Therefore, the students were convinced that it is inadequate to consider the electrons as particles, at least in a classical sense. This new way of considering the electrons drove us to introduce the concept of "quantum system".

7.2 *Analysis and Application of the Sum All Alternatives* (SAA) *Method for Free Electrons*

We started by declaring that there exists a set of laws that describe objects behavior from macroscopic to atomic scale. They are called generically Quantum Mechanics laws. They predict only the probability of an event. That is to say, given an initial state, what is the probability of arriving at a final state? In the case of the DSE the question would be: what is the probability for an electron to arrive at a given point on the screen having started from the source?

Experimentally, this probability is measured as a ratio between the number of electrons that actually reach the point and the total number of electrons emitted by the source, when the latter is very large. It is with these types of measurements that QM predictions are checked.

We have designed a sequence that emphasizes the probabilistic character of the predictions as a central aspect of the quantum theory. We adopted the Feynman's method for the QM and adapted it to the students' mathematic level, calling it SAA formulation. We replaced complex numbers by two-dimensional vectors. Moreover, integrals were approximated by sums and derivatives by finite increment ratios. The method is presented as follows.

1. Suppose as initial state (I) a particle at $x(t=0)=0$ and as final state (F) the particle at $x(T)=x_f$. We consider here one-dimensional paths for simplicity. Of course there are multiple forms (paths) to connect the initial state I with the final state F; some of them are shown in the following figure with straight sections (the only functions that the software used by students allows modeling).

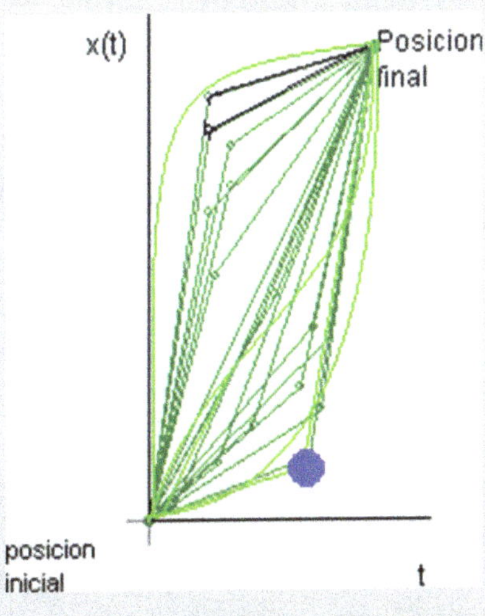

Then, with each possible path $x(t)$ we associate a numerical value called action, represented by "S". The action is the average difference between kinetic E_k and potential E_p energy times T.

$S = <E_k - E_p> T,$

where <> denotes temporal average. If the particle is "free", thus it is not in the presence of forces and E_p=0. Then, in this case the action is simply $S = <E_k> T$, i.e

$S = ½m<v^2> T$

2. With the action S, we construct a unitary two-dimensional vector, forming an angle $S/\hbar$ with respect to the positive x-axis. This vector is called "Probability amplitude"($A(x)$) associated with the path $x(t)$. The denominator of this quotient is $\hbar = h/2p$, where $h = 6.625x10^{-34}\ Js$ is the so called Planck's constant. That is to say:

   Every path $x(t)$ connecting I with F has a corresponding S, which is used to construct the Amplitude of probability vector associated to $x(t)$, whose components are:

   $\cos\frac{S}{\hbar}; \sin\frac{S}{\hbar}$
3. All amplitude of probability vectors associated to different paths connecting I with F are added. We call the resulting vector "total probability amplitude"($A_{tot}\ (x)$)
4. The square module of total probability amplitude gives the probability of arriving at final state F, having started at initial state I.

In the DSE particles can be considered as free (except on screens). We can also suppose they are sent at time intervals as long as there is no interaction with each other. The analysis of the free particle allows: (a) to validate the technique, (b) predict the distribution pattern on the screen, obtained in the first simulation.

To help the students apply the technique SAA to the free electron, a simulation using Modellus[3] was developed. Figure 5.1 shows an output screen of this simulation.

The use of the simulation software Modellus made it possible to evaluate the contributions of different paths to the probability amplitude. It allowed students to formulate the following conclusions:

- The classical path x_{class}(t) (a straight line from I to F) has the least action S.
- For atomic masses (e.g. electron mass), the angles of the amplitude vectors associated with those paths $x(t)$ near the classical path x_{class}(t) are very similar. However, the angles of the vectors associated to paths $x(t)$ which are

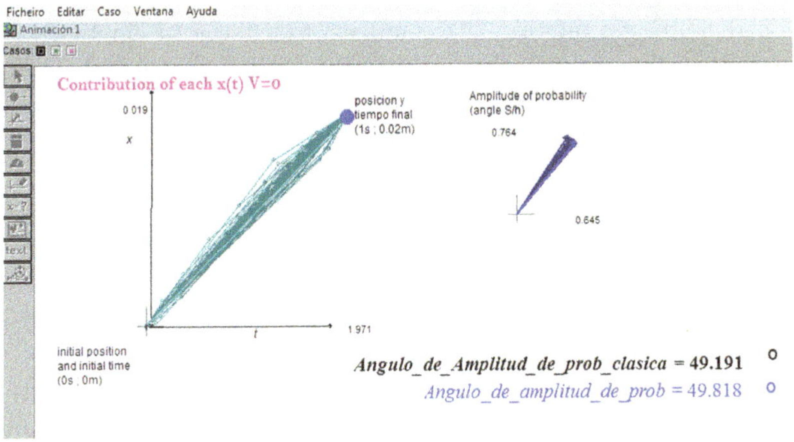

FIGURE 5.1 Selecting different functions $x(t)$ that connect initial and final states, the simulation shows the angles on the Cartesian plane and the angle value of this vector in sexagesimal degrees. The probability amplitude vectors are drawn simultaneously for each function $x(t)$ selected.

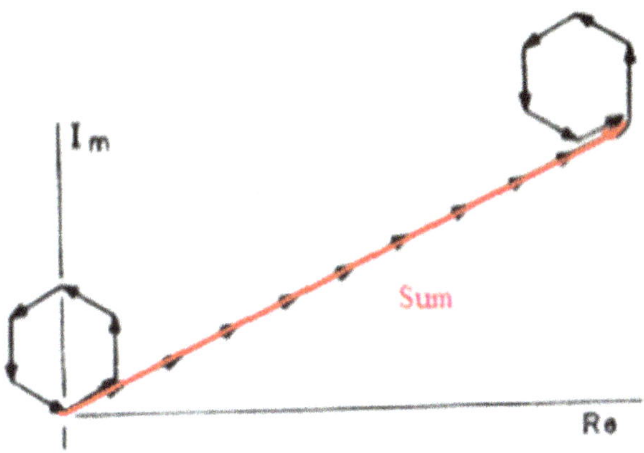

FIGURE 5.2 Schematic representation of the sum of amplitude of a finite set of $x(t)$ possible connecting the initial and final points in the case V=0. Each vector is unitary with an angle $S/\hbar$, and individual corresponds to the amplitude associated with one $x(t)$ of connecting the initial and final points.

far from the classical path are different from each other. This means that only a set of paths "around" the classical path contributes to the sum. The vectors associated to the paths that are far from the classical one have very different directions. They cancel each other in the sum. At this point it was emphasized that this is due to the fact that the electron is free, and that, in general, in a quantum context all paths contribute to the sum. Figure 5.2 is a schematic representation of the sum for V=0.

- As particle masses increase, the contribution to the total amplitude is reduced to paths extremely close to the classical path $x_{class}(t)$. In the limiting case of a macroscopic object only the classical path contributes to the sum. In this case QM predicts the same results as classical Physics, i.e., there is only one trajectory, the one of least action.
- The transition from quantum to classical behavior can be understood in terms of the small value of Planck's constant in a macroscopic context.

7.3 *Applying the SAA Method to Reconstruct the Interference Diagram with Electrons*

As mentioned in the previous section, in the case of the free particle, the classical trajectory plays a central role (even at the quantum level). In fact, it can be shown that the sum over all paths can be performed analytically in this case (Sankar, 1980).

Applying these analytical results to the DSE, the probability for a particle of mass m to arrive at a distance x of second screen center, starting from the source, yields the following result (Arlego, 2008):

$$P(x)\infty 4cos^2\left(\frac{md}{2\hbar T}x\right)$$

where d is the distance between screens. Of course we immediately recognize in the previous formula the interference phenomenon.

The derivation of this formula for $P(x)$ was made on the basis of the mathematical level of students. However, it was emphasized that it is a direct consequence of the quantum mechanics laws presented and the special role that classical paths play in the case of free particles, as they observed in the previous simulation.

The students discussed and analyzed in groups the functional form of the expression $P(x)$ above. Using this expression with typical values of the variables provided by the teacher, they made approximate graphical representations of $P(x)$, and located maxima and minima. As a result of this analysis students recognized that these graphs adopt a similar form to graphs representing the distribution of electrons obtained in the first simulation.

7.4 *Classic-Quantum Transition in the DSE*

A simulation with Modellus was generated to show that the ratio between the mass and Planck's constant generates, or not, the interference diagram. Fixing the rest of the parameters, it was observed how every larger value of the mass affected the $P(x)$ curve. The software also draws the associated vector to each

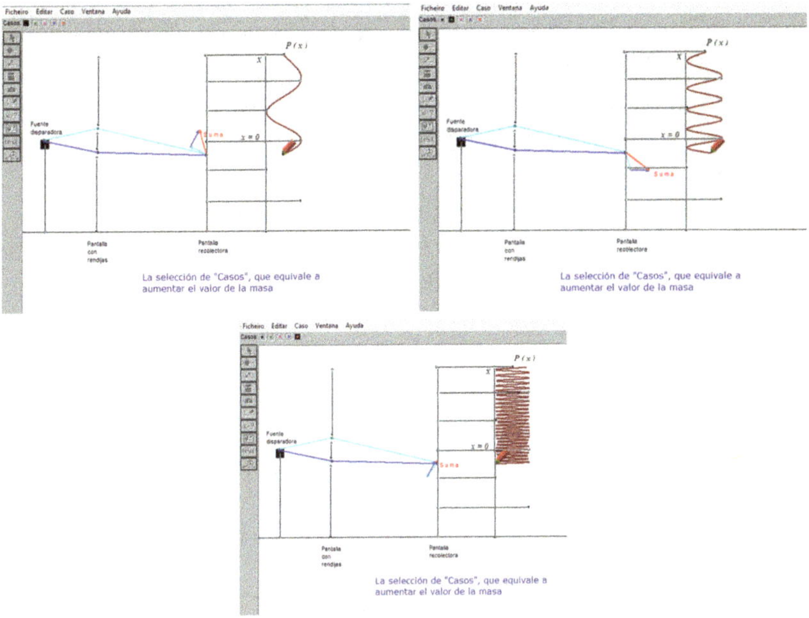

FIGURE 5.3 Screens showing double-slit experience simulations (obtained with the software Modellus)

alternative – starting with one slit or the other – the extreme vector and the curve. In Figure 5.3 we can see the interference diagram disappearing when the mass increases, making evident the transition between the QM and the classical mechanics.

Coming back to the DSE phenomenon and according to the electrons arriving at the screen one at a time, the students analyzed the results of the DSE obtained by Tonomura in 1974. They looked at a series of successive photographs of a collector screen.

From previous observations the students identified a wavelength (the distance between successive maxima) to be dependent on the ratio $h/(mv)$, where $v \approx d/T$ is necessary in order to give the correct units. Now, this wavelength depends only on the properties of the particle, so it makes sense to associate this wavelength to the particle itself. In this way we arrive at the concept of wavelength λ associated to the particle. It is called De Broglie wavelength, in honor to its discoverer and it is given by

$$\lambda = h/p,$$

where $p = mv$ is the momentum of the particle. From this formula it is clear that it is the ratio between m (more precisely, p) and h that determines the quantum or classical behavior of the system. For macroscopic objects λ it is so small that ondulatory characteristics are imperceptible. In turn, for electrons, for instance, where p is of the order of h, λ is large enough to exhibit interference effects. Matter behavior is actually well described by the Quantum Theory.

Immediately the question was put to the students why quantum interference is not detected if the experiment is realized with small balls. The students were invited to analyze the relationship between the associated wavelength and the interference diagram. Why it does not happen with the small balls while it is possible to detect it with electrons. In this case of the balls, the quotient between Planck's constant and the mass is extremely small, due to the value of h; therefore, the associated wavelength is too small, and the maximums and minimums on the curve $P(x)$ are indistinguishable, obtaining an average curve similar to the classical curve. The sequence finished by analyzing the role of Planck's constant as a fundamental constant in nature to establish if the quantum behavior is evident or not.

8 Obtained Results and Comments

The main tasks entailed in the PCST design were designing teaching situations, anticipating possible questions and answers, selecting the available software, creating simulations to visualize the SAA technique and the effects of mass increasing that were simulated with Modellus. We have described and analyzed the conceptualization and its related affective aspects. It has been a very complex process to reduce and manage the knowledge of Physics in this conceptual field to make it teachable at school. It was complicated to decide which concepts and principles should be studied and how a PCST should be designed, carried out and adjusted. We consider the PCST outlined as just the beginning to discuss, modify and talk to physicists, Physics teaching researchers, and teachers. Without consultation with these three groups of actors, it would be impossible to bring knowledge alive and bridge the gap between the school and the scientific community.

The OI like theorems-in-action, the inferences and the whole activity that were identified are described in Chapter 6 of this book. "Conceptualization" is a long-term process for it does not finish in the years of schooling. The students found out that the electrons had a special and characteristic behavior that allows us to think about them as quantum systems. Most of the students were unable to accept the impossibility of knowing which function would describe the electron movement. After the sequence, the students still thought: "*Finally, the electron must take some path or other*". The students agreed that the SAA technique was a suitable mechanism to explain the interference pattern in the DSE, which in other ways would remain inexplicable. Furthermore, they understood that the wave behavior allowed to associate a wavelength not only to the macroscopic particles but also to the microscopic ones. The students related the shape and detection of the interference pattern in the macroscopic and microscopic particle cases, giving a new meaning to Planck's constant. They understood its role in the quantum-classic limit.

The sequence implementation demanded a great effort of the teacher and students alike. We analyzed the affective aspects in two moments: during the classes and at the end of the sequence. We used as data source the protocols of the students situation by situation and the replies in a test, respectively. The sequence has been carried out according to the predicted steps in the estimated time. The emotional dynamic of coexistence in the CG is a necessary condition for quantum concepts construction. It requires the students to make the necessary cognitive effort. They expressed on many occasions they had made an intense but possible effort. The students were not surpassed by the proposed situations and they accepted the challenges.

On the one hand, a cognitive effort was required since the students had to relate the new concepts with the previous ones, and if it was necessary, to change their usual ideas. On the other hand, an affective effort had to be carried out, feeling well while facing challenges and questions, and accepting that the usual ideas are wrong. Moreover, the teacher had to do a big effort, making room to the students, without taking their responsibility in learning, allowing them to make mistakes and reconsider their ideas, accepting students' ideas, and waiting for students to solve problems.

Regarding software aspects, students recognized the advantages to support the understanding of new concepts, but they also emphasized the effort required, since that was a new tool for them. Although the designed tools tried to lighten certain difficult aspects of calculations, they did not suppose a passive use. They represented an indispensable part of situation conceptualization. Continuing with the topics discussed here, we plan to address the quantum aspects of light, in particular the emergence of the concept of photon.

Notes

1 We use the word 'didactic' based upon the theoretical framework provided by Chevallard's anthropological theory of didactics. Just as there are physical facts, there are didactic facts. In a slightly simplified way, one can say that they are made

up of the motley host of social situations in which some person does something – or even manifests an intention to do so – in order that some person may "study" – and "learn" – something (Chevallard, 2012).

2 "Doppelspalversuch" (2003). By Muthsam, K (Version 3.3, translated to Spanish by Wolfamann y Brickmann) Physics Education Research Group of the University of Munich. http://www.didaktik.physik.uni-muenchen.de/materialien/inhalt_materialien/doppelspalt/index.html

3 MODELLUS™ versión 2.5 Developed by Victor Duarte Teodoro, Joao Paulo Duque Viera; Filipe Costa Clérigo Faculty of Sciences and Technology Nova University, Lisbon, Portugal. The simulations crated are available in https://sites.google.com/site/mariaangelesfanaro/simulaciones-con-modellus

References

Arlego, M. (2008). Los fundamentos de la mecánica cuántica en la escuela secundaria utilizando el concepto de integral de camino [The foundations of quantum mechanics in secondary school using the Paths Integral method]. *Revista Electrónica de Investigación en Educación en Ciencias, 3*(1), 59–66.

Cabral de Paulo, I. J., & Moreira, M. A. (2005). Um estudo sobre a captação do significado do conceito de dualidade onda-partícula por alunos do ensino médio [A study on appropriation of the meaning of the wave-particle duality concept by high school students]. *Enseñanza de las Ciencias, Special issue*, 1–5.

Chevallard, Y. (1992). Concepts fondamentaux de la didactique: Perspectives apportées par une approche anthropologique [Basic concepts of didactics: Perspectives offered by an anthropological approach]. *Recherches en Didactique des Mathématiques, 12*(1), 73–112.

Chevallard, Y. (1997). Familière et problématique, la figure du professeur [Familiar and problematic, the figure of the teacher]. *Recherches en Didactique des Mathématiques, 17*(3), 17–54.

Chevallard, Y. (1999). L'analyse des practiques enseignantes en théorie anthropologique du didactique [The analysis of teaching practices in anthropological theory of didatics]. *Recherches en Didactique des Mathématiques, 19*(2), 221–265.

Cuppari, A., Rinaudo, G., Robutti, O., & Violino, P. (1997). Gradual introduction of some aspects of quantum mechanics in a high school curriculum. *Physics Education, 32*, 302–308.

Damasio, A. (2005). *En busca de Spinoza: Neurobiología de la emoción y los sentimientos* [Looking for Spinoza: Joy, Sorrow, and the Feeling Brain]. Barcelona: Editorial Crítica.

Fanaro, M. (2009). *La enseñanza de la mecánica cuántica en la escuela media* [The teaching of Quantum Mechanics in high school] (Doctoral dissertation). Programa Internacional de Doctorado en Enseñanza de las Ciencias (UBU-UFRGS), Burgos.

Fanaro, M., Arlego, M., & Otero M. R. (2006). Los caminos múltiples de Feynman y la mecánica cuántica en la escuela media [The multiple paths of Feynman and Quantum Mechanics in high school]. In *Actas del 8º. Simposio de Investigación en Educación en Física* (pp. 161–169). Buenos Aires: Asociación de Profesores de Física de la Argentina (APFA).

Fanaro, M., Arlego, M., & Otero, M. R. (2007). El método de caminos múltiples de Feynman para enseñar los conceptos fundamentales de la Mecánica Quántica en la escuela secundaria [Feynman's multiple paths method to teach fundamental concepts of Quantum Mechanics in secondary school]. *Caderno Catarinense de Ensino de Física, 22*, 233–260.

Fanaro, M., & Otero, M. R. (2008). Basics Quantum Mechanics teaching in secondary school: One conceptual structure based on Paths Integrals method. *Latin-American Journal of Physics Education, 2*(2), 103–112.

Fanaro, M., & Otero, M. R. (2009). Teoremas en acto y situaciones de Mecánica Cuántica en la Escuela Media [Theorems-in-action and Quantum Mechanics situations in high school]. *Latin-American Journal of Physics Education, 2*(2), 24–33.

Fanaro, M., Otero, M. R., & Arlego, M. A. (2009). Teaching the foundations of quantum mechanics in secondary school: A proposed conceptual structure. *Investigações em Ensino de Ciências, 14*(1), 37–64.

Feynman, R., & Hibbs, A. (1965). *Quantum mechanics and path integrals*. New York, NY: McGraw-Hill.

Fischler, H., & Lichtfeldt, M. (1992). Modern physics and students' conceptions. *International Journal of Science Education, 14*(2), 181–190.

Freire Jr., O., de Carvalho Neto, R. A., Rocha, J. F. M., Vasconcelos, M. J. L., Socorro, M., & Anjos, E. L. (1995). *Introducing quantum physics in secondary school*. Salvador: Instituto de Física/UFBA.

Garcia, R. (2000). *El Conocimiento en construcción, de las formulaciones de Jean Piaget a la Teoría de Sistemas Complejos*. Barcelona: Editorial Gedisa.

González, E., Fernández, P., & Solbes, J. (2000). Dificultades de docentes de ciencia en la conceptualización de temas de física actual [Science teachers' difficulties in conceptualizing contemporary physics themes]. *Actas del 5º. Simposio de Investigación en Educación en Física* (pp. 138–147). Buenos Aires: Asociación de Profesores de Física de la Argentina (APFA).

Greca, I. (2000). *Construindo significados em Mecânica Quântica: Resultados de uma proposta didática aplicada a estudantes de física geral* (Doctoral dissertation). Doutorado em Física/UFRGS, Porto Alegre.

Greca, I., & Moreira, M. A. (2004). Obstáculos representacionales mentales en el aprendizaje de conceptos cuánticos [Mental representational obstacles in learning quantum concepts]. In M. A. Moreira (Ed.), *Sobre el cambio conceptual, obstáculos representacionales, modelos mentales, esquemas de asimilación y campos conceptuales* (pp. 26–40) Porto Alegre: Instituto de Física/UFRGS.

Greca, I., Moreira, M. A., & Herscovitz, V. (2001). Uma proposta para o ensino de Mecânica Quântica [A proposal for teaching Quantum Mechanics]. *Revista Brasileira de Ensino de Física, 23*(4), 444–457.

Hanc, J., & Tuleja, S. (2005). The Feynman Quantum Mechanics with the help of Java applets and physlets in Slovakia. In *Actas del 10th workshop on multimedia in physics. Teaching and learning*. Berlin. Retrieved from http://pen.physik.uni-kl.de/w_jodl/MPTL/MPTL10/contributions/hanc/Hanc-Tuleja.pdf

Inhelder, B., & Piaget, J. (1955/1976). De la lógica del niño a la lógica del adolescente [The growth of logical thinking from childhood to adolescence]. Madrid: Paidos.

Lobato, T., & Greca, I. (2005). Análise da inserção de conteúdos de teoria quântica nos currículos de física do ensino médio [Analysis of the introduction of quantum theory concepts in high school Physics curricula]. *Ciencia & Educação, 11*(1), 119–132.

Maturana, H. R. (1991). *El Sentido de lo Humano*. España: Editorial Dolmen.

Maturana, H. R. (1995). *La realidad ¿objetiva o construida? I y II. Fundamentos biológicos de la realidad* [Reality: objective or constructed? I and II. Biological foundations of reality]. México: Ed. Anthoropos/Universidad Iberoamericano/Iteso.

Maturana, H. R. (2001). *Cognição, ciência e vida cotidiana* [Cognition, science, and everyday life]. Belo Horizonte: UFMG.

Montenegro, R. L., & Pessoa Jr., O. (2002). Interpretações da teoria quântica e as concepções dos alunos do curso de física. *Investigações em Ensino de Ciências, 7*(2), 107–126.

Moreira, M. A. (2002). Teoria dos campos conceituais de Vergnaud, o ensino de ciências e a pesquisa nesta área [Vergnaud's conceptual fields theory, science teaching, and research on this area]. *Investigações em Ensino de Ciências, 7*(1), 7–30.

Moreira, M. A. (2005). *Aprendizagem significativa crítica* [Critical meaningful learning]. Porto Alegre: Instituto de Física/UFRGS. Retrieved from http://www.if.ufrgs.br/~moreira/apsigcritport.pdf

Moreira, M. A., & Greca, I. (2000). Introdução a Mecânica Quântica: seria o caso de evitar a aprendizagem significativa (subordinada)? [Introduction to Quantum Mechanics: would it be the case of avoiding (subordinated) meaningful learning?]. In *Atas do III Encontro Internacional sobre Aprendizagem Significativa*. Peniche, Portugal.

Osterman, F., & Moreira, M. A. (2000). Uma revisão bibliográfica sobre a área de pesquisa "física moderna e contemporânea no ensino médio" [A bibliographical review about the area of research "modern and contemporary Physics in high school"]. *Investigações em Ensino de Ciências, 5*(1), 23–48.

Osterman, F., Prado, S., & Ricci, T. (2006). Desenvolvimento de um software para o ensino de fundamentos de Física Quântica [Development of a software to teach foundatins of Quantum Physics]. *A Física na Escola, 7*(1), 22–25.

Osterman, F., Prado, S., & Ricci, T. (2008). Investigando a aprendizagem de professores de Física acerca do fenômeno da interferência quântica [Investigating Physics teachers' learning about the quantum interference phenomenon]. *Ciência & Educação, 14*(1), 35–54.

Osterman, F., & Ricci, T. (2004). Construindo uma unidade didática conceitual sobre mecânica quântica: Um estudo na formação de professores de Física [Building a conceptual didactic unit about Quantum Mechanics: A study in Physics teacher education]. *Ciência & Educação, 10*(2), 235–257.

Otero, M. R. (2006). Emociones, sentimientos y razonamientos en didáctica de las ciencias [Emotions, feeling and reasonings in the didactics of sciences]. *Revista Electrónica de Investigación en Educación en Ciencias, 1*(1), 24–53.

Otero, M. R., Fanaro, M., Sureda, P., Llanos, V. C., & Arlego, M. (2014). *La Teoría de los Campos Conceptuales y la conceptualización en el aula de Matemática y Física.* Buenos Aires: Editorial Dunken.

Pessoa Jr., O. (1997). Interferometria, interpretação e intuição: Uma introdução conceitual à Física Quântica [Interferometry, interpretation and intuition: A conceptual introduction to Quantum Physics]. *Revista Brasileira de Ensino de Física, 19*(1), 27–47.

Pinto, A. C., & Zanetic, J. (1999). É possível levar a Física Quântica para o ensino médio? [Is it possible to bring Quantum Physics to high school?]. *Caderno Catarinense de Ensino de Física, 16*(1), 7–34.

Shankar, R. (1980). *Quantum mechanics.* New York, NY: Plenum Press.

Strnad, J. (1981). Pitfalls in the teaching of introductory quantum physics. *European Journal of Physics, 2*, 250–254.

Taylor, E. (2003). A call to action. *American Journal of Physics, 71*(5), 423–425.

Taylor, E. F., Vokos, S., & O'Meara, J. M. (1998). Teaching Feynman's sum-over-paths quantum theory. *Computers in Physics, 12*(2), 190–199.

Vergnaud, G. (1990). La théorie des champs conceptuels [The theory of conceptual fields]. *Recherches en Didactique des Mathématiques, 10*(2–3), 133–170.

Vergnaud, G. (1994). Le rôle de l'enseignant à la lumière des concepts de schème et de champ conceptuel [The role of the teacher in the light of the concepts of schema and conceptual field]. In M. Artigue, R. Gras, C. Laborde, & P. Tavignot (Eds.), *Vingt ans de Didactique des Mathématiques en France.* Grenoble: La Pensée Sauvage.

Vergnaud, G. (2000). *Lev Vigotsky, pédagogue et penseur de notre temps* [Lev Vygotsky, pedagogue and thinker of our times]. Paris: Hachette.

Vergnaud, G. (2013). Pourquoi la théorie des champs conceptuels? [Why the theory of conceptual fields ?]. *Infancia y Aprendizaje, 36*(2), 131–161.

Vokos, S., Shaffer, P. S., Ambrose, B. S., & McDermott, L. C. (2000). Student understanding of the wave nature of matter: Diffraction and interference of particles. *American Journal of Physics, 68*(7), S42–S51.

CHAPTER 6

Teaching the Foundations of Quantum Mechanics in High School: A Didactic and Cognitive Analysis of a Sequence of Situations

Maria de los Angeles Fanaro and María Rita Otero

Abstract

This chapter deals with the problem of teaching the fundamental concepts (QM) in high school. Many investigations in this area recognize the importance of the treatment of the quantum concepts at secondary school. Nevertheless, the usual way of teaching QM follows a strictly historical line. This prevents from approaching QM's fundamental aspects.First we ask which approximation to the "quantum world" is possible to teach at school. We have conceived a conceptual structure of reference related to the viewpoint of the Quantum Mechanics of Feynman "Path Integrals", which is alternative and complementary to the canonical method. Our design allows to avoid the strictly historical and traditional development that is usually adopted in QM teaching. We begin by the Classical Physics – using concepts familiar to the students- and we analyze the limit QM-classical. Thus, the ways of teaching the concept of quantum system and the Principles of Superposition and Correspondence are studied. Using a geometric-vector frame the mathematical formulation of the Path Integral is adapted to the student's mathematical knowledge. This sample allows the emergence of student's ideas: electrons like "small balls". Moreover, it shows how the concept of quantum system associated to the Path Integrals technique explains the probability curve of the electrons. A previous didactic analysis was made to anticipate as much as possible the actions of the students and the teacher. We have implemented the didactic sequence, and the results related to the concepts reached by the students, situation by situation, are presented here.

Keywords

Quantum Mechanics – high school situations – cognitive analysis

 | DOI: 10.1163/9789004409088_006

1 Introduction

This chapter deals with the problem of the teaching of the fundamental concepts of Quantum Mechanics (QM) in high school. Many investigations in this area recognize the importance of the treatment of the quantum concepts at this educational level (Cabral de Paulo & Moreira, 2004; Cuppari et al., 1997; Fischler & Lichtfeldt, 1992; González, Fernández, & Solbes, 2000; Greca, Moreira, & Herscovitz, 2001; Hanc & Tuleja, 2005; Montenegro & Pessoa, 2002; Moreira & Greca, 2000; Müller & Wiesner, 2002; Niedderer, 1996; Olsen, 2002; Osterman & Moreira, 2000; Ostermann & Ricci, 2004; Pessoa Jr., 1997; Pinto & Zanetic, 1999; Taylor et al., 1998; Taylor, 2003; Zollman, 1999; Lobato & Greca, 2005). Nevertheless, the usual way of teaching QM follows a strictly historical line. This prevents from approaching QM's fundamental aspects.

First, we ask which approximation to the "quantum world" is possible to teach at high school. We have conceived a conceptual structure of reference related to the viewpoint of the Quantum Mechanics of Feynman's "Path Integrals", which is alternative and complementary to the canonical method (developed by Dirac, Von Neumann, Schrödinger). Our design allows us to avoid the strictly historical and traditional development that is usually adopted in QM teaching. We begin by the Classical Physics – using concepts familiar to the students – and then we analyze the limit between the Quantum and Classical Mechanics. Thus, the ways of teaching the concept of quantum system and the Principles of Superposition and Correspondence are studied. Using a geometric-vector frame, the mathematical formulation of the Path Integral is adapted to the students' mathematical knowledge. This case allows the emergence of students' ideas[1]: electrons like "the small balls". Moreover, it shows how the concept of quantum system associated to the Path Integrals technique explains the probability curve of the electrons.

A previous didactic analysis was made to anticipate as much as possible the actions of the students and the teacher. We have implemented the didactic sequence and the results related to the concepts reached by the students, situation by situation, are presented here.

2 Theoretical Framework

In Physics, there are a lot of conceptual fields (Vergnaud, 1990) in which at least one Conceptual Structure of Reference (CSR) can be distinguished and recognized (Otero, 2006, 2007, 2008). When a Physics teacher invites her students to study a specific conceptual field, she adopts more or less explicitly a particular Conceptual Structure of Reference (CSR). A CSR is a set of concepts, the relationships among them, the principles, knowledge claims and the explanations relative to a conceptual field accepted by the scientific community of reference.

This investigation rebuilds a CSR based on Feynman's Paths Integral method (Feynman & Hibbs, 1965). A detailed analysis of this CSR can be consulted in Arlego (2008). The full proposal adapting a conceptual organization for students at the high school level can be found in Fanaro and Otero (2008) and Fanaro, Arlego, and Otero (2007). The CSR adopted should be partially or fully reconstructed by a class group, or by someone who tries to study it in high school, or in the basic or advanced courses in the university.

Any attempt at reconstruction originates a different conceptual structure, as much for the components as for the relationships among them. In a more or less explicit way, each teacher of a certain group will reconstruct or select – based on an existing structure – one conceptual structure to be taught, and, in the best of cases, she will invite her class to study it. We name this other structure Proposed Conceptual Structure for Teaching (PCST) (Otero, 2006, 2007, 2008). It is a set of concepts, the relationships among them, knowledge claims, principles, situations and explanations related to a certain conceptual field, which the teacher must reconstruct based on a CSR. The teacher aims at transforming the scientific knowledge and reconstructing it in a certain context in a given Institution (Otero, 2006, 2007, 2008).

The PCST design is a complex process that requires multiple actions and decisions related to the concepts and principles that could be studied. Which questions and problems would be more suitable for the conceptualization? What kind of inferences could the students make? What actions and activities we expect our students to do? The PCST design involves:
- Analyzing and selecting the key concepts of the conceptual field that will be reconstructed in the class group (CG);
- Creating the appropriate situations to use the software simulation.
- Choosing the suitable parameters to avoid actions that could mislead the study.
- Two simulations using *Modellus*™ are specially developed for the didactic sequence.[2]

The possible results for each simulation are analyzed beforehand and the actions of the students are anticipated.

The teacher and her class group will rebuild the PCST in a certain and specific institution, generating the Conceptual Structure Effective Reconstructed (CSER). A CSER is the set of concepts, relationships among them, principles, knowledge claims relative to a certain conceptual field that are reconstructed by the class group. The teacher and the students interact in conversations characterized by an emotional dynamic of coexistence.

Every member of the class group constructs a personal conceptual structure and a unique network of meanings – personal and private. Simultaneously, the class group conversations drive the meaning network construction, which is shared and made public. This meaning network is a consensual product. Its negotiation process can be both more or less explicit and self-conscious, depending on the professional practice of the teacher, and the distance between the CSR, the PCST and the CSER.

We adopt Vergnaud's ideas about concepts and conceptualization. We are interested in the students' activity. Activity must not to be confused with behavior, which is only the visible part of the activity. The most important aspect of knowledge is implicit; it cannot be easily verbalized. The implicit character of the activity does not mean that knowledge is not operational. Both the operational (which allows to act in a situation) and the predicative form of knowledge (which consists of enunciating the relationships among the objects) are important. The didactic sequence involves a set of situations where the students' ideas are accepted at the beginning, and afterwards they are unbalanced to reach a new balance state.

3 Methodological Aspects

The PCST was designed for a Physics course of the last year of high school of the Tandil city (Southeast of Buenos Aires Province, in Argentina). The group had thirty (30) 17–18 year-old students. It was a hard working group. The curriculum establishes two one-hour periods of Physics a week. The students had the required physical and mathematical knowledge: Classical mechanics, vectors, and trigonometrical functions. The habitual class work style of these students – who had been working in groups – was maintained.

The teaching sequence had thirteen lessons. The material was given period by period, regulating appropriately the new features and problem introduction. The conversations in each work groups were recorded in audio.

We made a previous analysis of each didactic situation. We considered the key questions arising in the situation, the required actions of the students, the key concepts, and the theorems-in-action and concepts-in-action that the students might use. After the implementation, the students' activity was analyzed in order to describe their learning. When analyzing the students' activity, the selection of information, the operative invariants, the control mechanisms, and the inferences had to be considered.

4 Analysis and Results

4.1 *Situation 1 – Imagining the Double Slit Experience (DSE)*

The students had to imagine and anticipate the results of DSE using small balls as projectiles. They had to predict the distribution of the small balls in the wooden wall and the distribution of frequencies depending on the distance to the center. The DSE was presented as follows:

Imagine an experience as shown schematically in the following figure.

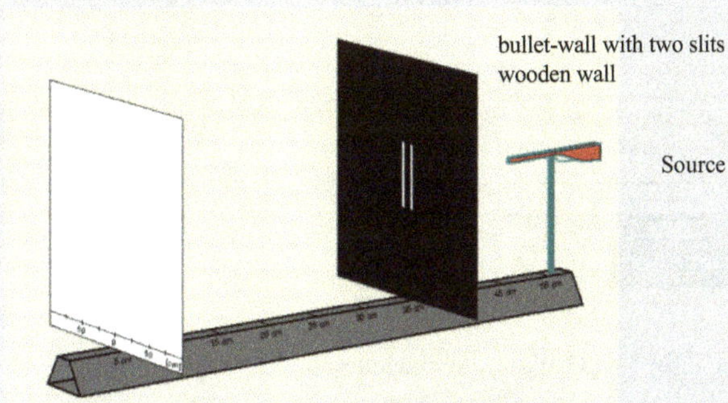

FIGURE 6.1 Scheme of the experimental disposition of the DSE with small balls

On the right, a source that shoots small balls is represented. It shoots off at random inside an angle big enough because, let's suppose, the source is not well fastened, and moves in all directions. There are two very important considerations:

1. The small balls that go out of the source are indestructible, and at the same time they come in entire units to the wooden wall.
2. They are shot one by one, at equal intervals of time, and with the same quickness.

15 cm from the source, there is a bullet-wall with two slits through which the small balls pass non stop. Let's suppose that the slits have a width of 10 mm and both slits have a separation of 10 mm. On the left, there is a wooden wall in which the small balls remain incrusted. This wooden wall has in its base a scale perpendicular to the gray ruler, in which the 0 represents the center of the screen, in that direction.

1. a) How will the small balls that manage to pass through the slits and come to the wooden wall be distributed? Will there be more balls in some place? Why?

b) Make a sketch of what you think the wooden wall will look like in the following figure:

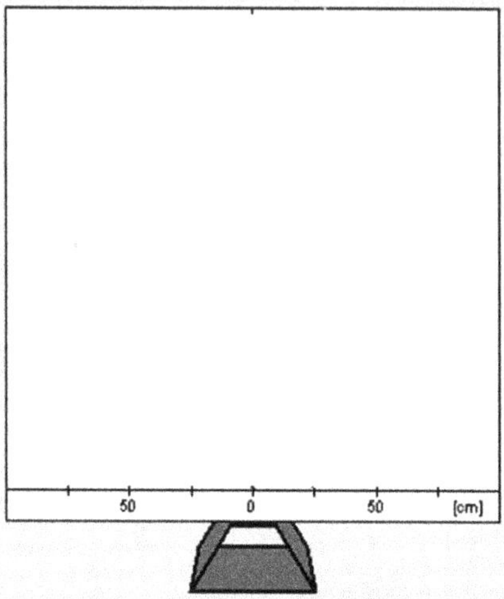

FIGURE 6.2 Distribution of the small balls in the wooden wall

The probabilistic ideas were introduced through the consideration of the following expression:

$$F = \frac{N^{o} \textit{ of balls that impact to a distance "x" to wall center}}{\textit{Total of balls shot } (N)}$$

It was proposed to the students to think in the limit of the frequency, when N is very big. Thus, the expression "curve of probability" was introduced. The students had to make a graph of $P(x)$:

2. Now let's think about the curve of probability according to the distance to the center of the slits. If to a certain distance x with the center o (without considering the vertical direction) many small balls are incrusted, the probability will be high. On the contrary, if there are few small balls, we will say that in that x the probability is low.
   a) Could you draw the approximated shape of the curve?

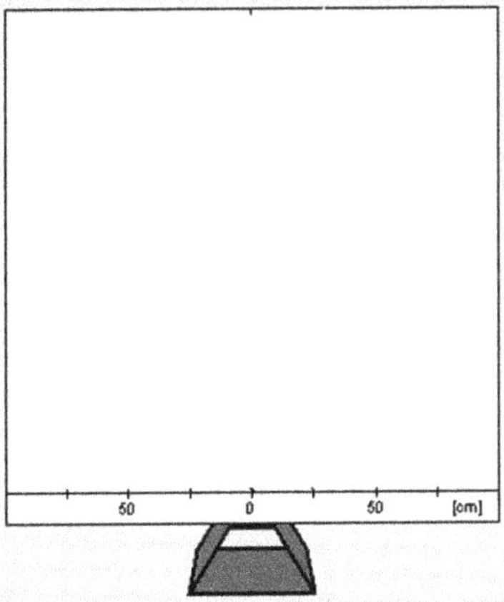

FIGURE 6.3 Probability curve depending on x

b) Describe the shape of the curve.

Aiming at the students' construction of the concept of sum of probabilities, it was proposed to analyze the changes both of the distribution and the curve of $P(x)$ when one or both slits are open:

3. Now let's imagine that we cover one of the slits. Then, we cover this slit and open the other one. Could you draw how the small balls will be distributed and how the curve of probability will change for each case? Now let's imagine that we cover one of the slits and make the experience. Then, we repeat the procedure with the other slit ... Could you draw how the small balls will be distributed in the wooden wall and how the curve of probability will be?

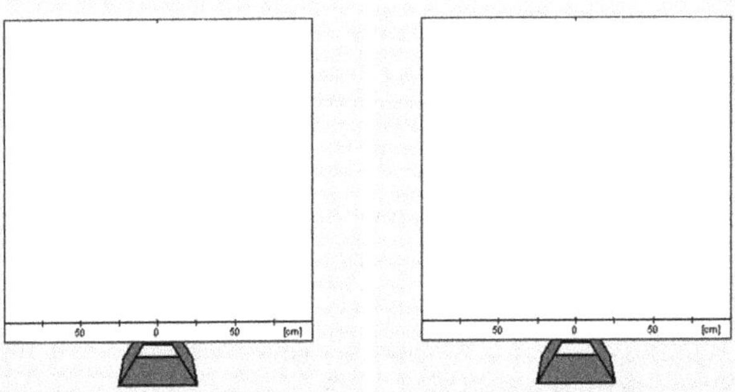

FIGURE 6.4 Blocking R_1. Left side: Distribution of the small balls that came to the screen. Right side: Curve of probability according to x

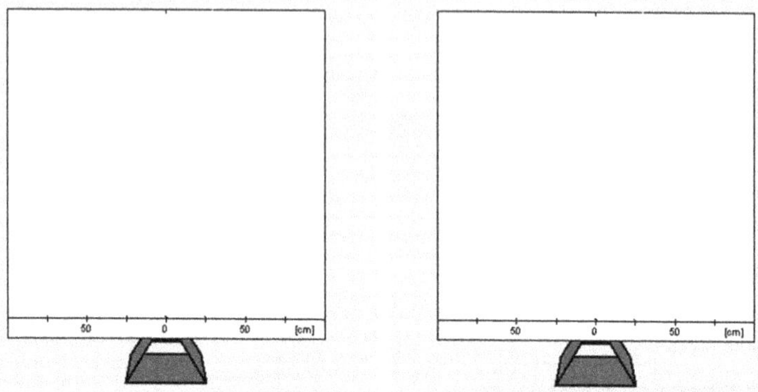

FIGURE 6.5 Blocking R_2. Left side: Distribution of the small balls that came to the screen. Right side: Curve of probability according to x

4. Explain the result of the experience with both open slits, in relation to the result obtained in each of the situations where one covers only one of the slits?

Discussion:
Regarding the students' activity, they selected the information focusing on the expression "random" and they evoked the theorem-in-action T_1 many times. It seems the students ignored that the slits affect the probability. The students had to leave the theorem-in-action T_1 to understand that the slits avoid a uniform distribution of the impacts on the wall. Only without slits a uniform distribution would arise. In addition, they have to understand the probability law and, after that, they will necessarily conclude that this law will be very different when electrons are used.

Some students had not taken into account the influence of randomness. They used the theorem-in-action T_2 at the beginning, tending towards the probabilistic ideas. The students thought that the small balls would be concentrated in certain places of the wall and not distributed over the whole screen. As soon as the students accepted that the slits impose a distribution of probability, they could approximately draw the curve of probability $P(x)$ using the theorems-in-action T_3 or T_4. Finally, we can say that the students were near the sum of probability curves idea in the case of the small balls. It is important to emphasize that the students answered and drew only based on their imagination.

TABLE 6.1 Previous analysis of the situation

Key question	Required actions	Key concepts	Expected theorems-in-action	Expected explanations and inferences
How is the distribution of the small balls that reach the wooden wall?	Imagine the experience and anticipate the results.	Distribution of impacts.	Even if the small balls go off at random, there is a zone where it is more probable to find them; this zone corresponds to the projection of the slits in the wall.	There will be two curves corresponding to every slit.
How is the distribution represented in x-axes?	To draw the impacts of the small balls in the wall.	Curve of probability.		
What is the relation between this distribution and the one obtained by closing each slit at a time?	To graph the curve of probability depending on the center of the wall. To compare the drawn curves when each slit is open at a time and when both are opened simultaneously.	Highest of the curve P (x).		The curve of probability is the overlapping of the individual curves.

TABLE 6.2 Posterior analysis of the situation

Selection of information	Operative invariants used by the students	Mechanisms of control	Inferences
Focusing on the random	T_1: "If the processes are at random, there is no law of probability"		The small balls will be distributed uniformly on the screen.
The fraction N° of small balls that arrived at a distance "x" of the center of the wall/N° of small balls shot in total	T_2: "Although the process begins at random, it follows a law of probability"	Comparing the pictorial external representation, its description and the imagined results	Since there are two slits, the small balls copy the shape in the wooden wall and they form "two columns of concentrated impacts"
The slits are more separated so that the effects are not superimposed in the center.	T_3: The number of highest is proportional to the number of slits	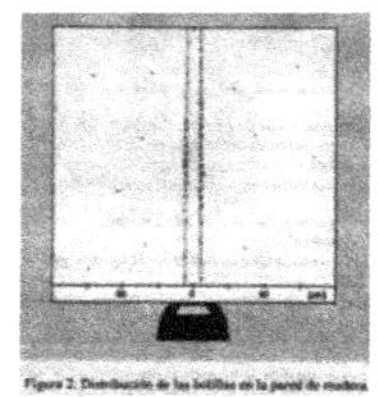	
The slits are very close.	T_4: There is an overlapping of individual effects in the center	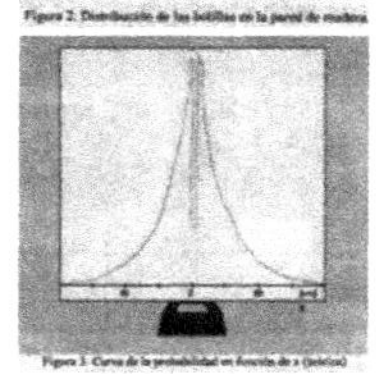	There is a highest of probability in the center

Selection of information	Operative invariants used by the students	Mechanisms of control	Inferences
		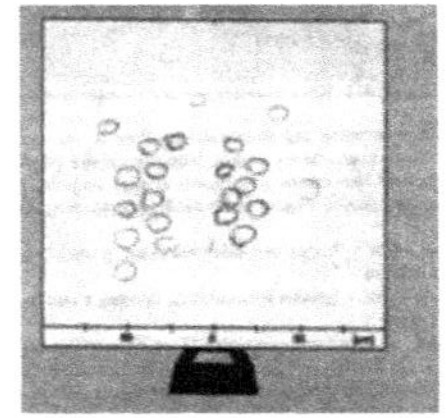 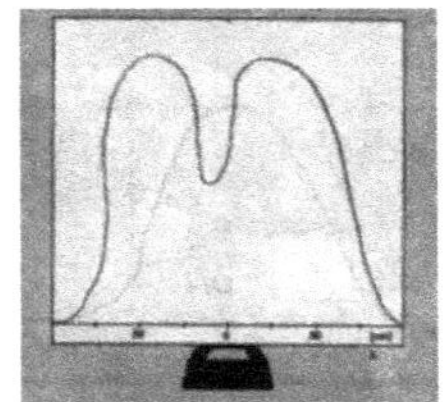 Figura 3: Curva de la probabilidad en función de x (teórica) 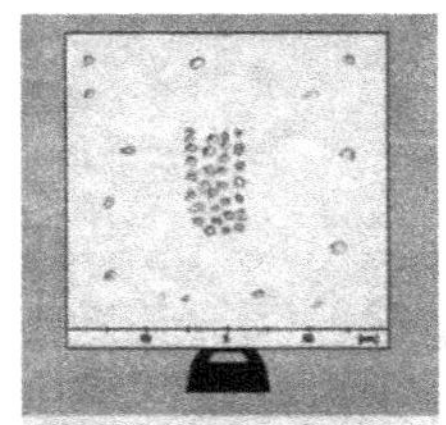 Figura 2: Distribución de las bolillas en la pared de madera	

4.2 *Situation 2 – Simulating the DSE Experiment*

It was proposed to use the software "Doppelspalt"[3] to simulate the DSE. This tool allows to estimate the individual impacts of the projectiles on the screen and the evolution in time, and to generate the histograms of frequencies, and the curve of the frequencies distribution aspect or "probability curve" $P(x)$. Figure 6.6 shows screens of the software with the kinds of outputs produced by it.

When the separation of the slits is the same (or smaller) than the width, the software shows a central highest (Figure 6.7).

A challenge appears to the students when they have to explain the highest in the center of the slits, where the small balls did not pass. Hence, the construction of the idea of sum or overlapping curve can occur.

The students were invited to use the software for simulating DSE with certain experimental conditions, and to answer the following questions:

a) How were your predictions regarding the results shown in the simulation?
b) Draw the shown curve $P(x)$ in your folder.

To analyze the changes in the curve of probability when the width of the slits and their separation are modified, it was proposed to the students to fix the width and to reduce gradually the separation. These questions allow to explain the differences between the graphs drawn by the students and the one shown by the software. We asked the following questions:

a) Describe the impacts distribution shown by the simulation
b) Describe and interpret the shape of the $P(x)$
c) Try to explain the shape of $P(x)$ in $x = 0$

In order to find the relation between the obtained curve when both slits are open and the individual curve, the students were proposed to close one slit at a time:

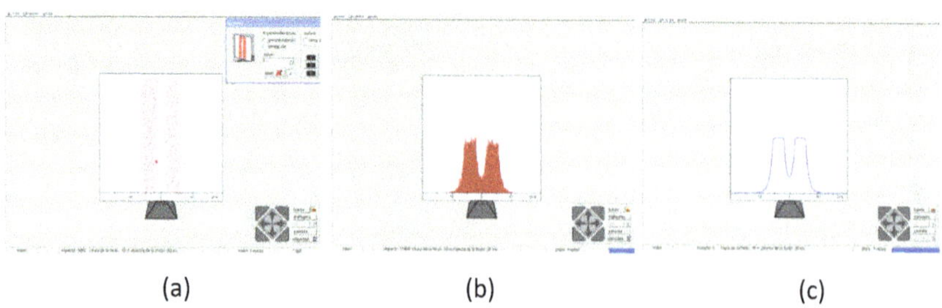

FIGURE 6.6 Simulations of (a) "individual impacts", (b) "histogram" and (c) "theoretical curve" by the software Doppelspalt

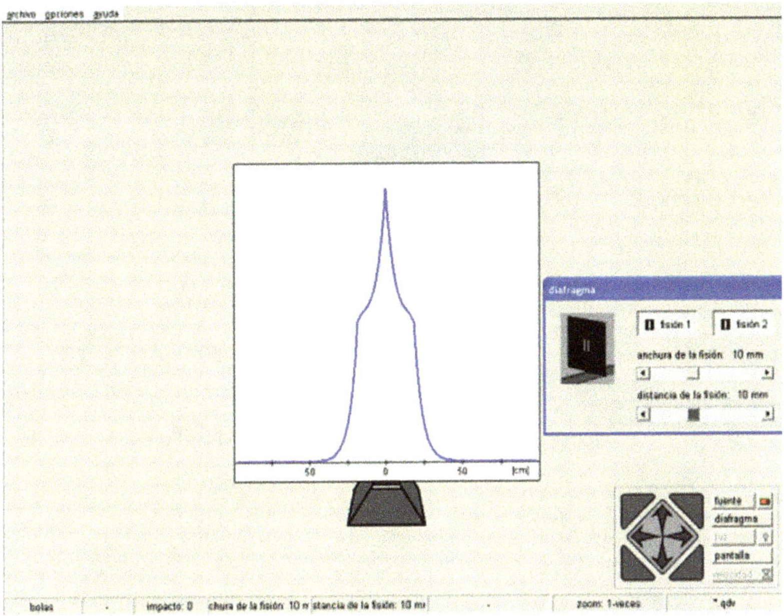

FIGURE 6.7 Curve obtained setting the same width and separation of the slits

a) Draw each curve of probability when each slit is closed separately
b) How might the curve of probability be related when both slits are open with each of the curves separately?

Before simulating the experience with electrons, the following questions were formulated, for allowing the students to express their ideas about the electrons and to anticipate the results that the software will show then:

Are the electrons like little electric balls? Will they behave in the same way?

Then it was proposed to simulate the DSE choosing electrons instead of small balls. The disagreement between the students' prediction and the software results should provoke the need to look for an explanation for the unexpected behavior of the electrons, because the electrons' distribution would not show the corpuscular behavior of the balls. Figure 6.8 represents the simulation for the electrons.

The following tasks were given:

a) Describe the distribution of the electrons on the screen.
b) Reproduce the graph of $P(x)$, and describe its shape.
c) How do you interpret this graph?

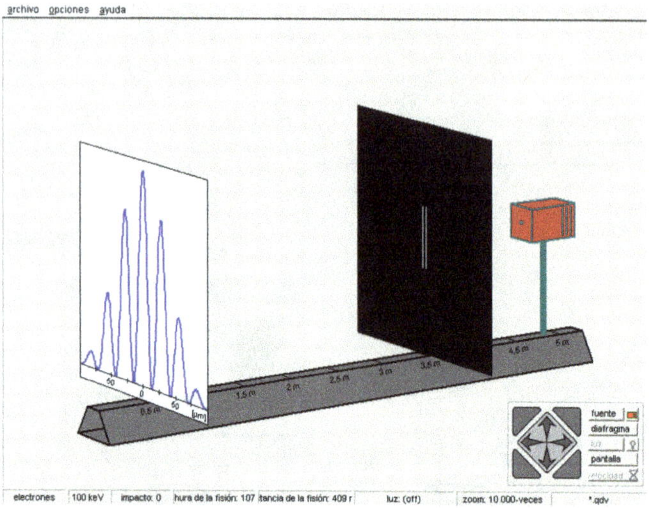

FIGURE 6.8 P(x) for electrons with both slits open

The prearranged values were looked in such a way that the students perceived the difference between the graph of the function $P(x)$ for electrons and for small balls, considering the highest and minimums of the graph as the highest or minimum concentration of electrons.

For the students to estimate that in the case of the electrons the curve is not the sum of individual curves, it was proposed to close one slit to notice its shape, and to compare it with the one obtained having both slits opened. The given tasks were:

a) Describe the distribution of the electrons on the screen.
b) How are the curves of probability in each case? Draw the curves.
c) Do these curves have any relation with the curves obtained with the small balls?
d) Discuss this statement: With the electrons the curve of probability when both slits are open is the sum of each curve of probability separately.

Discussion:
The students who used T_6 also predicted the uniformity of the distribution. When they saw the simulation, they were confused because the used theorem-in-action could not explain what was shown by the software. On having noticed the striping impacts in the whole wall, they resorted to the explanation that the "electrons – little balls" cross the screen and are distributed in this way. They neither noticed that there were zones of the screen that did not have any impact, nor related this pattern with the wave concepts, which evidently were not available at that moment.

TABLE 6.3 Previous analysis of the situation

Key question	Required actions	Key concepts	Expected theorems-in-action	Expected explanations and inferences
How can we explain the form of P(x)?	To contrast the predictions with the simulation.	Histogram of frequencies	If the slits are near each other, in the center there will be a highest.	For the small balls, the curve obtained with both slits open has a highest in the center because the effects of both slits are superimposed.
What is the relation between the individual curves and the curve with both slits open?	To close a slit at a time and to analyze what happens.	Curve of probability Sum of curves	If the slits are remote from each other, there will be two highests in the curve.	
	To describe the curves obtained with electrons.	Highest and Minimums of P (x)		
What is obtained in the DSE when it is made with electrons?	To differentiate the curves obtained with electrons and with small balls.		Electrons are like small balls.	The experience must give results similar to the one made with the small balls.

TABLE 6.4 Posterior analysis of the situation

Selection of information	Operative invariants used by the students	Mechanisms of control	Inferences
The electron is so small that the slits do not modify its trajectory.	T_5: "The electrons are very tiny balls"	The students observed a central highest of the curve P(x) (They did not notice the presence of other relative highest)	The results are the same for electrons and for small balls.
Being a question about electrons, the knowledge learned in the chemistry classes must be used.	T_6: "The electrons are special particles"	On the collecting screen they interpreted the distribution of the electrons as uniform, since they imagined the experience with small balls.	The electron can "cross" the armored wall, because they have a very special quality: "crossing the wall". It is as if they could move along the "prohibited": Because of that the collecting screen will show a uniform distribution of electrons.

According to the students, as the electron is considered a "tiny particle" (T_5), this experience must be framed in classical mechanics. There is no reason for them to update knowledge of the ondulatory phenomena. They did not perceive the minimums in the curve $P(x)$. Therefore, the teacher had to point out this information. He also insisted on the following ideas:

- The slits affect the distribution of the electrons, and they no longer "copy" the form of the slits in the screen since they are not small balls.
- The electrons are named as "quantum system" indicating their special and proper behavior, which was very different from the expected behavior for them in the DSE. From this it was settled the necessity to look for an explanation of the form of the curve that "does not add".

Although they had limited knowledge of the algebra of functions (in this case the sum), they could make relevant inferences for the conceptualization.

4.3 *Situation 3 – Analyzing and Using the Sum of All Alternatives* (SAA) *Technique with Free Electrons*

On having accepted that the electrons arrive at the screen in entire units but they are distributed in a different way from the small balls (performing the curve of probability with several highests and minimums), the following information was presented:

How is it possible to obtain the probability curve of the DSE with electrons?

Let's leave for a moment the DSE and let's consider a free particle moving at a certain speed in one direction in order to simplify the situation. Let's try to answer an even more basic question:

What is the probability P of reaching final state F having left initial state I?

$$P[(t_i;\, x_i) \dashrightarrow (t_f; x_f)]$$

Every "state" is described with the position and the time of the particle. The initial state I is characterized by an initial time, $t= t_i$, at a certain initial position $x = x_i$ and the final state F is characterized at a final time t = tf and a final position $x = x_f$. To simplify, let's consider that the initial state, characterized by initial time (t = 0*s*) and initial position (x = 0*m*) and expressed as a couple where its first component is the time and its second component is the position, is (0; 0). The physicists have developed a method to explain these results, proposing a mathematical model that describes the behavior of both small balls and electrons that pass through the double slit.

This method is universal, because it allows to describe with the same mathematical model the behavior of both macroscopic objects and quantum systems (like electrons). And this works for any value of mass!

To help the students use the "Feyman's Method of multiple ways for the QM", the complex numbers were replaced by a vectorial representation. The method can be applied to any physical system. To start with the free particle case to later deduce the properties of quantum systems was a key didactic decision. This case joins the most general properties of these systems. The method to calculate the probability was named "Sum of all the Alternatives" (SAA) and it was presented to the students in the following steps:

1. Suppose an initial state (I) at $x(t=0)=0$ and a final state (F) at $x(T)=x_f$. We consider here one-dimensional paths for simplicity.

Of course paths connecting the initial state I with the final state F can show multiple forms. Some of them are shown in the following figure with straight sections (the only functions that the software used by the students allows modeling).

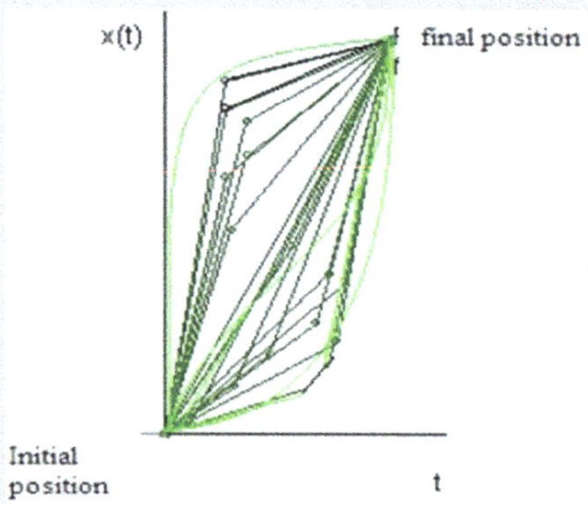

Then, with each possible path $x(t)$ we associate a numerical value called action, represented by "S". The action is the average difference between kinetic Ek and potential Ep energy times T.

$$S = \langle E_k - E_p \rangle T,$$

where <> denotes temporal average. If the particle is "free", thus it is not in the presence of forces and $E_p=0$. Then, in this case the action is simply $S = \langle E_k \rangle$ T, i.e.,

$$S = \frac{1}{2} m < v^2 > T$$

2. With the action S, we construct a unitary two-dimensional vector, forming an angle $S/\hbar$ (with respect to the positive x-axis). This vector is called "Probability amplitude" associated with the path $x(t)$. The denominator of this quotient is $\hbar = h/2\pi$, where $h = 6.625x10^{-34}$ Js is the so called Planck's constant. That is to say:

Every path $x(t)$ connecting I with F has a corresponding S, which is used to construct the Amplitude of probability vector associated to $x(t)$, whose components are:

$$\cos\frac{S}{\hbar}; \sin\frac{S}{\hbar}$$

3. All amplitude of probability vectors associated to different paths connecting I with F are added. We call the resulting vector "total probability amplitude".
4. The square module of total probability amplitude gives the probability of arriving at final state F, having started at initial state I.

The students were helped to apply the technique SAA to the free electron, using a simulation that was specifically developed for this situation in *Modellus*™. To acquaint the students with the software, the screen with initial and final conditions was presented before running the simulation. The following questions were presented:

a) From the linear function, what type of movement is possible to deduce?
b) How can the angle of probability vector associated to $x_{class(t)}$ be obtained?

Figure 6.9 shows the software displays when some functions $x(t)$ alternatives that connect the initial and final states are selected in the case of the electron. It only allows to construct functions in two linear sections. On the right, the simulation "draws" the vectors associated with every chosen function. Some of the selected functions remain drawn with its respective associate vector. Simultaneously, it shows the numerical value of the angle of the drawn vector. As the students are not acquainted with radial system, it was decided that the software should show the angle in the sexagesimal system.

The proposed questions to use the Simulation 1 were:

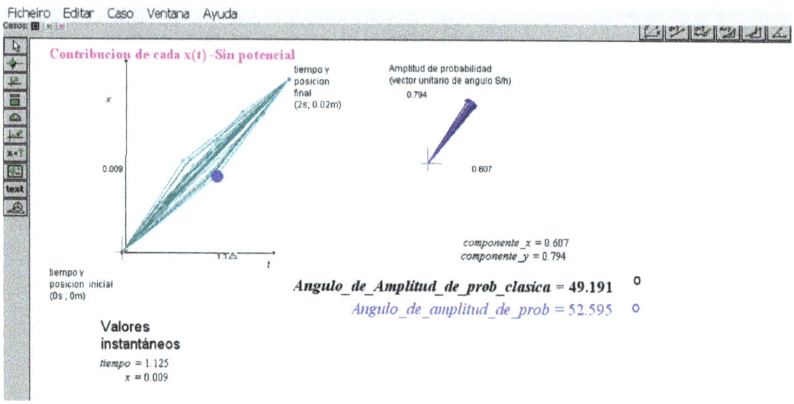

FIGURE 6.9 Screen simulation with Modellus™

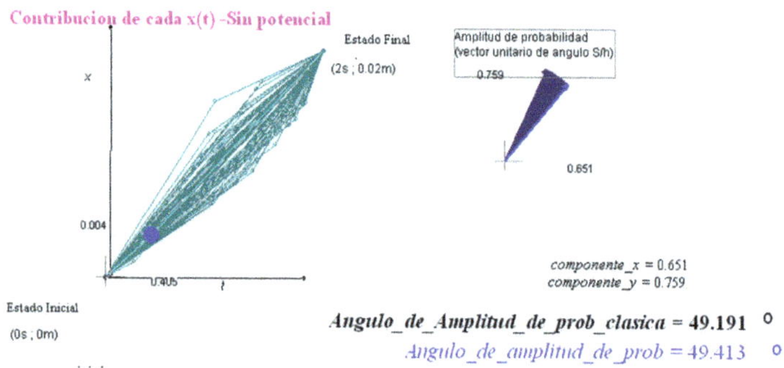

FIGURE 6.10 Simulating the SAA for a free electron, on having selected x(t) "near" the xclass (t)

Select possible different $x(t)$ that connect the initial state to the final state.

a) Compare the value of the action of each one $x(t)$ shown with regard to the action of $x_{class}(t)$ that you calculated previously. What can you conclude?

b) How are the directions of associated vectors of each $x(t)$ that are close to $x_{class}(t)$? What happened with the functions removed to $x_{class}(t)$?

The students had to select different functions and compare the value S of each with the corresponding action to $x$class(t). In this way, the students could conclude that the action for any function, S(x (t)), is always greater than the action of the classical function S(xclass (t)). Figure 6.10 shows the simulation on having selected functions "near" to the classical function:

Using the SAA technique to calculate probabilities, all the vectors associated with each of the alternative functions must be added. The key question was:

How can we consider "all" the x(t)?

To lead the students to conclude that the sum is finally reduced to a set of vectors which are associated to the functions near the classical function, sixteen ordered pairs ($t;x$) were given (see Table 6.5). These pairs were extracted after the simulation. The students were invited to draw the vectors, placing each one after the other to obtain the geometric sum.

The graphs obtained are like the one shown in Figure 6.11.

TABLE 6.5 Ordered pairs ($t;x$) (first and second column) and the angle amplitude of probability (third column), obtained from the simulation

0.38	0.011	90
0.243	0.011	140
0.279	0.019	230
0.303	0.018	270
0.018	0.015	320
0.126	0.02	50
0.486	0.03	51
1.296	0.01	53
1	0.01	49.5
0.882	0.011	52
1.775	0.002	320
1.746	0.003	270
1.854	0.006	230
1.422	0.002	140
1.03	0.001	90

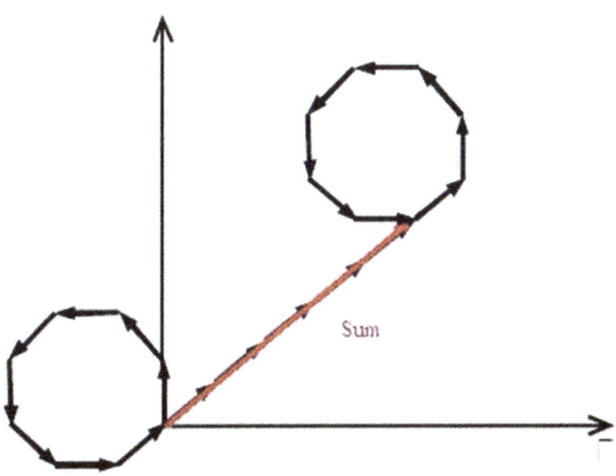

FIGURE 6.11 Schematic sum of the vectors

TABLE 6.6 Previous analysis of the situation

Key question	Required actions	Key concepts	Expected theorems-in-action	Expected explanations and inferences
How is the calculus of the probability for a free electron that goes from an initial to a final state?	To interpret the technique SAA relating to the classical concepts.	Quantum system Electron	It is necessary to make other calculations for the probability of the electrons, since they did not behave as the small balls.	The action S is minimum for the classical functional relation x(t) – a straight line – if it is compared with other arbitrary functional relations x(t).
	To simulate selecting alternative functions x(t).	Initial and final State		
	To draw the vectors geometrically.	Functions x(t) near to the classical function.		The angles of the amplitude vectors associated with those paths x(t) near the classical path xclass(t) are similar. However, the angles of the vectors associated to the x(t) placed far from the classical path are different from each other.
	To analyze the contribution of the nearby and distant vectors to the classical one.	Functions x(t) distant to the classical function.		
				Only a set of paths "around" the classical path contributes to the sum.

inquiry learning,[2] we can mention the Workshop Physics at the Dickinson College,[3] which is a method of teaching introductory physics without formal lectures. The central focus of the program is concentrated in active learning, in questioning, critical thought and problem-solving, all in collaborative working groups. The observation leads to build operational definitions and its relations, and empirical models, before the introduction of formal definitions and theoretical relationships. Computational tools are used to facilitate these processes. This project has been seen with great interest among physics teachers in several Latin-American countries.[4]

We agree with the approaches of the LW for the physics courses as a scientific research activity, triggered by a problematic situation, in which the theoretical and experimental contents associated with it are in permanent interdependence. However, we believe that the resolution of the LW implies a complex cognitive activity that requires analyzing and understanding by us, in order to mediate learning. We now present a proposal that we have recently developed in our research group.

4 Cognitive Process in Solving Physics Lab Work

We consider Laboratory Work as an activity that arises from problem-situations and shows an indissoluble link between the theoretical and the methodological aspects throughout the process of solution. The grouping of the experimental tasks in phases (Figure 8.1) implies that we have general goals of the LW, and specific goals for the phases. Each of them requires the mobilization of concepts, procedures and representations, both in theoretical and methodological domains. If the student does not understand them, the phases will result into sub-problems.

In order to mediate learning in each phase, we have modeled its dynamics during the resolution of the LW, taking as reference the theory of conceptual fields (TCF) (Vergnaud, 1990) and mental models (MM) (Greca & Moreira, 2008).

The TCF constitutes a cognitive referential that addresses the development and learning of complex skills to explain and predict behavior, in attention to conceptual problems in a specific field of knowledge. It is assumed that the student perceives and integrates new knowledge into their cognitive structure through an adaptive process. Cognitive development is shaped by the actions of individuals in specific situations through the used schemes (Vergnaud, 1990). An effective learning process must be founded upon the knowledge of the difficulties related to the cognitive tasks, the procedures and the representations available for the student, and the evaluation of its distance from the expected development (Vergnaud, 2007).

From these theoretical grounds, it is assumed that to dominate a class of situations implies the development of a general scheme that allows the student to organize sequences of specific actions. In such a sense, we ask ourselves: How do the schemes guide the actions when dealing with a situation (Vergnaud, 1998)?

The schemes include four elements: (a) Goals and expectations, (b) Readiness to use knowledge, (c) Rules of action, and (d) Possibilities and inferences. The interaction between them allows to generate arguments, to evaluate the rules and anticipations about the situation, and to make decisions on how to approach the situation.

The concepts-in-action and the theorems-in-action (C&TA) include one part of the concepts and theorems that the person has into her knowledge structure, therefore making them relevant to guide learning. In order to account for the C&TA, external representations will be necessary: linguistic, graphics or gestural.

The theorems-in-action are propositions constructed with the concepts to approach the situation, which can be true or false from a scientific perspective, and they allow to make inferences or operations to reach the goal (Vergnaud, 1990).

For the student to increase the significance and complexity of C&TA, it is necessary that she operates with them in a variety of classes of situations, and that she may be able to represent them.

According to Vergnaud (1990, 2007), the concept of situation is restricted to the cognitive processes and to the subject's responses as a function of the situation. For this reason, the sense is not in the situations themselves, but in the relation of the subject with the situations and with the meaning (Vergnaud, 2007). This does not discard the importance of other dimensions to give an integral meaning to the situation, like the affective and the social ones (Moreira, 2008).

A situation is a complex task that can be analyzed as a combination of subtasks. Its difficulty does not derive from the form of the statement, nor from the amount of tasks or the elements of these, but from the complexity of the conceptualizations required to approach each task, and to relate them with one another. It is then relevant for teaching to establish the sets of concepts and theorems that allow to solve a situation, and to investigate the C&TA that the subjects use in it.

This theory allows establishing a gradient of the complexity of the situations in terms of the concepts, theorems and abstraction of the representations that are involved in the appropriate rules of action. These have to do with the area of knowledge. This is relevant to the organization of teaching, concerning the students in action.

The schemes can be classified according to the person's functionality when facing a situation, including:

Known Situations, when the schemes have been developed to approach a class of situations and are activated in a relatively immediate way in front of a similar one, turning out to be efficient and, in some cases, effective to the individual.

Unknown Situations, when there are no ready schemes to dominate them. In these cases of uncertainty, reflection and failed attempts occur, as well as other actions that lead to the achievement of the goal or to its abandonment. In addition, the schemes begin to enter into a process of competition and accommodation, and may permit the emergence of new schemes for new situations (Vergnaud, 1990). Nevertheless, for these cases it is not clear in the TCF how some elements of the schemes in the long-term memory are activated.

To this end, we have considered the mental models (MM) (Johnson Laird, 1983), a proposal derived from cognitive psychology. The MM is built to perceive the real world and operate in the short-term memory like working models, giving functional account of the cognition before the unknown (Moreira, 2008). In addition, the MMs are conceived like analogous structure of reality and the schemes include some elements (C&TA) of the same nature (Vergnaud, 1990); therefore, the MMs can be considered as mediators between the new situation and schemes. Thus, it is possible that the MMs mobilize to the working memory the C&TA of schemes that have some relation with the situation, in order to enable to form the precursors of new schemes, in the context of the action itself.

In summary, the TCF shows some relevant aspects to the research in didactics, as the following ones:

- The *action before the situations* allows the mobilization of schemes. If we can mediate so that the students make explicit the C&TA, discuss and evaluate their relevance and validity in the conceptual domain established from science, then we can facilitate the development of knowledge closer to the scientific one.
- The *actions for responding to new situations*. These constitute the way for the incorporation of new schemes to the cognitive structure. What situations within a conceptual field are new for the students?
- The *variety of situations*. It is possible to organize this variety in classes of situations within a given conceptual field. What are the kinds of situations giving meaning to the concepts and procedures that we want to teach in a conceptual field?

- *History of learning*: The students progressively model conceptualizations (knowledge) dealing with situations they master, in a slow and complex process, with forward and backward steps, which raises questions such as: What sequence of *situations* promotes the conceptual development in a field of knowledge? How to formulate the problems from one area of knowledge like *situations* that permit the conceptual development?

We can say that these theoretical referents have allowed us to account for the process of learning before new situations regarding *laboratory tasks*. To do this, we have described and represented the process of the LW in the *Model of Learning in the Physics LabWork* (MLePLab) (Figure 8.2).

The laboratory work begins by placing students in a problem-situation; they perceive and identify the overall goal. If it is new, they do not find a scheme to approach it, and then they enter in a stage of uncertainty and reflection (Vergnaud, 1990). They advance to the first specific demand of the LW (Phase I) that implies the conceptual analysis of the problem, restate it and formulate questions about the physical world according to pertinent variables, interpretive models and working hypotheses, which may result partially new. The students will build MMs as mediators, which recursively evolve until reaching their functionality (Moreira, 2008). These are transitory and contain elements of the C&TA type that remain and determine the identification of the pertinent information to the problem, and also allow producing operational predictions and inferences for the LW. Consequently, it becomes necessary to explicit the C&TA, referred to theoretical and experimental contents, for the construction of new knowledge from them.

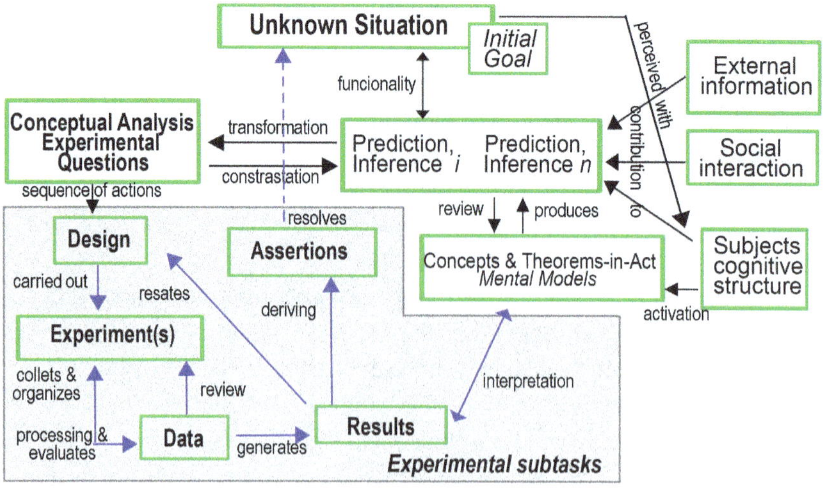

FIGURE 8.2 Model of learning in the Physics LabWork (MLePLab)

In this process we can intervene through: social interaction among peers and with the teachers, searching information from external sources, comments, guidance and modeling by the teacher actions, among others. All these can progressively mould the C&TA, making them closer to the scientific knowledge and incorporating new meanings. One assumes that this mental activity stops after the developed tasks.

In addressing the other tasks of the LW (Phases II to V, Figure 8.1), each one results in a situation, whose solution requires a sequence of actions organized according to scheme(s) that may not be in the student knowledge structure.

Thus, the goal of LW is achieved when the student resolves the sub-problems (phases) by means of gradual adjustments of the assimilation schemes in each phase, interrelated to each other. The C&TA contain elements of theoretical, methodological and epistemological domains, varying the emphasis depending on the phase. For science education is one hope that the C&TA may come closer each time more to the knowledge of science.

In summary, the LW can be seen as cognitive cycles of resolution of partial tasks (Phases), which then together allow solving the problem concerning the experiment. The phases that result new for the students implies a cognitive processing similar to the one shown in the MLePLab for Phase I (shaded area, Figure 8.2).

For each experimental work, we expressed learning objectives that guide the teaching and evaluation processes, associated with: i) the conceptual, understood as concepts and theorems, and representations, referred to the methodological and theoretical domains, according to the particular situation, and ii) the epistemological aspects.

From these theoretical grounds, we believe that if labwork starts from problem-situations, learning during the different phases (Figure 8.1) can be understood and facilitated by the MLePLab model. Using this hypothesis, we designed a methodological proposal to assess the effectiveness of the labwork in physics teaching. The results have shown to be reasonably acceptable; therefore, we have derived and tested a methodological approach to promote the learning of physics in relation to the LW (Andrés, Pesa & Meneses, 2006a; 2008; Andrés, 2009; Miranda, 2010).

5 The Physics Labwork Developed from the MLePLab Model and the Laboratory "V"

The potentiality of the MLePLab model to guide and provide an understanding of the cognitive process during a LW was evaluated in the context of a laboratory course with students of the teaching career, at UPEL-IPC. We applied a sequence of three LW focused on problem-situations. In the first LW (oscillations), we assessed the feasibility of the model and the students' initial conceptual development. Also, we presented the overall plan of action with the grouping of experimental tasks in phases and their representation in the 'V'.

The other two LW (mechanical waves) were organized to produce explicit learning in the methodological, epistemological and theoretical domains. Each LW had duration of 4–5 weeks (20 hrs.). The teacher mediated the cognitive cycles in each sub-task, towards the established goals of learning. Students worked in pairs and the teacher assisted to mobilize the initial C&TA, provoke debates and critical reflection, and promote awareness about the experiment-theory and the process interrelationship (Andrés et al., 2006a, 2008).

In this first study, we were able to promote the learning expected in most of the students in the three domains (Andrés et al., 2006a, 2008), and validated the MLePLab model with the research methodology. Furthermore, this study allowed establishing a methodology for didactic intervention in the LW.

The learning process during the LW, oriented by this model, requires setting up the conceptual field of the laboratory (CFL), constituted by:

1. Classes of situations that we suppose are novel to the student, which are ranked, taking into account the conceptual complexity.
2. The concepts and theorems from the discipline that allow solving them.

We have identified typical contents for the phases, from: the analysis of texts referring to the methodological content of physics laboratory (Gil & Rodríguez, 2001; Taylor, 1997); some classifications of the knowledge to be learned with the LW (De Pro Bueno, 1998; Gott & Duggan, 1996; Millar, Le Maréchal, & Tiberghien, 1999); a proposal of CFL,[5] and the description of the process by means of the "V" (Table 8.3).

Under this approach, we restructured the laboratory courses for physics teacher education (UPEL-IPC). To do this, we selected the specific problem-situations for each laboratory course, considering the previous content of the theoretical course and a sequence of learning objectives. Furthermore, we developed criteria for the curricular organization of the experiments.

On the other hand, to conduct the teaching process it is necessary to know the history of student learning in laboratory situations, that is to say, to make explicit the C&TA that she mobilizes in the LW, in order to promote her progress. To this end, we used tests with situations, individual or collective interviews, laboratory reports, and anecdotal records of the process.

In order to *visualize* and guide the LW from its beginning to the final communication, it is used the representation of the process in the laboratory "V" (Figure 8.1), which has been a favorable meta-cognitive tool (Andrés, Pesa, &

TABLE 8.3 Some contents concerning the experimental task for each phase

Phase	Experimental task	Contents
I	*Conceptual analysis*	Sources information. Concepts of the theoretical domain of the situation. Relations between concepts. Argumentation. Modeling.
	Formulation of key questions	Key questions. Prediction. Hypothesis.
II	*Experimental Design*	Experiment. Variables and classification according to the role. Control. Approaching from the model. Real model.
III	*Data collection*	Estimation of measures. Classification of measurements. Measuring range. Uncertainties. Classification of Errors. Sources of error. Significant figures. Data. Anomalous data. Appreciation. Precision. Accuracy. Sensitivity. Accepted value. Repeatability. Measurement techniques.
	Data, processing and transformations	Classification, organization. Tables, graphs. Measures of central tendency (mean, median) Measures of variation (standard deviation of the measure, deviation from the mean ...) Distributions. Error propagation.
IV	*Analysis and interpretation*	Patterns. Empirical equation. Fitting. Discrepancies. Trend of data. Contrast-data model. Modeling. Rejection results. Reliability. Significance.
V	*Conclusions*	Inferences. Generalizations. Conclusions. Explications.
	Communication to peers	Oral and written report. Report types. Writing scientific papers.

Meneses, 2007; Caraballo, 2009; García, Insausti, & Merino, 2003). Furthermore, the critical reflection about the relations between the domains in each phase and between the phases has allowed to promote and to infer the learning about the epistemological aspects. We estimate that with the domain of sequences of classes of situations, the students can achieve in the long term a vision about the nature of the experimental activity in science.

This proposal, aiming at critical investigative learning, requires a coherent assessment. Therefore, we have considered that the written and oral reports are pertinent means for this purpose, since they are an important activity in scientific communication. Furthermore, the report allows to make evident the meanings in action and their relationships, as well as the level of students' symbolic representation.

To this end, we have designed a procedure and an instrument in the context of the laboratory courses. They were developed by three professors, who used the approach described above. A set of rubrics and criteria were established for the qualification in each phase through an instrument (Andrés, 2009), which can be tailored by selecting the rubrics according to the type of experiment (exploratory, contrast or application) and the learning goals. The students know the criteria and the instrument since the beginning of the course, using them as a guide.

6 Perspectives of Research and Conclusions in Relation to the Laboratory Work in Physics Teacher Education

We think that our proposal for changing the experimental activities in the physics courses has yielded favorable results for future teachers' learning. At present, we continue with new investigations, as described below. Considering that in the LW the theory is used in the practice, the employed theoretical knowledge will be reconstructed in the process to promote their development. Therefore, it is important to assess the problem-situations in the various fields of the theoretical domain, in attention to the C&TA that the students have activate. This should be done in order to the identify those situations where the C&TA showed to be different to what was agreed in science or showed little conceptual development.

In addition, we are developing new directed studies to evaluate the effectiveness of problems with real context and with different types of experiment, in various fields of the theoretical knowledge (Miranda & Andrés, 2009).

Furthermore, the experimental work in both science and learning constitute social activities that require the negotiation of meanings between peers. Therefore, we have initiated a new line of study, where the LW centerd on problems is combined with the social organization of students in cooperative groups. In a first work, situated in high school, we obtained some progress in the vision of the students about science with the cooperation taking place in the social interactions between students, with the teacher's mediation (Caraballo, 2009).

So far, we have identified that the students face difficulties when solving the phases with greater demand of the theoretical-experimental interrelation. In phase I the theoretical knowledge is applied to the problem to model it and to approach the experiment. This demands the following tasks: restating the problem in response to a theoretical framework, formulating relevant questions and hypotheses or possible solutions in response to a feasible model, deductive reasoning, generating predictions according to experimental conditions. In addition, in phase IV, the experimental results are interpreted and explained using theoretical ideas. All these actions are not easy for the students. Therefore, these tasks deserve a more exhaustive analysis of the implicit cognitive processes, in order to propose and evaluate more efficient forms of didactic intervention.

In the research program, we have incorporated diverse technologies (real-time measurements, digital videos of phenomena, simulations and software) in order to improve the experiments and to make its development more productive (Andrés, 2007; Miranda & Andrés, 2009). In order to mediate the learning processes, it is relevant to evaluate the impact that we have achieved with the inclusion of technology, and to generate educational criteria for the implementation of each of these tools. We are also designing proposals to develop and improve the implementation of phases I and IV, supported by simulations, animations, concept map software and platforms for distance learning environments.

We have found some achievements with the introduced changes in the laboratory courses (in UPEL-IPC), which will lead to the study about the impact that these can have in the scholar practice. We think that the learning that can be generated using the proposals of laboratory work that were presented in this chapter will contribute to one of the goals of science education: scientific literacy[6] since it has proved to be an important component in the critical and reflective formation, necessary to exert citizenship and reach a protagonist role in the development of our region.

In Latin America, research over the past five years[7] on the subject has focused on the study of the cognitive processes involved in the inquiry activity of the LW, the reasons why teachers do not use LW in teaching, or proposals to incorporate technological resources or new experiments in the LW. However, it seems that it has had little impact in the educational institutions, although recently some pilot assays[8] at primary school in Latin America have been reported. For example, two approaches[9] based on the so-called guided inquiry are being studied in two Argentinean provinces, in schools located in poor communities, since the academic year 2009. Scientific reasoning, experimentation, group work and dialogue characterize the used approaches. They constitute a sharp departure from simple transmission of information. Some relevant data about the context from this study are: Fewer than 38 percent of the participant teachers expresses to have confidence in their ability to carry out experiments in their lessons. In addition, only 5.4 percent of schools have science laboratories (Näslund-Hadley, Cabrol, & Ibarrará, 2009).

In secondary school institutions in Latin America the rupture between theory and practice still predominates in science teaching. As we discussed before, among the few teachers that implement the LW, the majority uses an instrumental approach and the classical experiments of the available laboratory manuals. This is partly due to lack of knowledge (Laburú, 2007) and the conception about the experimental activities in science that they have developed during their teacher education (Andrés et al., 2006b). Consequently, the region still requires science teachers with an adequate preparation about a topic like this, so valuable and rich, so that they may be able to design a more effective laboratory work and have the disposition to do this, in attention to the context and school reality. All of this is in order to contribute with a humanistic science education.

Notes

1 Laburú (2005) makes a summary of the research in this field (pp. 161–162).

2 Discovery learning also promoted inquiry but from an inductive view of science.

3 See http://physics.dickinson.edu/~wp_web/wp_homepage.html.
This project, coordinated by Priscilla Laws, has been established with support materials in this century, but has its roots in research dating from the end of the past century.

4 Since 2008, there have been three regional conferences and workshops of the Southern Cone under the auspices of the Universities of Córdoba and of San Luis, Argentina. See http://www.famaf.unc.edu.ar/congresos/aaeym/index.html

5 Established for the physics laboratory courses for physical teacher education, UPEL-IPC.

6 Scientific literacy, as characterized by Sasseron and Pessoa (2008, p. 334), involves- "the ability to organize their thoughts logically, and the help in building a more critical view about the world around him", based on the idea of literacy designed by Paulo Freire.

7 In journals like *Ciência & Educação* (http://www.scielo.br/scielo.php?script=sci_serial&pid=1516-7313&lng=en&nrm=iso), *Investigações em Ensino de Ciências* (http://www.if.ufrgs.br/ienci/), *Enseñanza de las Ciencias* (http://ensciencias.uab.es/), among others.

8 Some of them have been supported by the Inter-American Development Bank.

9 Science and Technology through Creativity (CTC) and the Scientific Literacy Program (PAC) has also been used in Brazil.

References

Abd-El Khalick, F., & Lederman, N. (2000). Improving science teachers' conceptions of nature of science: a critical review of the literature. *International Journal of Science Education, 22*(7), 665–701.

Andrés, Ma. M. (2003). El docente de física en servicio. Concepciones y desempeño en el aula [The in-service physics teacher. Conceptions and performance in the classroom]. *Paradigma, XXIV*(2), 57–86.

Andrés, Ma. M. (2007). Desarrollo conceptual en un experimento con videos: velocidad de propagación de pulsos en cuerdas [Conceptual development in an experiment with videos: Speed of propagation of pulses in strings]. *Ciencia, 15*(1), Article 16.

Andrés, Ma. M. (2009). *Evaluación del aprendizaje en trabajo de laboratorio centrado en resolver situaciones problema* [Assessment of learning in labwork centered on solving problem-situations] (pp. 366–370). *Enseñanza de las Ciencias, Número Extra,* VIII International Congress on Research in Science Teaching, Barcelona.

Andrés, Ma. M., Pesa, M., & Meneses, J. (2006a). Desarrollo conceptual acerca de ondas mecánicas en un laboratorio guiado por el modelo MATLaF [Conceptual development about mechanical waves, in a labwork guided by the MATLaF Model]. *Revista Electrónica de Enseñanza de las Ciencias, 5*(2), 260–289.

Andrés, Ma. M., Pesa, M., & Meneses, J. (2006b). La actividad experimental en Física: visión de estudiantes universitarios [Experimental activity in Physics: view of university students). *Paradigma, XXVI*(1), 349–363.

Andrés Z, Ma. M., Pesa, M. A., & Meneses, J. V. (2007). Efectividad metacognitiva de la heurística V de Gowin en trabajos de laboratorio centrados en la resolución de situaciones problemáticas [Metacognitive effectiveness of Gowin's V heuristics in labwork centerd on the resolution of problematic situations]. *Indivisa. Boletín de Estudios e Investigación Monografía VIII. V International Conference on Meaningful Learning* (pp. 203–215). Madrid: La Salle Centro Universitario.

Andrés, Ma. M., Pesa, M., & Meneses, J. (2008). Efectividad de un laboratorio guiado por el modelo de aprendizaje MATLaF para el desarrollo conceptual asociado a tareas experimentales [The effectiveness of a laboratory guided by the MATLaF model for the conceptual development associated with experimental tasks]. *Enseñanza de las Ciencias, 26*(3), 343–358.

Andrés Z, Ma. M., Pesa, M. A., & Moreira, M. A. (2006). El trabajo de laboratorio en cursos de física desde la teoría de campos conceptuales [Labwork in physics courses from the perspective of conceptual fields theory]. *Ciência & Educação, XXII*(2), 129–142.

Brown, J., Collins, A., & Duguid, P. (1989). Situated cognition and the culture of learning *Educational Researcher, 18*(1), 32–42.

Bunge, M. (1975). *La investigación científica* [Scientific research]. Barcelona: Ariel.

Caraballo, D. (2009). *Desarrollo del pensamiento científico en los alumnos mediante la realización de trabajos de laboratorio tipo investigación y la V de Gowin como herramienta orientadora en el proceso* [Development of scientific thinking in students by conducting inquiry labwork and the V Gowin as guiding tool in the process] (Master's thesis). Universidade de Oriente, Cumaná, Venezuela.

Chalmers, A. (2000). *¿Qué es esa cosa llamada Ciencia?* [What is that thing called Science?] (3rd ed. translated, revised and enlarged). Madrid: Siglo XXI.

Cudmani, L., Salinas, J., & Jaén, M. (sf). (2000). *Epistemología de la Física. Tópicos introductorios.* [Epistemology of Physics. Introductory Topics]. Fac. Ciencias Exactas y Tecnología, Argentina: Universidad Nacional de Tucumán.

De Pro Bueno, A. (1998). ¿Se pueden enseñar contenidos procedimentales en las clases de ciencias? [Can you teach procedural contents in science classes?]. *Enseñanza de las Ciencias, 16*(1), 21–42.

Duschl, R. (2000). Making the nature of science explicit. In R. Millar, J. Leach, & J. Osborne (Eds.), *Improving science education* (pp. 187–206). Philadelphia, PA: Open University Press.

Etkina, E., Van Heuvelen, A., Brookes, D. T., & Mills, D. (2002). Role of experiments in physics instruction – A process approach. *The Physics Teacher, 40*(6), 351–355.

Franklin, A. (2002). Experiment in physics. *Stanford encyclopedia of philosophy* (First published October 5, 1998; substantive revision October 8, 2002). Retrieved from http://plato.stanford.edu/entries/physics-experiment/

Garcia, P., Insausti, Ma. J., & Merino, M. (2003). Evaluación de los trabajos prácticos mediante diagramas *"V"* [Assessment of practical work with diagrams *"V"*]. *Revista Electrónica de Enseñanza de las Ciencias, 2*(1), Article 3.

Gil, D., Carrascosa, J., Furió, C., & Martínez-Torregosa, J. (1991). *La Enseñanza de las Ciencias en la Educación Secundaria. Cuadernos de Educación 5* [Teaching Science in Secondary Education. Education Papers 5]. Barcelona: Horsori.

Gil, S., & Rodríguez, E. (2001). *Física re-creativa. Experimentos de Física usando nuevas tecnologías* [Re-creative Physics. Physics Experiments using new technologies]. Buenos Aires, Argentina: Prentice-Hall.

Gott, R., & Duggan, S. (1996). Practical work: Its role in the understanding of evidence in science. *International Journal of Science Teaching, 18*(7), 755–760.

Greca, I., & Moreira, M. A. (2008). La integración de modelos mentales y esquemas de asimilación para la comprensión de procesos de asimilación [The integration of mental models and patterns of assimilation for understanding processes of assimilation]. In Ma. L. Rodríguez-Palmero (Ed.), *La teoría de aprendizaje significativo en perspectiva de la Psicología cognitiva* [The theory of meaningful learning in cognitive psychology perspective] (pp. 133–161), Spain: Edc Octaedro.

Hodson, D. (1994). Hacia un enfoque más crítico del trabajo de laboratorio [Towards a more critical approach of the labwork]. *Enseñanza de las Ciencias, 12*(3), 299–313.

Johnson-Laird, P. (1983). *Mental models*. Cambridge, MA: Harvard University Press.

Laburú, C. (2005). Seleção de experimentos de Física no ensino médio: uma investigação a partir da fala de professores [Selection of Physics experiments in high school: An investigation from teachers' speech]. *Investigações em Ensino de Ciências, 10*(2), 161–178.

Laburú, C., Alves B., M., & Kanbach, B. G. (2007). A relação com o saber profissional do professor de Física e o fracasso da implementação de atividades experimentais no ensino médio [The relationship with the physics teacher's professional knowledge and the failure of the implementation of experimental activities in high school]. *Investigações em Ensino de Ciências, 12*(3), 305–320.

Laudan, L. (1986). *El progreso y sus problemas* [Progress and its problems]. Madrid: Encuentro.

Millar, R. H., Le Maréchal, L.-F., & Tiberghien, A. (1999). "Mapping" the domain – Varieties of practical work. In J. Leach & A. Paulsen (Eds.), *Practical work in science education – Recent research studies* (pp. 33–59). Dordrecht/Roskilde: Kluwer Academic Publishers/Roskilde University Press.

Miranda, C. (2010). *Propuesta para la enseñanza de la física en el contexto de la ruta del chocolate, basada en el diálogo de saberes* [Proposal for physics teaching in the context of the chocolate route, based on the dialogue of knowledge] (Master's thesis). UPEL-IPC, Caracas, Venezuela.

Miranda, C., & Andrés, M. (2009). *Propuesta didáctica para el aprendizaje en el laboratorio basado en resolución de problemas reales* [Didactic approach to learning in the lab based on real problem solving] (pp. 377–380). *Enseñanza de las Ciencias, Número Extra*, VIII International Congress on Research in Science Teaching. Barcelona.

Moreira, M. A. (2008). Aprendizaje significativo: la asimilación ausubeliana desde una visión cognitiva contemporánea [Meaningful learning: Ausubelian assimilation from a contemporary cognitive view]. In Ma. L. Rodriguez-Palmero (Ed.), *La teoría de aprendizaje significativo en perspectiva de la Psicología cognitiva* [The theory of meaningful learning from the cognitive psychology perspective] (pp. 198–221). Barcelona: Octaedro.

Murcia, K., & Shibeci, R. (1999). Primary students teachers' conceptions of the nature of science. *International Journal of Science Education, 21*(11), 1123–1140.

Näslund-Hadley, E., Cabrol, M., & Ibarrará, P. (2009). *Experimental math and science education in Argentina* (Technical notes. Nov. IDB education). Retrieved from http://idbdocs.iadb.org/wsdocs/getdocument.aspx?docnum=35018947

Novak, J., & Gowin, D. B. (1984). *Learning how to learn*. Cambridge: Cambridge University Press.

Peixoto, M. A. N., & da Silva, F. W. (2003). *Os laboratórios de ensino de Física nas escolas estaduais de nível médio de Belo Horizonte* [The physics teaching labs in state high schools of Belo Horizonte]. *Proceedings of the XV National Symposium of Physics Teaching, Physics Education: Present and Future*. Curitiba, Brasil: CEFEP-PR.

Pesa, M. (2002). La concepción estándar de las ciencias y las propuestas superadoras – Algunas implicancias para la educación en ciencias [The standard conception of science and the overcoming proposals – Some Implications for science education]. In M. Andrés (Ed.), *Investigación en enseñanza de la física. Memorias de la IV Escuela Latinoamericana* [Research in physics education. Proceedings of the IV Latin-American School] (pp. 120–124). Caracas, Venezuela: Universidad Pedagógica Experimental Libertador.

Ryder, J., & Leach, J. (1999). University science students' experiences of investigative project work and their images of science. *International Journal of Science Education, 21*(9), 945–956.

Sasseron, L. H., & Carvalho, A. M. P. (2008). Almejando a alfabetização científica no ensino fundamental: a proposição e a procura de indicadores do processo [Aiming at scientific literacy in basic education: the proposition and the search for indicators of the process]. *Investigações em Ensino de Ciências, 13*(3), 333–352.

Séré, M.-G. (2002). Towards renewed research questions from the outcomes of the European Project Labwork in Science Education. *Science Education, 86*(1), 624–644.

Taylor, J. R. (1997). *An introduction to error analysis. The study of uncertainties in physical measurements* (2nd ed.). Sauselito, CA: University Science Books.

Tobin, K., Tippins, D., & Gallard, A. J. (1994). Research on instructional strategies for teaching science. In D. Gabel (Ed.), *Handbook of research on science teaching and learning* (pp. 45–93). New York, NY: MacMillan.

Vergnaud, G. (1990). La théorie des champs conceptuels [The conceptual fields theory]. *Recherches en Didáctique des Mathématiques, 10*(2), 133–170.

Vergnaud, G. (1998). A comprehensive theory of representation for Mathematics Education. *Journal of Mathematical Behavior, 17*(2), 167–181.

Vergnaud, G. (2007). ¿En qué sentido la teoría de los campos conceptuales puede ayudarnos para facilitar aprendizaje significativo? [In what sense the conceptual fields theory might help us to facilitate meaningful learning?]. *Investigações em Ensino de Ciências, 12*(2), 285–302.

Wong, S. L., & Hodson, D. (2008). From the horse's mouth: What scientists say about scientific investigation and scientific knowledge. *Science & Education, 93*(1), 109–130.

CHAPTER 9

Research on Colombian Science Teachers' Education: A Review

Rómulo Gallego Badillo, Royman Pérez Miranda,
Adriana Patricia Gallego Torres and Deisy Baracaldo Guzmán

Abstract

In this chapter the main characteristics of science teacher education in Colombia, Brazil and Argentina are described as well as the historical-epistemological and pedagogical transformations that occurred during the second half of the twentieth century. The authors present an interpretation of the history and the current state of the field of curriculum studies in these countries, based on the works of researchers and practitioners published during the last five decades. To do so, the development of the curriculum and the meanings attached to this notion by educators and educational researchers are examined. It is observable that different educational reforms around the world have emphasized scientific literacy as the main objective of science education, but the curricular approach guided by governmental agencies, along with value-committed, ideological interpretations made by teachers and researchers, have made it difficult for science education research to take root and flourish in the three countries. In addition, there are few opportunities for research on practice, as these opportunities tend to occur at the end of the undergraduate program. It can be said from this review that in Colombia, Brazil and Argentina, there are very few studies that deal with the systematic study of the concrete results obtained with the official curricula or programs of science teacher education. In this sense, the new generations of science teachers as well as the colleagues dedicated to science education research have the responsibility to overcome these shortcomings in the future and to ensure that the programs are maximally useful for science teachers.

Keywords

science – teachers education – research

 | DOI: 10.1163/9789004409088_009

1 Introduction

This chapter is a review of the research that has been carried out on science teachers' education in Colombia, Brazil, and Argentina. The results discussed here have been published in research reports and articles in some of the most widely circulated printed and electronic journals in the continent, as well as in Spain. We discuss here science teachers' education as a field of research divided into subfields that are addressed in the chapter.

2 Origins of Colombian Teachers' Education

The first institution established in the late seventeenth century for teachers' education was the "Seminari Scolastici". This "Normaleschule" was founded in Vienna in 1770. With the "French Revolution" and within a free public secular education project, "Normal Schools" were created. T. Husen (1988) argues that education as a field of study, but not of research, was introduced into German universities in the eighteenth century. The first program in education was established at the University of Halle, in 1718. It implemented courses for secondary teachers, which dealt with the philosophical foundations of the art of teaching.

L. Mumford (2006, p. 195), in his study originally published in 1932, points out that the large-scale organization of the factory made workers learn to read. In 1832, in England, there was a reform to provide education to the children of these workers. The author notes that as these students would also be in the future workers of the factory system, teachers opted for a repetitive method that made students learn by heart and listen to the teachings in silence. In the late nineteenth and early twentieth century, the movement to bring science to primary and secondary education appeared (Calatayud et al., 1986, p. 12). Therefore, the conversion of science in school content followed the model established by the English education reform of 1932.

In Colombia, the late nineteenth century saw the arrival of Normal Schools to educate teachers, usually run by Catholic religious communities, with men and women taught separately. With the "Second German Mission" in 1926, the German pedagogue J. Sieber came to Colombia to be the principal of the Normal School for Boys in Tunja city, capital of the Department of Boyacá. An additional "Supplemental Specialization Course" was created in this institution with a duration of two years. Boyacá Departmental Assembly legalized it by the Ordinance 38 of 1929. It was the first training course for science teachers in Colombia.

Liberal President Enrique Olaya Herrera (1880–1937) and his Minister of Education, Luis Carrizosa Valenzuela, converted the "Supplemental Specialization Course" into the College of Education and its duration was extended to four years. In 1934 through Decree 1379, it was renamed as the College of Education for Boys, affiliated to the National University of Colombia (UN) in Bogota. Through Decree 2178 of 1930, the "Graduate Course" for women was established at the National Pedagogical Institute (IPN). The first course was to train science teachers. With Decree 850 of 1934, it was transformed into the College of Education for Women, depending on the IPN and being a part of the UN, with a duration of three years. During the government of Olaya Herrera it was decreed that the title of the graduates should be that of "Licenciados".

In the first presidential term of Alfonso Lopez (between 1934 and 1938), the three Colleges of Education were unified by Decree 1917 of 1935, as a result of the administrative difficulties in the management of the separate Colleges. With Law 39 of 1936, the Unified School of Education became the "Escuela Normal Superior de Colombia". Among its objectives was the training of teachers for the secondary level of education, including natural sciences teachers. The respective academic programs were Biology, Chemistry, Physics, and Mathematics (Gallego Badillo, Pérez Miranda, & Rincón Pabón, 2008, p. 8).

The "Escuela Normal Superior de Colombia" (ENSC), structured similarly to the "Ecole Normale Superieure de Paris" and under the direction of the psychiatrist José Francisco Socarrás, was the first institution to give Colombian women a university education, against the policies that governed at that time (Socarrás, 1987, pp. 22–28). This educational project was the subject of political and religious persecution during the second presidential term of Alfonso López (between 1942 and 1945), under the justification that they could not educate young ladies with young men. The conservative government of Laureano Gómez (between 1950 and 1953), with Decree 1955 of 1951, dissolved the ENSC and created the "Escuela Normal Universitaria para Varones", in Tunja city, and the "Escuela Superior Universitaria para Mujeres", in Bogotá. In 1953, under the dictatorship of Gustavo Rojas Pinilla (between 1953 and 1957), it was transformed into the Pedagogical University of Colombia, which later became the Pedagogical and Technological University of Colombia (UPTC). In 1955, the second gave rise to the Women's Pedagogical University. This institution became the National Pedagogical University (UPN) in 1968. In the 1970s, in the capitals of various departments that make up the Colombian administrative political structure, universities created Colleges of Education. Thus, the prevalence of UPTC and the UPN in the teachers' education for secondary level was lost.

In 1969, sponsored by the Ford Foundation, the UPN adopted the North-American scheme with the credit system. To graduate, each student had to take a total of 128 credits and had the opportunity to opt for a major (one main area, and one basic area) or two majors. The title was the Bachelor of Science Education, with majors in Chemistry, Physics, Biology, and other academic disciplines.

With Decree 080 of 1980, it became mandatory for the title to be in a specific discipline, for example, BSc in Chemistry, Biology, Physics, or another discipline. The UPN was reformed, creating three areas of teaching: disciplinary, humanistic, and research area, and an area of pedagogical and didactic training. With the reform of the UPN three faculties were organized: Education, Humanities, and Science and Technology. The latter introduced three seminars in the research area: History of Science, Logic of Science and the Philosophy of Science.

With Decree 272 of 1998, the degree was extended to five academic years, specifying that Pedagogy and the Curriculum were the foundational knowledge of the teaching profession. The General Education Law 115 of 1994 organized the basic education system, creating the field of natural science and environmental education. Many colleges have opted to suppress science degrees with appropriate reforms that came from the ENSC and opted for a degree in Elementary Education with an Emphasis in Natural Sciences and Environmental Education. Only the UPN and the Distrital University continued with degrees in Biology, Chemistry, and Physics. The Technological University of Chocó also continued with its previous degree in Biology, Chemistry, Mathematics, and Physics (Gallego Badillo, Gallego Torres, Pérez Miranda, & Figueroa Molina, 2010).

In the case of Brazil, where science teachers are educated in the universities, the title of "Licenciados" has also been given to science teachers since the 1950s, with training programs for practitioners. Degree programs were imposed in an authoritarian manner by the military regime in the 1970s. It is said that, in general, with some exceptions in the 1940s or 1950s, "technical rationality" has prevailed in teacher education (Terrazzan, 2007, p. 146). Nardi (2003) argues that it was in that decade that science education research started, adding that teachers for primary and secondary education were previously formed by the Faculty of Philosophy.

In Argentina, E. González (2003) argues that teacher training for those levels of education has been promoted for years by the Teacher Training Institutes, mostly at the non-university tertiary level. These institutes have a tradition of excellence, even though they have many difficulties in carrying out their educational task due to structural problems. Among the problems, we can mention: not having dedicated teachers, a shortage of administrative resources and teaching materials, and restricted opening hours.

3 Establishment of a Research Field

One of the arguments justifying the conservation of outstanding academic programs to train science teachers is the answer to the question: "Are science teachers born or are they formed?" A positive response to the first part is called "essentialism" (Gil et al., 1991, p. 10), which means that people are born with the skills to perform professionally as teachers. Within this "essentialism" we should include the practice of improvisation when we accept teachers only because they have received a degree in science. These improvised teachers lack training in Pedagogy and Science teaching, in spite of their experience during their student days. First, they have developed a conception of the nature of the sciences which they teach. At the same time, in-service teachers have a conception based on empathy with those teachers who taught them, under the influence of which they created an idea of how to teach science. This is an idea that can even be kaleidoscopic, a non-reflected synthesis of several experiences.

Along with this reflection is the conception of the masters of the Middle Ages, where the art of giving public displays of mastery was taught. This meant that it was sufficient to know the discipline and to obtain the title to teach science. This position demands critical analysis as encompassed in the questions: "To what extent do the teachers know what science is?" and "what the specific science they teach is?"

In our context, researchers have been influenced by work initiated by well-known Spanish specialists. The new generations that are now investigating this discipline have learnt a great deal from those specialists. Research on the initial and continuing training of science teachers in this context began with systematic studies in the 1990s. Research was required due to changes and innovations that were proposed. In the same way, it was perceived as being necessary to characterize and identify the obstacles and difficulties presented when the initial training of teachers aims at transforming the learning model they had been developing previously (Porlán & Rivero, 1998, p. 10; Gallego Badillo & Pérez Miranda, 1999, p. 67). Thus, the presence of a spontaneous teaching is highlighted (Mellado & Gonzalez, 2000, p. 542).

4 Research Subfields

It is accepted that teaching, learning and teacher education are three related but different components, and that these components must be present in research on initial and continuing training of science teachers. In this regard, the growing importance of these components in classroom research should be noted (Carvalho, 2007, p. 192). Since teacher education is linked to curriculum projects and pre-service teachers' historical, epistemological, didactic and pedagogical conceptions, it is suggested that they take part as research constructs. If we consider the idea that the results of teaching and learning are valuable aspects, the problem of assessment models must also be investigated. As teacher education is based on the use of textbooks at the university level, these texts should be equally subjected to systematic study.

In accordance with the above, the field of research on initial and continuing training of science teachers should be looked at as being composed of several subfields. This means we have to cover the historical-epistemological and pedagogical conceptions that dominate in their respective areas. Dominance does not mean exclusivity, since each program is to allow the deployment of other positions. It would also be necessary to consider the history of science, the internalist the externalist approaches, and the possibility that science teachers teach from that history (Laudan, 2005, p. 68).

In this context it is important to specify the STS relationships (Fontes & Cardoso, 2006), which were raised in the social studies of science, and also the methodologies used in teaching science contents and the expected learning in interactions between in-service and pre-service teachers. It is also important to include the kind of evaluation (Ostermann & Trieste, 2004) which is mainly used to identify and characterize the changes in the conceptions with which they enter into the process (Alonso, Gil Perez, & Martinez-Torregrosa, 1996). This evaluation must consider the strengths and weaknesses of the curriculum projects and the teachers in these times of accreditation. In this literature review, the researchers did not find reports or articles in specialized journals which would allow discussions to expose this issue in the case of initial and continuing training of science teachers.

Regarding the epistemological aspect, besides identifying and characterizing the approach that guides the programs, it is important to define the extent to which they enroll in the non-dogmatic epistemology proposed by the end of the twentieth century (Estany, 2005), as there is no longer a scientific model taken as a reference to determine the scientific status of the others (as it happened until the 1960s when physics was the model). For this characterization, it would be useful to highlight the category of the scientific model which has come to replace the theory (Greca & dos Santos, 2005; Lombardi, 1998). Thus, it is a subfield which is the object of researchers' attention (Justi, 2002), to the extent that it has been put forward as Didactics of modeling (Gallego Torres, Gallego Badillo & Pérez Miranda, 2006).

In a research study on scientific modeling with physics students, the conclusions reached by the authors express the difficulties students face when engaging in this exercise (Islas & Pesas, 2003). Even teachers expressed this difficulty. The authors attributed this result to an education focused on textbooks, whose contents are limited to definitions and algorithms to solve pencil and paper exercises.

Reconstructions of the internal history of science for research in science education have caused, among communities of scholars in general, consensus between the perspectives of convergent or parallel methodologies (Batista, 2007). The proposals for the inclusion of history and philosophy of science in teacher education have been identified as a complex problem in a review and analysis of literature (El-Hani, 2007, p. 300).

A question that needs to be answered is that formulated by some experts regarding the initial and continuing training of science teachers, which has been linked to the texts edited in developed countries, and written according to their contexts. They give rise to a decontextualized teacher education and, therefore, strengthen the dependency status of people from peripheral countries (Kreimer, 2009, p. 33), thus, framing the teachers and educational programs to impart knowledge of the past (Kuhn, 1972, p. 3) and standardized laboratory techniques without a discussion (Peduzzi, 2007).

The revisions made by the authors of this chapter, in their attempt to contribute to this field, concluded that the textbooks had to be reviewed in order to guide theses of the Master's course in Chemistry teaching. After an internalist historical analysis, the results of these investigations showed that the analyzed texts conveyed a distinctly positivist version, as well as misconceptions (Camacho Gonzalez, Gallego Badillo & Pérez Miranda, 2007; Cuellar Fernandez, Gallego Badillo & Perez Miranda, 2008; Herreño Chaves, Gallego Badillo, & Pérez Miranda, 2010).

5 Curricular Programs

In a research project on science teacher education in Colombia, it was necessary to analyze official documents of the respective curriculum projects through matrix analysis and interviews with department directors, managers or coordinators of the programs, teachers and students. Twenty-two out of a total of fifty one Colombian programs were investigated as they had already obtained prior approval from the Ministry of Education (ordered by the decree 272 of 1998). The results showed the dominance of Comte's positivist epistemological approach, which sought to replace the 'brainpower approach' of Rationalism by leveraging the principles of the natural sciences (such as Physics, Chemistry and Biology). They also showed a lack of knowledge about the Didactics of science as recorded in contemporary journals (Gallego Badillo,

Pérez Miranda, & Torres de Gallego, 2004; Gallego Badillo, Pérez Miranda, Torres de Gallego, & Amador Rodriguez, 2004, p. 137).

Here, the authors highlight the comparative study about the epistemological, didactic and pedagogical conceptions of two academic programs for Chemistry teachers from two public universities in Bogotá. The methodology used was document analysis of the syllabi. For the analysis, three matrices were designed, one for the epistemological aspects, one for the didactic aspects, and one for the pedagogical issues of each program. It was concluded that in the documents of both programs we can perceive that they follow a communitarian construction. One of them states that Didactics is concerned with the problems related to teaching and learning to read and write. In the other, Didactics is conceived as the result of a set of assumptions and theoretical and methodological frameworks with epistemological identity. In both programs it is stated that there are methods available to analyze and interpret education as an object of study (Gallego Badillo, Pérez Miranda, Torres de Gallego, & Amador Rodriguez, 2010).

A similar study which used the same methodology and had a comparative nature was carried out with academic programs for science teachers at some universities in Argentina (Universidad de Buenos Aires/UBA and Universidad Nacional de San Martin/UNSAM), Colombia (Universidad Pedagogica Nacional/UPN and Universidad Distrital Francisco José de Caldas/UDFJC), Chile (Pontificia Universidad Católica de Chile/PUC and Pontificia Universidad Católica de Valparaíso/PUCV), Brazil (Universidade Estadual de São Paulo/UNESP, campus of Bauru) and Venezuela (Universidade de los Andes/ULA). The results obtained in this analysis stress that education programs in UBA, PUC, PUCV, and UNESP/Bauru differ from those from the other universities. Their programs are not working from positivist approaches or the constructivist version of science teaching. However, it is emphasized that, in short, it is the positivist version which continues to dominate among teacher educators who train new science teachers and the technique employed is the usual verbal transmission of curriculum content (Gallego Badillo, Pérez Miranda, Torres de Gallego, & Amador Rodríguez, 2010).

In a project sponsored by the Colombian National Administrative Department for Science and Development (COLCIENCIAS), conducted between 2008 and 2009, which involved researchers from three Colombian universities (Universidad Pedagogica Nacional/UPN, Universidad Distrital Francisco José de Caldas/UDFJC and Universidad del Atlántico/UA), the conceptual foundations took into account studies of the sociology of scientific knowledge (Vessuri, 1992; Restivo, 1992) and the need for a plural conception of the history of each one of the sciences (Estany, 2005). Furthermore, the relations between external and internal history were taken into account (Shapin, 2005, p. 95). From this perspective, the project established that the social history of science education must be treated as a special component of the social history of science. In this sense, science teacher education is part of every scientific community, since it involves the younger generations through their teachings in the task of developing the potential of each scientific model.

As part of the methodological design, an interview was applied to Heads of department, teachers, and students. Twelve programs, thirty-six teachers, and sixty graduates were selected to participate. We conducted a total of 420 interviews. Regarding the question "How do science teachers teach?", almost all the interviewees, with a few exceptions, answered that they rely on verbal transmission of curriculum content. With respect to the second question, "Is history involved in teaching?" all of them, without exception, responded negatively. When asked "What is the relationship between science and technology?", the majority of students reduced the technological aspect to computers, adding that it was still uncommon to use information technology. Among the students, some said there was no relation. Regarding the question "What is the analysis you can make from the texts recommended by teachers in different science subjects?", everyone agreed that those were the ones which teachers followed in their classes. When asked "How did education impact on your professional performance?", a few replied that it has been meaningful and the majority said that there had been very little impact (Gallego Badillo, Pérez Miranda, Torres de Gallego, & Amador Rodríguez, 2010).

As it can be seen, the contents transmitted in science education was historically decontextualized, a science without history. In the case of technology, in general the role of instruments in the laboratory practice was not even contemplated, a science without any relation to technology, precisely because of the decontextualization mentioned above. As a consequence, we have a version of science from Comte's positivist approach, which reflects cumulative linear version of the texts of their own teaching. While these texts have been investigated in the case of the academic programs of science teacher education and deserve a subtitle in this chapter, what was deduced from the results is that these teachers and graduates did not have a critical position with respect to those texts.

In a piece of research that aims at determining the revision that was being done to the Brazilian National Curricular Parameters in the educational programs for science, mathematics and technology, Ricardo and Zylberstajn (2007) collected by means of interviews points of view on the concepts of skills, interdisciplinarity, and contextualization of 17 methodology and teaching practice teachers who belonged to Biology, Physics, Mathematics, and Chemistry

Departments in three public universities of three different regions of Brazil. The results showed that the majority of the teachers in this study held uninformed conceptions about these constructs and that further research was required to better understand how teachers apply these concepts in their classes and how the pedagogies of teacher education inform this conceptualization.

6 Didactic and Pedagogical Conceptions

With respect to didactic and pedagogical conceptions, the conceptions elaborated on Didactics and Pedagogy by 400 Colombian teachers with Bachelor and Master degrees, were identified and characterized. The instruments used in the study were a test with a combination of ten closed questions and ten open-ended questions validated by academic peers and a Likert questionnaire with twenty items for the respondents to specify their level of agreement or disagreement with a series of statements on a symmetric agree-disagree scale. After the analysis of the results, it was concluded that most of them used the habitual definitions, i.e., simple and basic definitions that teachers repeat to each other. It also showed misunderstandings regarding Didactics and Pedagogy definitions, i.e., Pedagogy was defined as a teaching method that follows a consistent scientific approach or educational style and was also reduced to a list of teaching techniques (Gallego Torres, Gallego Badillo, & Pérez Miranda, 2006).

In a case study aimed at identifying the evolution of didactic models regarding the concepts developed by pre-service chemistry teachers, the focus was on the analysis of teaching units. The intervention was centered on the conception of a didactic research model. The results showed that in the end the subjects had differences between their initial and intermediate concepts which showed significant changes in didactic models conception (Predebon & Del Pino, 2009).

7 Historical and Epistemological Conceptions

With respect to historical and epistemological conceptions, it was essential to identify and characterize the philosophical approaches which beginning teachers employ in their work (Ghisolfi & Pacheco, 2000). It was seen that graduates from education programs for science teachers tend to take these approaches in a non-linear way. It was therefore necessary to investigate newly graduated teachers to know how they express their version of the nature of science in their lessons (Bianchini et al., 2003). The results concerning the conceptions of the nature of science in a group of Chilean biology teachers showed that these teachers were rational, conservative, dogmatic, and empiricists (Ravonal Moreno & Quintanilla Gatica, 2010).

Epistemological education for physics teachers remains a concern of researchers. Thus, certain specific aspects of communication between scientists could contribute to destabilize the misconceptions that often underlie teaching performance. The information gathered about epistemological contents can give support to the feasibility of incorporating these issues, which require necessary renovations in the dynamics of education programs (Islas, Sgro, & Pesa, 2009).

Here, we can refer to an investigation of pre-service chemistry teachers that aimed at identifying and characterizing the mental models with which students entered the teacher education process and the changes that resulted from the educational work done. The course where the sample taken was called Chemical Theory I, which belonged to the curricular program of the UPN Chemistry Department. The analysis of the results led to the conclusion that the chemistry teachers' initial mental models were modified. Before the intervention, nearly all pre-service teachers considered a scientific model as a simplified or schematic representation of reality. After the intervention, 70% of the pre-service teachers described a scientific model as a conceptual representation of some phenomenon (Amador Rodriguez, Gallego Badillo, & Pérez Miranda, 2008).

Considering the role of the History of Science in initial and continuing education of science teachers, particularly in physics, it deserves attention a research aiming at integrating the history of science into physics teaching on the basis of the historical development of the subject of gravitational attraction. According to the methodology used, there were changes in the concepts held by physics teachers in their initial education (Gatti, Nardi, & da Silva, 2010).

8 Professional Identity of a Science Teacher

This aspect is a problem in the field of science teacher education, because of the misconception held by most of the students and teachers: "To teach science, it is enough to know or dominate it". This statement has been challenged with the question: What is the source of this knowledge? Maybe, the textbooks and the education policies. Some studies have focused on the problem of the professional identity of science teachers, which is one of the reasons for science teaching not being considered as a scientific activity in some of our countries.

Thus, one educational strategy aimed at reviewing and constructing models in Didactics and models in each of the sciences that are the subject of education (Justi, 2002). The results obtained from the survey carried out in a course for undergraduate students in Chemistry and a Master course in Chemistry teaching at the UPN showed the difficulty involved in working in the teaching of modeling, because the students were not able to develop the expected models. It was difficult for the pre-service teachers to adapt themselves to the process of building scientific models in the classroom. The Master program students who performed the role of advisors used verbal transmission of content and algorithms (Pérez Miranda, Gallego Badillo, & Garay Garay, 2006). The role that Didactics must fulfill in science teacher education programs (Craven III & Penick, 2001) is also mentioned, as the component that allows teachers to prepare and apply the target description to the planning and execution of teaching.

In this line of research it is necessary to mention a project that investigated the career prospects of students in a degree program in Biology from a public university of the state of São Paulo, in Brazil (UNESP/Bauru). Through the results obtained with semi-structured interviews, it was concluded that the academic program contributed little to build the students' identities as biology or science teachers and, by contrast, generated in these prospective teachers the idea that they would be biological researchers (Brando & Caldeira, 2009).

9 Teaching Internships

By teaching internships we refer to the space where pre-service science teachers, after meeting certain requirements, experience what their future as professionals in science education might be in educational institutions where they hypothetically could be hired once they graduate. The results of an investigation carried out in Colombia in this regard allow us to infer that such teaching internships are characterized by strengthening the tradition, since the teachers of the institutions that collaborate with science teacher education programs require practitioners to continue behaving professionally in accordance with their own tradition. In addition, there is no agreement between educational institutions and the education faculties that requires the commitment of these parties with the universities (Gallego Badillo, Pérez Miranda, & Torres de Gallego, 2004).

Similarly, this seems to happen in another process of science teacher education in relation to these internships. Based on the information obtained through structured and semi-structured interviews, Harres et al. (2010) concluded that there are contradictions between what we think about teaching practice and the pre-service teachers' behavior in the educational institutions. There is no consistency between what is being worked in academic programs and the context of these educational institutions.

In turn, the results of another investigation revealed that some practitioners tend to reproduce the style of physics teaching experienced by them in their basic education and university studies. It was also showed in the same study that they assume the view that educational work in this science could lead to a less rigorous learning and fewer scientific concepts if related to social and environmental dimensions (Silva & de Carvalho, 2009).

10 Conclusion

After the literature review performed, we agree with Terrazzan (2007, p. 1) when he says that this specialist field is not structured as such, since although there are a growing number of projects in the graduate programs, it still lacks a necessary link between research and agreed conceptualizations, something that derives from the idea that if students read enough of the research in an area of interest, they will somehow magically be able to produce sensible ideas for further research. Indeed, it is usual to include in this field studies on epistemological, didactic and pedagogical conceptions of pre-service and in-service science teachers in initial training or in practice, although there are a few that systematically and continually assess the actual results of the goals or intentions proposed for the curricular projects.

It can be said from this review that in Colombia, Brazil and Argentina, there are very few studies that deal with the systematic investigation of the concrete results obtained with the official curriculum or program of science teacher education. In this sense, an assessment of the impact generated by the graduates in terms of the socialization of the historical-epistemological and pedagogical transformations that occurred during the second half of the twentieth century among the school population is necessary. It could be stated that the medieval conception still prevails: "To teach a science, it is enough to know it". This holds despite the educational research conducted in the field of science teachers' initial training.

Similarly, in teaching the content of science subjects included in the science teacher education programs, the focus on historical reconstructions is still incipient. In this regard, it should be noted that in the absence of this historical approach, something similar happens with the introduction of social studies of science or sociology of scientific knowledge in science teaching.

These education programs focus on monitoring textbooks, as well as transmitting a science out of context. First, because they neglect the knowledge problems raised in each of the groups that produced the scientific models made to work in the classroom. Second, the cultural, political and economical context in which each group received the recognition of the importance of their work is usually unknown in those education programs.

We conclude that the programs for science teacher education should give pre-service and in-service teachers an opportunity to experience science authentically, free of the misconceptions and idealizations inherent in the myths about the nature of the scientific enterprise. There must be increased opportunity for both pre-service and in-service teachers to learn about the social history of science education and the epistemological and didactic foundations by means of careful review of textbooks, in order to remove the misconceptions that have helped in constructing an inaccurate view of the nature of science teaching.

References

Adúriz-Bravo, A. (2001). Integración de la epistemología en la formación del profesorado de ciencias [Integration of epistemology in science teacher education] (Doctoral dissertation). Universitat Autónoma de Barcelona, Barcelona, Spain.

Alonso, M., Gil Pérez, D., & Martínez-Torregrosa, J. (1996). Evaluar no es calificar. La evaluación y la calificación en la enseñanza constructivista de las Ciencias [To evaluate is not to qualify. Evaluation and qualification in constructivist science teaching]. *Investigación en la Escuela, 30*, 15–26.

Amador Rodríguez, R. Y., & Gallego Badillo, R. (2004). Estudio inicial de dos programas para la formación de Licenciados en Química [Initial study of two chemistry teacher education programs]. *Revista TEΔ, 16*, 64–83.

Amador Rodríguez, R. Y., Gallego Badillo, R., & Pérez Miranda, R. (2008). Desde qué versiones epistemológicas construyen modelos mentales los profesores em formación inicial: una investigación didáctica [From which epistemological versions do pre-service teachers build mental models: a didactic investigation]. *Revista TEΔ, 24*, 8–22.

Batista, I. L. (2007). Reconstruções histórico-filosóficas e a pesquisa em Educação científica e matemática [Historical-philosophical reconstructions and research in science and mathematics education]. In R. Nardi (Ed.), *A pesquisa em ensino de ciências no Brasil* [Research in science teaching in Brazil]. (pp. 257–272). São Paulo: Escrituras.

Bianchini, J. A., Johnston, C. C., Oram, S., & Cavazos, L. M. (2003). Learning to teach science in contemporary and equitable ways: The successes and struggles of first-year science teachers. *Science Education, 87*(3), 419–443.

Brando, F. da R., & Caldeira, A. M. de A. (2009). Investigação sobre a identidade profissional em alunos de Licenciatura em ciências biológicas [Investigation on the professional identity of biological sciences pre-service teachers]. *Ciência & Educação, 15*(1), 155–173.

Calatayud, M., Carbonell, F., Carrascosa, J., Furió Mas, C., Gil Pérez, D., Grima, J., Hernández, J., Yopis, R., Martínez Torregrosa, J., Payá, J., Ribó, J., Satoca, J., Solbes, J., Soler, J., & Vilches, A. (1986). *Iniciación a las ciencias físico-químicas en la enseñanza media* [Initiation in physical-chemical sciences in secondary education] (Vol. 1). I. C. E. de la Universidad de Valencia. Valencia: Soler.

Camacho González, J. P., Gallego Badillo, R., & Pérez Miranda, R. (2007). La ley periódica. Un análisis histórico, epistemológico y didáctico de algunos textos de enseñanza [The periodic law. An historical, epistemological, and didactic analysis of some textbooks]. *Educación Química, 18*(4), 278–288.

Carvalho, A. M. P. (2007). A pesquisa em sala de aula e a formação de professores [Classroom research and teacher education]. In R. Nardi (Ed.), *A pesquisa em ensino de ciências no Brasil* [Research in science teaching in Brazil] (pp. 193–218). São Paulo: Escrituras.

Cortela, B. S. C., & Nardi, R. (2004). Formadores de professores de Física: uma análise de seus discursos e como podem influenciar na implantação de novos currículos [Physics teacher education: an analysis of their discourses and how they can influence the implementation of new curricula]. *Revista TEΔ, 16*, 102–112.

Craven III, J. A., & Penick, J. (2001). Preparing new teachers to teach science: The role of the science teacher educator. *Electronic Journal of Science Education, 6*(1). Retrieved July 28, 2015, from http://ejse.southwestern.edu/article/view/7670/5437

Cuellar Fernández, L. H., Gallego Badillo, R., & Pérez Miranda, R. (2008). El modelo atómico de E. Rutherford. Del saber científico al conocimiento escolar [E. Rutherford's atomic model. From scientific to school knowledge]. *Enseñanza de las Ciencias, 26*(1), 43–52.

El-Hani, C. N. (2007). Notas sobre o ensino de história e filosofia da Biologia na educação superior [Notes on the teaching of history and philosophy of biology in higher education]. In R. Nardi (Ed.), *A pesquisa em ensino de ciências no Brasil* [Research in science teaching in Brazil] (pp. 293–315). São Paulo: Escrituras.

Estany, A. (2005). El papel de la historia de la ciencia en los estudios interdisciplinares de la ciencia [The role of the history of science in interdisciplinary studies on science]. In S. F. Martínez & G. Guillaumin (Orgs.), *Historia, filosofía y enseñanza de la ciencia* [History, philosophy, and science teaching] (pp. 291–303). México: UNAM.

Fontes, A., & Cardoso, A. (2006). Formação de professores de acordo com a abordagem Ciência/Tecnologia/Sociedade [Teacher education according to the Science/Technology/Society approach]. *Revista Electrónica de Enseñanza de las Ciencias, 5*(*1*), Article 2.

Galagovsky, L., & Adúriz-Bravo, A. (2001). Modelos y analogías en la enseñanza de las ciencias naturales. El concepto de modelo didáctico analógico [Models and analogies in natural sciences teaching. The concept of analogical didactic model]. *Enseñanza de las Ciencias, 19*(2), 231–242.

Gallego Badillo, R., & Pérez Miranda, R. (1999). *El problema del cambio en las concepciones epistemológicas, pedagógicas y didácticas* [The problema of the change in epistemological, pedagogical, and didactic conceptions]. Bogotá: Universidad Pedagógica Nacional.

Gallego Badillo, R., & Pérez Miranda, R. (2002). El problema del cambio en las concepciones de estudiantes de formación avanzada [The problema of the change in the conceptions of students in advanced education]. *Enseñanza de las Ciencias, 20*(*3*), 401–414.

Gallego Badillo, R., Pérez Miranda, R., & Rincón Pabón, L. I. (2008). *La formación de profesores de ciencias en la Escuela Normal Superior de Colombia. Desde sus inicios hasta 1953* [Science teacher education in the *Escuela Normal Superior de Colombia*. From its beginnings until 1953]. Bogotá: Universidad Pedagógica Nacional-CIUP.

Gallego Badillo, R., Pérez Miranda, R., & Torres de Gallego, L. N. (2004). Formación inicial de profesores de ciencias en Colombia: un estudio a partir de programas acreditados [Pre-service teacher education in Colombia: a study from accredited programs]. *Ciência & Educação, 10*(2), 197–218.

Gallego Badillo, R., Gallego Torres, A. P., Pérez Miranda, R., & Figueroa Molina, R. (2010). *Historia social de la educación en ciencias en Colombia. La segunda mitad del Siglo XX* [Social history of science education in Colombia. The second half of the 20th century]. Bogotá: Universidad Pedagógica Nacional.

Gallego Badillo, R., Pérez Miranda, R., Torres de Gallego L. N., & Amador Rodríguez, R. Y. (2004). *La formación inicial de profesores de ciencias en Colombia. Contrastación de fundamentos* [Pre-service science teacher education in Colombia: Comparison of foundations]. Bogotá: Universidad Pedagógica Nacional.

Gallego Badillo, R., Pérez Miranda, R., Torres de Gallego, L. N., & Amador Rodríguez, R. Y. (2010). *Comparación de programas de formación de docentes de ciencias de nueve universidades Sudamericanas* [Comparison of science teacher education programs in nine South-American universities]. Bogotá: Universidad Pedagógica Nacional.

Gallego Badillo, R., Pérez Miranda, R., Torres de Gallego, L. N., & Gallego Torres, A. P. (2007). El papel de la "prácticas docentes" en la formación inicial de profesores de ciencias [The role of "teaching practices" in pre-service science teacher education]. *Revista Electrónica de Enseñanza de la Ciencia, 5*(*3*), 481–504.

Gallego Torres, A. P., Gallego Badillo, R., & Pérez Miranda, R. (2006). ¿Qué versión de ciencia se enseña en el aula? Sobre los modelos científicos y la didáctica de la modelación [What version of science is taught in the class? On scientific models and the didactics of modeling]. *Educación y Educadores, 9*(1), 105–115.

Gatti, S. R. T., Nardi, R., & Da Silva, D. (2010). História da ciência no ensino de Física: um estudo sobre o ensino de atração gravitacional desenvolvido com futuros professores [History of science in physics teaching: a study about the teaching of gravitational attraction developed with future teachers]. *Investigações em Ensino de Ciências, 15*(1), 7–59.

Gianotto, D. E. P., & Diniz, R. E. S. (2009). Formação inicial de professores de biologia: a prática colaborativa e o uso pedagógico do computador [Biology teachers' pre-service education: collaborative practice and the pedagogical use of the computer]. *Revista Electrónica de Enseñanza de las Ciencias, 8*(2), 422–439.

Gil, D., Carrascosa, J., Furió, C., & Martínez-Torregrosa, J. (1991). *La enseñanza de las ciencias en la educación secundaria* [Science teaching in secondary education]. Barcelona: Horsori.

González, E. (2003). *La formación de los docentes de ciencias en la Argentina: un caso no tan extraño* [Science teacher education in Argentina: a case that is not so strange]. *Revista TEΔ, Número Extra,* 85–97.

González, E., Arena, L., Budde, C., De Longhi, A., Ferreira, A., & Re, M. (1996). Cinco ejes para la discusión sobre la formación inicial y la capacitación de los docentes de ciencias. Aportes a un tratamiento interdisciplinario [Five axes for the discussion about science teacher education and training: contributions to an interdisciplinary treatment]. *Revista de Enseñanza de la Física, 9*(2), 75–83.

Greca, I. M., & dos Santos, F. M. T. (2005). Dificuldades da generalização das estratégias de modelação em ciências: o caso da física e da química [Difficulties in the generalization of modeling strategies in the sciences: the case of physics and chemistry]. *Investigações em Ensino de Ciências, 10*(1), 31–46.

Harres, J. B. S., Pizzoto, M. C., Sebastiany, A. P., Diehl, I., F., & Fonseca, M. C. (2010). La práctica docente declarada y realizada por profesores en formación inicial [The declared and performed teaching practice of pre-service teachers]. *Revista Electrónica de Enseñanza de las Ciencias, 9*(1), 1–17.

Herreño Chaves, J. I., Gallego Badillo, R., & Pérez Miranda, R. (2010). Transposición didáctica del modelo científico de Lewis-Langmuir [Didactic transposition of the scientific model of Lewis-Langmuir]. *Revista Eureka. Enseñanza y Divulgación de la Ciencia, 7*(2), 527–543.

Humphreys, P. (2007). Epistemología del siglo XXI [21st century epistemology]. *Revista Ántropos. Huellas del conocimiento, 214*, 65–70.

Husen, T. (1988). Research paradigms in education. *Interchange, 19*(3), 2–13.

Islas, S. M., & Pesa, M. A. (2003). Concepciones de los profesores sobre el rol de los modelos científicos en clases de física [Teachers' conceptions about the role of scientific models in physics clases]. *Revista de Enseñanza de la Física, 17*(1), 43–50.

Islas, S. M., Sgro, M. R., & Pesa, M. A. (2009). La argumentación en la comunidad científica y en la formación de profesores de Física [Argumentation in scientific community and physics teacher education]. *Ciencia & Educação, 15*(2), 291–304.

Justi, R. S. (2002). Modeling, teacher's views on the nature of modeling, and implications for the education of modelers. *International Journal of Science Education, 24*(4), 369–387.

Kreimer, P. (2009). *El científico también es un ser humano. La ciencia bajo la lupa* [The scientist is also a human being. Science under the microscope]. Buenos Aires: Siglo Veintiuno.

Kuhn, T. S. (1972). *La estructura de las revoluciones científicas* [The structure of scientific revolutions]. México: Fondo de Cultura Económica.

Lombardi, O. (1998). La noción de modelo en ciencias [The notion of models in the sciences]. *Educación en Ciencias, II*(4), 5–13.

Machado, D. I., & Nardi, R. (2004). Uma proposta de software hipermedia para o ensino de Física moderna e contemporânea [A proposal of hypermedia software for teaching modern and contemporary physics]. *Revista TEΔ, 16*, 84–101.

Medeiros, A. (2007). A história da ciência e o ensino de Física moderna [History of science and the teaching of modern physics]. In R. Nardi (Ed.), *A pesquisa em ensino de ciências no Brasil* [Research in science teaching in Brazil] (pp. 273–292). São Paulo: Escrituras.

Mellado, V., & González, T. (2000). La formación inicial del profesorado de ciencias [Pre-service science teacher education]. In J. Perales Palacio & P. Cañal de León (Eds.), *Didáctica de las ciencias experimentales* [Didactics of experimental sciences] (pp. 535–556). Alcoy (España): Marfil.

Mumford, L. (2006). *Técnica y civilización* [Technics and civilization]. Madrid: Alianza.

Nardi, R. (2003). A educação em ciências, a pesquisa em ensino de ciências e a formação de professores no Brasil [Science education, research in science teaching, and teacher education in Brasil]. *Revista TEΔ, Número Extra*, 12–19.

Ostermann, F., & Ricci, S. F. S (2004). Construindo uma unidade didática conceitual sobre mecânica quântica: um estudo na formação de professores de física [Building a conceptual didactic unit about quantum mechanics: a study in physics teacher education]. *Ciência & Educação, 10*(2), 235–258.

Peduzzi, L. O. Q. (2007). Do átomo grego ao átomo de Bohr: receptividade inicial e perspectiva de pesquisa em um texto voltado para uma disciplina de evolução dos conceitos da Física [From the Greek atom to Bohr's atom: initial reception and research perspective in a text for a discipline on the evolution of physics concepts]. In

R. Nardi (Ed.), *A pesquisa em ensino de ciências no Brasil* [Research in science teaching in Brazil] (pp. 317–338). São Paulo: Escrituras.

Pérez, A., & Gimeno, J. (1992). El pensamiento pedagógico de los profesores: un estudio empírico sobre la incidencia de los cursos de aptitud pedagógica (CAP) y de la experiencia profesional en el pensamiento de los profesores [Teachers' pedagogical thinking: an empirical study about the impact of pedagogical ability courses and the professional experience on teachers' thinking]. *Investigación en la Escuela, 39*, 51–73.

Pérez Miranda, R., & Gallego Badillo, R. (2006). Concepciones sobre pedagogía y didáctica de un grupo de docentes. Informe de investigación [Conceptions about pedagogy and didactics of a group of teachers. Research communication]. *Revista Educación y Pedagogía, XVIII*(44), 129–137.

Pérez Miranda, R., Gallego Badillo, R., & Garay Garay F. (2006). A construção de modelos na formação inicial e continuada de professores de Química [The construction of models in pre-service and in-service chemistry teacher education. In R. Nardi & M. J. P. M. de Almeida (Orgs.), *Analogias, leituras e modelos no ensino da ciência* [Analogies, readings and models in science teaching] (pp. 141–159). São Paulo: Escrituras.

Porlán, R., & Rivero, A. (1998). *El conocimiento de los profesores* [Teachers' knowledge]. Sevilla: Diada.

Predebon, F., & Del Pino, J. C. (2009). Uma análise evolutiva de modelos didáticos associados às concepções didáticas de futuros professores de Química envolvidos em um processo de intervenção formativa [An evolutionary analysis of didactic models associated with didactic conceptions of future chemistry teachers involved in a process of educational intervention]. *Investigações em Ensino de Ciências, 14*(2), 237–254.

Ravonal Moreno, E., & Quintanilla Gatica, M. (2010). Caracterización de las concepciones epistemológicas del profesorado de Biología en ejercicio sobre la naturaleza de la ciencia [Characterization of in-service biology teachers' epistemological conceptions about the nature of science]. *Revista Electrónica de Enseñanza de la Ciencia, 9*(1), 111–124.

Restivo, S. (1992). La ciencia moderna como problema social [Modern science as social problema]. *Fin de Siglo, 3*, 20–39.

Ricardo, E. C., & Zylbersztajn, A. (2007). Os Parâmetros Curriculares Nacionais na formação inicial dos professores das ciências da natureza e matemática do Ensino Médio [The National Curricular Parameters in pre-service natural sciences and mathematics high school teacher education]. *Investigações em Ensino de Ciências, 12*(3), 339–355.

Seixas das Neves, L., Beltrán Núñez, I., Leite Ramalho, B., Luna da Silveira, G. C., & Pascoal Diniz, A. L. (2002). O conhecimento pedagógico do conteúdo: Lei e tabela periódica. Uma reflexão para a formação do Licenciado em Química [Pedagogical content knowledge: periodic law and table. A reflection for pre-service chemistry teacher education]. *Revista Brasileira de Pesquisa em Educação em Ciências, 1*(2), a9.

Shapin, S. (2005). Disciplina y delimitación: la historia y la sociología de la ciencia, a la luz del debate externismo-internismo [Discipline and bounding: the history and sociology of science as seen through the externalism-internalism debate]. In S. F. Martínez & G. Guillaumin (Orgs.), *Historia, filosofía y enseñanza de la ciencia* [History, Philosophy, and Science Teaching] (pp. 67–119). México: UNAM.

Silva, L. F., & de Carvalho, L. M. (2009). Professores de Física em formação inicial: o ensino de Física, a abordagem CTS e os temas controversos [Pre-service physics teacher education: physics teaching, STS approach, and controversial themes]. *Investigações em Ensino de Ciências, 14*(1), 135–148.

Silva, R. M. G., & Schnetzler, R. P. (2000). Bases epistemológicas e enfoques didáticos implicados na formação do educador [Epistemological grounds and didactic approaches involved in teacher education]. *Enseñanza de las Ciencias, Número Extra*, 421–422.

Socarrás, J. F. (1987). *Facultades de Educación y Escuela Normal Superior* [Schools of Education and the Higher Normal School]. Tunja: La Rana y el Águila.

Terrazzan, E. A. (2007). Inovação escolar e pesquisa sobre formação de professores [School innovation and research on teacher education]. In R. Nardi (Ed.), *A pesquisa em ensino de ciências no Brasil* [Research in science teaching in Brazil] (pp. 145–192). São Paulo: Escrituras.

Vessuri, H. M. C. (1992). Perspectivas recientes en el estudio social de las ciencias [Recent perspectives in the social studies of science]. *Fin de Siglo, 3*, 40–52.

PART 4

Discourse Analysis and Argumentation in Science Education

CHAPTER 10

Student Participation in Science Classroom Discourse: Research in Latin America

Antonia Candela

Abstract

This is a review focused on spontaneous forms of students' participation in science classroom discourse in Latin America done in 2010. Some historical elements are analyzed in order to look for possible reasons for scarce works around the topic. It has to be taken into account that since then some advances must have happened in this field that are not shown here, however, the general orientations are still the same. Some researches are presented, as examples of how different theoretical-methodological perspectives contribute to the process of approximation to the comprehension of students as actors of the educational processes. It is assumed that every theoretical perspective entails a different interpretation of the social logic that underlies the interaction. My point of view in this work is that in order to better understand learning processes in educational context it is important to focus the research on children's ways of participation in classroom discourse at everyday lessons. A section is included in which I illustrate, with my own work and some others, what research focusing on student participation in science classrooms can bestow. I conclude with some discussion points such as the importance of more work on students' participation on classroom discourse in order to understand learning and scientific knowledge as a social construction in educational contexts.

Keywords

participation – students – science – classroom discourse – Latin America discourse

 | DOI: 10.1163/9789004409088_010

1 Introduction

Studies focused on spontaneous forms of students' participation in science classroom discourse are very unusual in Latin America. I will describe some historical elements from the international context, in order to analyze the possible reasons for this fact. Then I will review some researches conducted in this region as examples of the different theoretical-methodological perspectives that contribute to the comprehension of students as actors of the educational processes. I assume that every theoretical perspective entails a different interpretation of the social logic that underlies the interaction and every one contributes in different forms to our knowledge of what are and what the students do at the classrooms, as well as the educational impact of these actions. My point of view in this work is that in order to better understand learning processes in educational context it is important to focus the research on the ways children participate in classroom discourse. In this sense I include a section in which I illustrate with some works, including my own, what researches focusing on student participation in science classrooms from an "emic" approach can contribute, especially to understand students' points of view in relation to the construction of science knowledge at the classroom. I will conclude with some discussion points.

2 A Bit of History

Research on science education emerged in the 1960s under the pressure to improve scientific education in Western schools in order to compete with the USSR in the space race. First, attention was given to learning, so that we began with innovations based on psychological orientations, such as Bruner's "learning by discovery" (1963). Later, proposals based on so-called "conceptual change" (Posner, Strike, Hewson, & Gertzog, 1982; Driver et al., 2000) gained importance. The latter were based initially on genetic psychology (Piaget, 1967), which focuses on children as subjects who interpret the physical world from their own conceptions and change them through individual processes of construction (constructivism). In order to understand children representation of the natural world, beginning in the 1970s, a great deal of experimental research was developed (through clinical interviews and questionnaires) on the different conceptual notions of students of diverse ages and contexts (Flores & Gallegos, 1999). This research also studies the conditions that will allow the students to change from those notions to ideas closer to science.

The scarce impact that these proposals had in the classroom and the influence of research from other disciplines, such as educational anthropology, led to the recognition that conceptions are dependent on the social context in which they appear (Cicourel, 1974; Bruner, 1984). It is therefore necessary to study the construction of knowledge in the classroom.

At the same time, we saw the emergence of an awareness that the popularity of Piagetian constructivism has led to a conceptualization of the process of construction of knowledge as a self-structuring, basically individual activity that contrasts with the processes of shared understanding that are characteristic of the school settings (Edwards & Mercer, 1988). For Piaget, social interaction plays a role in learning only if the structures formed in interaction with the physical world already exist in the child.

For Vygotsky (1984), children representations of the physical world are strongly influenced by cultural categorizations they internalized from the social milieu. For him, and his sociocultural followers, knowledge and human thought are basically cultural. "I am increasingly aware that most learning is a community experience, a sharing of culture" (Bruner, 1984). In accordance with these conceptions, coming from psychological perspectives, we came to the idea that to make educational proposals it is necessary to understand the characteristics of the social construction of knowledge in the school as a cultural context.

Turning toward the classroom, research on science teaching also encountered the legacy of so-called new sociology (Young, 1971), in which descriptions are centered on the teacher as the one who holds power over the educational process. Studies on linguistics, such as those of Sinclair and Coulthard (1975), and from "constitutive ethnography", such as Mehan (1979), also reinforce the idea of teacher control exercised through the educational discourse defined by the Initiation-Response-Evaluation (IRE) structure. The teacher exercises power because he/she defines the questions and evaluates the responses, socializing pupils in the rules of educational discourse.

Sociology and anthropology also posited the need to link the educational experience with the social and cultural context outside the school, and approaches emerged, such as those of Bourdieu and Passeron (1977), which reiterate the role of the school as a reproducer of the dominant social relationships through the power that teachers exert over students.

An important legacy of anthropology, relevant to distinguish among studies of classroom cultures, is the difference between "etic" an "emic" approaches (Saville-Trioke, 1989). An "etic" description is based on concepts and categories "from the outside" of the analyzed context, useful when you want to compare similar settings, or evaluate them in terms of certain standards, as those developed for innovation in science education. In contrast, an "emic" approach attempts to describe a culture "from the inside", to understand members' ideas in terms of their own definitions, in the way they express them. This second perspective is better when you want to understand, before changing it, the complexity of classroom interaction, to study the habits, the culture or the meaning construction of participants from their own perspective (Elbers & Streefland, 2000).

All these conceptions represent references that influence the ways in which research on science teaching approaches the classroom. Studies generally focus on teacher participation and this orientation has made it very difficult to observe and examine student participation in the classroom as a central topic in education. Likewise, research on science teaching in the classroom usually assumes an "etic" approach because it remains focused on pedagogical experimentation with an educational proposal as a reference to see science construction, and then on the teachers as responsible for desirable educational changes. Among other factors, this is due to the pressure exerted by international funding agencies, which demand research to have impact on schooling. Fewer researchers have developed an "emic" perspective in order to study interaction in classrooms "from within", and less studies from this perspective are done in Latin America.

3 Students in Science Classroom Research

One of the first studies to place students as the object of research in the classroom are that of Willis (1981), who introduced the concept of cultural resistance in the face of reproductionist conceptions. Paul Willis finds that young people of working-class origin show resistance to learning, as part of a "counter school culture". In 1986 Frederick Erickson postulated that only teachers have "legitimate" power in the classroom because of their institutional position and greater knowledge of the topic. However, he explains that students have the power to resist to learn what teachers want to teach them. Erickson states that this resistance can become a form of power when it changes the dynamics of interaction.

In studies of classroom discourse, student participation is often thought of as interaction among peers or as the result of communication between an expert and an apprentice (Vygotsky, 1984), in which the teacher, as an expert, controls the content and the structure of the discursive exchange. Most studies of classroom interaction focus on teachers; consequently, student participation is analyzed in terms of how well the guidance or scaffolding set up by the teacher is followed (Edwards & Mercer, 1987; Cazden, 1986). However, recent studies suggest that this perspective limits interpretation from the start and does not allow for adequate reconstruction of the importance and meaning of student contributions to the social organization and content of the discourse negotiated with the teacher. Hugh Mehan (1979) also refers to students as active participants who build their own participatory context and do not just react to the teacher's orientations; however, this participation is far from being shown empirically in his examples. Other researches (Cazden, 1986; Dorr-Bremme, 1982) on student participation in the structure of the class tend to conclude that when the students intervene spontaneously, they do so primarily to ask for help or to inquire about procedures.

Neil Mercer (1995) explicitly states the importance of student contributions to the process of classroom communication, and recognizes that they influence the dynamics of the social construction of knowledge. Nevertheless, he mentions that opportunities for studying this influence are limited, and the description of the ways in which this happens is a starting point in his work that, he says, should be continued. Mercer himself recognizes that the emphasis on the teacher's power and control has led to underrating the potential significance of student contributions to classroom discourse. Before starting empirical analysis, he mentions that being a student implies, among other things, "asking about organization of a lesson, but in general not about its content (and never suggesting that the teacher might be wrong)" (Mercer, 1995, p. 45).

Jay Lemke's work (1990), from a sociolinguistic perspective, is relevant to the topic of this chapter because he shows that, in order to learn science, students have to understand how scientific discourse describes the physical evidence. This assertion states discourse as an important feature of science education. Lemke describes situations in which secondary school students take the initiative in an ongoing interaction, and their interventions influence the dynamics of classroom discourse. Nonetheless, Lemke's study is centered on the analysis of the thematic patterns of the connections among the meanings of words in science lessons According to him, the thematic patterns are those offered by teachers around science topics, and he compares those patterns the students possess and acquire in relation to the teacher's ones. However, the characterization of thematic patterns of science as a reference for analyzing the students' construction in the classroom can introduce an "etic" approach, because it can constitute an external judgmental stance for the wealth of understanding this discursive construction.

Recent studies, such as those based on co-constructivist (Vygotsky, 1987) and "community of practice" perspectives (Lave & Wenger, 1992), among other orientations, emphasize student participation; however, most of them are based on educational reports of how certain proposed innovations work in classroom experiences (Wells, 1993), and very few contribute to our knowledge of student participation in the everyday practices of common teachers who do not have any especial support.

Derek Edwards (1993) provided some of the first studies focused on children discourse in classroom interaction and contributed to questioning those works that take students' talk as evidence of what they really think (Donaldson, 1978; Driver, 1983), "as a kind of windows upon the mind". He, instead, takes children's discourse as descriptions, accounts, and explanations, constructed for the occasion, which can change in relation with the context.

This overview shows us that students seem to be the last ones to have been considered as the main point of reference of research in the classroom, especially in trying to understand their own views on the construction of scientific knowledge. However, in this chapter I will try to recover the analysis of everyday mechanisms of how educational practices develop, to capture the multifaceted educational experience, and focus on student participations. I particularly describe those studies that contribute to our understanding of the characteristics of student participation in "natural" classroom settings without previous categories to orient the perspective. The intention is to capture the logic of students' interventions, the presence of their interests in science classes and the way that they negotiate them with teachers, the resources they use to defend alternative ways to explain, their silences or sounds as forms of resistance or ways to defend versions that are not demanded in the classroom.

Since what is important is student participation in the discursive interaction at classrooms and the meanings that are interactively constructed, I will not take into account certain studies that, although undertaken in science classrooms, analyze teacher and student discursive participation as isolated utterances (Mares et al., 2004).

Classroom research is generally carried out through case studies. I understand that case studies cannot be taken as direct evidence of what happens in all science classrooms. Nevertheless, as Michael Billig states, "although a case study cannot proclaim its representativity, it is expected that an in-depth analysis may unveil characteristics and complexities that convey a wider generality" (1989, p. 2004).

4 Between "Natural" Research in the Classroom and Pedagogical Experimentation in Science Classes in Latin America

Pedagogical experimentation is the main orientation of research on science teaching in the classroom in our region. It involves designing a proposal to improve science teaching, based on some psychological theory or epistemological model regarding scientific construction (Adúriz-Bravo & Bonan, 2006; Quintanilla, 2007; Rojas-Drummond et al., 1998). This proposal is taken to the classroom to be tested in real teaching conditions, either by a teacher specifically trained to do this work or one who is not a teacher. Those teachers are frequently at the same time part of the research group in which they prepare themselves theoretically to understand the importance of conducting certain practices that will foster the scientific education of their students (Tobin, 1990).

These studies help demonstrate the potential and limits of a model when it comes into contact with the school context. However, in these cases, observation and analysis of the discursive interactions tend to focus on those aspects that the proposal wishes to highlight, and frequently other factors that are not previously conceived are not properly taken into account.

These investigations differ from research under natural conditions. The action-research perspective of these works and the narrow relationship that is sometimes established between researchers and teachers make it difficult to distinguish among what appears as a result of researchers' actions, what would have occurred if they had not participated, and what can be maintained after their direct influence.

Among the researchers with the broadest background and recognition in the field of science teaching in Latin America is the Brazilian Eduardo Fleury Mortimer. I will take up some of his works (Mortimer, 1998: Mortimer & Scott, 2003; Aguiar, Mortimer, & Scott, 2010) as examples of what can be obtained with regard to our topic, from studies that can be situated between pedagogical experimentation in the classroom and research on "natural" settings in science classrooms.

Mortimer and Scott's (2003) book is a representative text of Mortimer's contributions and his international collaborations. Its goal is to better understand science teaching and learning in actual classrooms and to comprehend how teachers introduce learners to the social languages of school science. The authors combine a sociocultural approach (Vygotsky, 1987) with a Bakhtinian use of social languages of science, speech genres of school science and the dialogic nature of understanding (Voloshinov, 1929), in order to analyze interactional episodes of secondary school chemistry classes in the north of England as well as in Minas Gerais, Brazil. An important theoretical reference for their perspective is the communicative approach in which they combine dialogical or authoritative communicative perspectives with interactive or non-interactive ones as an analytical framework to describe teacher orientation. In order to study a broader picture of the lessons in relation to scientific content they complement the micro-analysis with the analysis of what they named the "rhythm of discourse".

The book's analytical chapters examine two groups of science lessons. The first sequences are from the "natural" classes of a "good" teacher in the UK working on chemical reactions. They study the process of gradual decontextualization, where content and the rhythm of discourse are transformed under the teacher's guidance within a communicative cycle. Analysis shows that in some sequences there is a genuine exploration of students' ideas. The other group of sessions can be placed between natural sequences and pedagogical experimentation because the teacher, with a group of researchers, develops a proposal centered on debates among the students for the purpose of selecting one of their models of the structure of matter. In these interactions the authors find it relevant that some students "played the role of teacher". They also show the way students' talk construct science story going through a path of appropriation of scientific views with the help of teacher interventions after the debates.

Although it is not the main purpose of the research, the authors mention that student participation in classroom interaction can be place within one of four types of responses to the teacher and the content being learned: total *rejection* of scientific content and teacher orientations; *passive engagement* without participating in classroom interaction; *active engagement* in the lesson, asking questions and participating in the activities; and *taking the initiative* in searching for information and developing ideas beyond those required by the teacher, as reported by Arruda et al. (2001).

In a previous article Mortimer (1998) analyzes the discursive and conceptual dimensions of student patterns of evolution with respect to their ideas of matter (Newman et al., 1989). He describes "conceptual profiling" (the evolution of students' ideas) and movement from multivoicedness (several voices as versions of content) to a situation of univocality.

A recent work of this group of researchers (Aguiar, Mortimer, & Scott, 2010) focuses on the impact that students' questions have upon the teacher explanatory structures within innovative lessons. These lessons have been developed and tested by the group for secondary science classrooms in Brazil. They used their communicative approach (interactive/non-interactive vs dialogic/authoritative) and the patterns of interaction (IRE structures) to analyze spontaneous wonderment questions (Chin & Brown, 2002) of students within those lessons. The study is also based on some previous categories they developed in relation with Candela's work (1999) (continuation, extrapolation, and contestation). The analysis arrives at the conclusion that the wonderment questions of students can "continue" or "extrapolate" teachers' orientations, but they can also "contest" them, pushing the discourse dynamics towards a different direction than that chosen by the teachers. They found evidence of the dilemmas and tensions teachers have between orienting students to the "right" answer in order to "get through the lesson" using an authoritative approach, and staying at the dialogic communicative approach opened by the students' questions. The students' questions can be analyzed, then, in terms of these tensions. They also found that the communicative approach as well as the content and patterns of interaction emerge from the interaction between teacher and students, and not only from the teacher's discourse.

The relationship of sequences of micro-analysis with the categories of conceptual profile, communicative approaches, patterns of discourse and rhythm of discourse to obtain the evolution of students' conceptualizations and the impact of students have in classroom dynamics are important contributions from Mortimer's studies to the field, because different scales of analysis and different topics are articulated in order to describe the students' processes of construction of scientific representations, especially in relation to educational proposals.

This kind of work has an "etic" side as they construct an analytical framework to advance the study of the discoursive interaction. They are interesting in order to see students' contributions to science construction, but are not the best way to show students' meaning construction from their own logic or categories. The previous categories and frameworks tend to orient the interpretation towards what the researcher wants to see, rather than offering an understanding of other processes, resources and meanings the participants can use.

5 Research in "Natural" Settings of Classroom Interaction

The goal of investigating science education in "natural" settings of classroom interaction is to understand the complexity of the discursive interaction between teachers, students and educational knowledge. This type of research brings us closer to understanding the spontaneous participation of children in the classroom, although studies from different disciplines and theoretical approaches (psychology, anthropology, linguistics, pedagogy, history, sociocultural studies, and various orientations of research on discourse and modes of communication) show that we have yet been unable to comprehend, in all their complexity, the processes that occur in these educational contexts (Candela, Rockwell, & Coll, 2004).

Educational research in "natural" settings analyzes discursive interaction when the intervention of researchers is limited to different forms of observation and data gathering, attempting to influence as little as possible the everyday practices of teachers and students.

Ethnography is an "emic" research approach representative of investigation of "natural" contexts. It proposes to study the multifaceted school experience within specific times and spaces, recognizing the daily mechanisms and practices that operate in the classroom, and trying to understand why certain practices persist while others are modified (Rockwell, 1982) from the perspective of the educational actors: teachers and students. To achieve this, it is necessary to break with an evaluative viewpoint and avoid, as much as possible, the use of previous categories for the analysis. The dominant tendency of evaluating practice based on an ideal model is an obstacle for reconstructing processes and the meaning they have for the participants (Candela & Rockwell, 1991).

A pioneering study that opens the field of ethnographic research in the classroom in Latin America is that of Rockwell and Gálvez (1982), "Ways of Transmitting Scientific Knowledge: A Qualitative Analysis". This study provides some important results that were guides for later works. Following are some of its conclusions related to children's participation:

1. Children's responses constitute keys to their reasoning, but it is not possible to infer what children learned from what was transmitted in the school, since learning implies one's own cognitive processes not usually shown in classroom discourse.
2. The particular presentation of content in the classroom defines the limits of what is possible to learn from the school experience.
3. Students pay attention simultaneously to reasoning that follows the logic of the content presented by the teacher and reasoning that follows the logic of the interaction itself (Erickson, 1982).

These authors conclude that classroom knowledge is integrated and limited not only through a formal presentation scheme, but also through those elements that characterize the network of relations between teachers and students. In this way students are seen as relevant actors of knowledge construction in the classrooms whose participation must be analyzed within this network of relations.

Another important Latin American researcher in the field of science teaching is the Argentinean Ana Lía de Longhi. In her article "Teacher's and student's discourse: didactic analysis in science classes" (De Longhi, 2000), she declares to be more interested in carrying out a multidimensional analysis of "natural" classes in different secondary school science subjects than studying what is prescribed. She proposes to analyze the teaching-learning process as a complex system (Bertalanffy, 1976; Pérez Gómez, 1985). The analysis focuses on four levels (didactic context, analysis of interventions, didactic inferences, and conceptual synthesis). However, data are only presented for the second level, where she classify and quantify isolated questions or declarations from the teacher and students. It is interesting to observe that this study declares as one of its intentions the advisability of carrying out ethnographic research that would recover multidimensional discourse; however, discursive interaction is not analyzed. This work also concludes that students' participation is a response to the teacher's strategies and demands. Thus, the idea of teacher control is maintained. With regard to the scarce number of questions from students, they are requests to clarify information, confirm ideas or expand on a subject, such as Cazden posited in 1986.

This study represents another type of transition toward the need to incorporate theoretical orientations that distance themselves from proposals and recover spontaneous interventions. Nevertheless, the use of prior categorizations and the lack of analysis of the meanings constructed through the discursive interaction do not help to reach its proposed goals.

Among the few investigations in Mexico that have ventured into the analysis of "natural" science classes is the work of Miguel Ángel Campos H. and Sara Gaspar H. (2004) titled "Analysis of intertextuality and argumentation in the educational context". From a sociolinguistic point of view (van Dijk & Kintsch, 1983) the authors analyze the discourse, its intertextuality and argumentative form in an integrated manner. Following Caballero (2001), they understand intertextuality as a process constructed between the participants in the interaction, which produces common explicit and implicit discursive elements. They take Lo Cascio (1998) model to analyze the argumentative construction of the discourse, following its components: opinion, general rules (that shape the semiotic-logical connectors), counter-arguments, data, qualifiers, and reservations.

Several university-level science classes are analyzed regarding two topics of undergraduate biology (physiological adaptation of organisms and sexuality). The authors describe interactional sequences from the first topic in which the professor practically constructs the entire argumentation regarding the content. However, there is an analytical effort to recover the participation of the students in the process of constructing intertextuality. This is why they revisit aspects that have been the subject of very little study, such as students' affirmative gestures or taking notes, as signs of comprehension. In the second group of sessions the complex dynamic of constructing intertextuality is described through an open and participative conversational structure based on inverted dialogical pairs of I-R, since it is organized around students' questions and the professor's answers.

In their conclusions, Campos and Gaspar (2001) establish that the participation of the students in these classes is important for constructing intertextuality, both explicit and implicit, when revisiting the cases in which the students do not participate verbally. Likewise, they suggest that the argumentation in the conversational context is constructed in successive semantic blocks where the students also contribute to this joint construction of the argumentation. In fact, the conversational dynamics of science classes in higher education is considered, in this work, a discursive process of a predominantly argumentative genre. The interesting conclusion, then, is that "the lack of intertextuality is as important in the social process of constructing knowledge as is its presence, since it modulates the argumentative and cognitive process at a strategic level".

We can thus establish that these authors consider the participation of students fundamental for the construction of both scientific content as well as its argumentative forms, whether their participation can be explicit or implicit through less evident semiotic forms. In this way they refine the criteria between active and passive students and contribute to the understanding of their role in classroom discourse.

Argentinean researchers Ana G. Dumrauf and Silvina Cordero (2004) analyze "natural" science classroom sequences of a teacher considered an innovator, in a public high school of lower-middle class students, at La Plata, Argentina. The study is carried out from a multidimensional analysis perspective (Leander & Brown, 1999) that provides a variety of nuances and levels that can be recovered in the analysis. This framework includes six different levels of analysis: conceptual, focal, discursive/symbolic, institutional, social and affective, and for each one there is a description of a dynamic of movement and negotiation defined as (in)stabilities.

In the analyzed discursive extracts they show a process that starts out with the students' prior ideas about heat and change through reasoning and argumentation among the participants. Through the focal dimension the authors recover the way that the teacher takes up again the students' ideas and opinions and resorts to improvisation in order to make them change, showing him as a reflexive "teacher-artist". With regard to institutional instability, the manifestation of different "voices" in the classroom is analyzed. From the discursive-symbolic point of view, they find that the teacher shifts back and forth between macro- and microscopic models of heat and temperature, showing the content construction. In the social and affective dimension, as well as the conceptual one already described, the students' participation in this session appears more clearly. The dynamics introduced by students' different positions, sometimes manifest and other times as silences, but equally influential, results in a relaxed climate that allows them to express their moods and feelings.

In this study, the authors' theoretical approach, but also their openness to "what the data say", allow them to show high school students who spontaneously influence classroom dynamics with their points of view, with their arguments, and who do not always let themselves be convinced by the teacher.

The two last works, as well as Rockwell and Gálvez's, are also good examples of studies that contribute to the comprehension of students' participation in science classroom discourse. It is because, even though they do not focus the research in students' perspective, they carefully analyze the details of their interventions in order to study their influence on the teachers' discourse and on the classroom dynamics in relation with scientific knowledge construction.

6 From the Students' Perspective

Since the middle of the 1980s I got involved in investigating how the proposals for teaching science that we developed in the Educational Reform of the 1970s for primary education in Mexico entered the classroom.[1] I revisited the ethnographic approach initiated by Rockwell and Gálvez (1982). However, the purpose of my research was explicitly to analyze student interventions in experimental activities proposed in the new textbooks (Candela, 1997a). I was interested in understanding their participation, when and how students were discursively engaged, and the meaning of the content that they negotiated with teachers. However, I was trying to analyze them from the students' point of view, from an "emic" approach.

I carried out empirical research in several 4th and 5th grade classrooms in public primary schools in marginalized areas of Mexico City, in which the teachers were volunteers selected among those in the school community itself that were considered "good" teachers. These teachers had no special training nor were given any suggestions about how to do their work. In many cases, the teachers had been even described as traditional.

I found that the pedagogical proposals, and especially the problem solving that were posed in the textbook through experimental activities, were transformed when they entered the social space of the classroom. Teachers often modified them turning into demonstrations, possibly in an attempt to maintain more control over the constructed knowledge. However, students' questions and comments generally reopened inquiries that transformed the activity into a new problem, in a second transformative move (Candela, 1997b). For example, after children in one lesson on the solar system performed a representation in which they moved around the classroom simulating the way the planets revolve around the Sun, a series of questions arose spontaneously that opened new queries about the causes of planetary dynamics (Why don't the planets crash into each other? Why do they possess two movements? What would happen if they stopped moving?) (Candela, 1991). These results are important because they indicate that in everyday settings proposals are reinterpreted by the participants and are articulated with other traditions and experiences, from both the teachers and the students. That is, the classroom is a living space where content acquires a life of its own in relation to the social dynamics in which the ideas, concerns, interests, and tendencies of the participants, and especially those of the pupils, are expressed.

In these first studies I also found that non-solicited participations on the part of students can generate argumentative contexts in which the children defend versions of the content that differ from the one the teacher tries to propose (Candela, 1997c). Sometimes they even generate an alternative consensus to the teacher's and the textbook's position (Candela, 1995). It is in these moments, when conflict talk (Grimshaw, 1990) appears, that the students' reasoning can be best examined. I found that they follow a logic related to the content in order to describe what they think, while simultaneously they follow the dynamics of social interaction so as to know when and how to participate (Candela, 1990).

In later studies I found the perspective of discursive psychology (Edwards & Potter, 1992) very powerful for an in-depth analysis of the children interventions and its effect on classroom interaction. The study of the turn sequence contributes to the analysis of the meaning construction the actors themselves bestow on the utterances expressed (Sacks, Shegloff, & Jefferson, 1974), paying attention to the rules of social interaction and not only the grammatical structure of the utterances. However, I kept the ethnographic perspective that allowed me to incorporate into the analysis the social and cultural context beyond what was revealed by the study of the discursive sequentiality.

In studies of discourse analysis (Edwards & Potter, 1992), it is thought that by speaking one carries out a contextual construction of concepts, which may produce multiple versions according to the everyday situation in which they are produced. In discursive psychology (Edwards & Potter, 1992) and rhetoric studies (Billig, 1987), speech is an activity situated in a discursive context, which constructs meaning, reality, and even cognition itself. From this perspective language is not an instrument for transmitting information, but a dynamic means of social action. Conversational analysis and social studies of science (Latour & Woogar, 1986; Gilbert & Mulkay, 1984) allowed me to go into greater depth with regard to students' treatment of experimental activities in order to understand science as a socially constructed topic, and not as an objective description of reality. In several analyzed extracts I found that the students do not "see" in experiments what the teacher establishes as evident. On other occasions they interpret the data differently from the way the teacher explicitly requests (Candela, 2002). Their participation is sometimes of active and explicit refusal and in other times they reject the teacher's versions through their silence. Silence becomes a powerful means of social action in classroom interaction as is stated by Campos and Gaspar (2004).

I continued with the study of argumentation, finding new situations of conflict talk (Grimshaw, 1990) in which the students participate spontaneously with alternative versions, providing unsolicited explanations or asking questions that teachers find hard to answer. The discursive resources that students use in conflict settings, where they demonstrate their discursive competence in defending and negotiating their points of view, are especially analyzed. Although these examples show students who do not allow themselves to be easily convinced, I also found interventions of children oriented toward the construction of consensus regarding the content, or who explicitly demand that the teacher says which version is correct (Candela, 1995). In this way they show an understanding and shared responsibility, contributing to the school's social role in constructing legitimated knowledge. In the same vein, I found that students frequently take responsibility for the classroom dynamics and participate as co-authors of institutional practices, helping teachers to carry out the educational tasks (Candela, 2005). In my work, the pair of argumentation/consensus appears as a conceptualization that allows for an analysis of the dynamics and tensions of the construction of knowledge and its forms of negotiation in the classroom.

By paying attention to the students and trying to understand the meaning and impact of their participation in the classroom from their perspective, the phenomena of power acquire new features. Students exercise power in the classroom not only as a refusal to learn (Willis, 1981), but because of their interest in particular topics of academic knowledge that move them to actively defend alternative versions they believe in. Students can break away from teacher's control even when the discursive structure has the IRE form (Candela, 1998). They can do this by rejecting the teacher's orientations to certain answers as they sustain other versions of knowledge. In their responses, the children make use of their relative autonomy to decide whether or not they will follow the teacher's orientation, depending on the academic task and their opinion about the specific topic content they are working on. There are also various examples of situations where the IRE form is maintained, but the students take the role of asking evaluative questions and of evaluating others' turns, thus reversing interactive roles. In the process, they appropriate those functions through which teachers develop their power to influence the discourse of others, demonstrating their role as active and competent communicators who can use the available resources of discourse in the classroom in order to influence other versions.

As Derek Edwards (1999, p. 2) says in the foreword of my book "Science in the classroom: students between argumentation and consensus construction": "the children in Candela's classrooms engage in argument and dispute, just as scientists do, where the nature of the world is at issue – being the outcome of, rather than the input to, their talk about it".

In recent research of physics classes at the university level (Candela, 2010), I studied the impact of sociocultural contexts on the definition of students' trajectories. A comparative analysis of the paths of undergraduate students of physics in the U.S. and Mexico turned out to be very illustrative. I found that while in the U.S. university life narrowed undergraduate student's participation in classroom discourse, in academic spaces and in diverse activities and relationships, in Mexico the opposite occurs and new possibilities of relationships and activities appear to broaden more during undergraduate studies. Some possible explanations are the differences in culture and history of the two university environments, in the features of the academic programs, and in the role that this disciplinary field plays in the economic structure of the two countries. In this work the cultural and historical topics show their relevance to understand the students' participation not only in classroom interaction but also in the constitution of social networks more extended than those of the academic disciplines, but yet related to them.

With an ethnographic perspective, the analysis made by Naranjo and Candela (2006) describes the knowledge used by a primary school teacher with no special training, in a normal classroom, when teaching science to a fourth-grade group that includes a blind student. The study shows the centrality of the students' participation in interactional practices. The teacher's activities are oriented by his everyday knowledge related to main aspects as how to ensure that students with heterogeneous characteristics make homogeneous progress in curricular content and how to resolve the tension between individual needs and the whole group progress. For example, the teacher had to discursively explain every physical activity in more detail and carefully described the textbook images and their meanings for the blind boy. These activities also benefited the rest of the group. From this work we may infer that the presence of students with special needs can diversify and enrich classroom dynamics for all children when the attitude, teaching everyday knowledge, and the teacher's interests are focused on students' understandings.

María Eugenia de la Chaussée is another Mexican researcher who has made multiple contributions to research regarding the role of students in the discourse in chemistry classes. She studies "natural" chemistry classes in three public universities in Mexico. From an ethnographic approach, De la Chaussée analyzes the ways in which students participate in the construction of academic knowledge as well as with regard to its discursive structure. In a study published in 1999 she reviews the type of knowledge and mental operations that university students put into play to predict different reactions in organic chemistry. She also analyzes their hypotheses, factual judgments and proofs regarding the conductivity of electricity of different chemical substances (De la Chaussée, 2010) and the argumentative strategies (examples, analogies, comparisons) used by them and the professors in chemistry learning and teaching (De la Chaussée, 2009; De la Chaussée & Rugarcia, 2007).

7 Final Remarks

Through this review I have attempted to describe the process through which Latin American studies on discourse in science classrooms at different educational levels and from different theoretical-methodological perspectives contribute to the understanding of students' participation in the classroom.

I found that pedagogical experimentation in the classroom is being articulated in different ways, including the analysis of spontaneous, unprogramd and unmodelated situations in which students' interventions appear on their own initiative. However, in this kind of "etic" research, the emphasis is placed on teacher participation, and on knowledge construction, and, thus, the conclusions indicate that students participate in content and forms basically at the behest of teachers. Very few of their analytical descriptions show the special influence of students' spontaneous participations on the scientific knowledge construction. One exception is the work of Aguiar, Mortimer, and Scott (2010), in which they elaborate an analytical framework in order to study the impact of students' spontaneous questions on the dynamics of scientific knowledge construction guided by the teacher.

We also found a variety of theoretical approaches to "natural" investigations of discursive interaction. Those analyses that, regardless of their expressed orientation, limit themselves to examining isolated utterances, such as that of De Longhi (2000), cannot interpret classroom dynamics and easily direct their conclusions toward the teacher's control over the participation of the students. We can say that the analysis of the relationships among the interventions (or the turn sequence) is fundamental in order to interpret meaning as socially constructed in context.

School contexts are constructed by networks of relations among teachers, students, and the academic content. Students' participations have to be understood within the context of these networks. The verbal part of these communicative networks can be studied by the analysis of the discursive interaction of the participants. From several discourse analysis perspectives as discursive psychology (Edwards & Potter, 1992), ethnomethodology, conversation analysis, and the theory of speech acts (Searle, 1969), speech is understood to be an action situated in a discursive context, and, thus, to be a situational construct that can change according to the context of the interaction. Then, the meaning of words in an utterance must be understood in reference to the previous turn, because the signification of that version is constructed for the occasion. In the same sense, next utterances must be analyzed in relation to this turn that conditions and give signification to the next utterances in the context of the discursive interaction (Sacks, Schegloff, & Jefferson, 1974). The meaning of the discursive actions is defined by the set of activities in the interactive situation that is being created (Wittgenstein, 1953). That is why the meaning of the utterances out of context is ambiguous and its definition can better reflect the researchers' than the participants' orientations.

The approaches used in studies such as those carried out by Campos and Gaspar (2004) and Dumrauf and Cordero (2004) (intertextuality and multidimensionality through (in)stabilities) represent ways of arriving at different outlooks, points of view that gather the dynamics of educational processes. These perspectives, without the commitment to come up with their own proposals, allowed the authors to recover the meaning of student participation in science classroom discourse. The authors' sensitivity in getting closer to the point of view of the educational actors is also a factor that helps understanding the classroom dynamics and approaching to a dense description (Geertz, 1981).

The non-verbal and implicit forms of participation encountered by Campos and Gaspar at the university level are important. This coincides with some statements I have gathered in interviews with university professors of physics for whom it is fundamental to pay attention to students' gestures in order to realize if they are following the arguments and explanations. One could suggest the hypothesis that, according to the educational level and cultural context, students' participation tend to follow different forms and modes of communication, which seem to be less explicit as the students advance in the educational levels.

In ethnographic studies carried out from the students' perspectives, such as my own and De la Chaussée's, we see active students with their own positions, which they defend and argue for. Naranjo and Candela (2006), like Dumrauf and Cordero, find that in everyday situations of classroom interaction, teacher improvisation is necessary in order to respond to unexpected and unsolicited students' demands. Our work shows that social participation dynamics do not only depend on the different interactional signals (as discourse timing, rhythm, tone intensity, emphasis, or sequential positioning) (Erickson, 1982) and on the various contextualizing cues (Gumpertz, 1977). We find the important influence of content significance for students in order for them to participate.

In the foreword of my book (Candela, 1999), Derek Edwards states: "This notion of scientific knowledge, as not just influenced, but constituted by a nexus of social-discursive practices, in which knowledge claims have to be worked up, defended, made robust, undermined, and so on, provides us with an understanding of how science 'proper' works, that can be applied to studies and practices of science education". And he concludes that the children presented in my analyses "seem to be participating in what Bruno Latour has called 'science in action' where knowledge is fluid and at issue, the product of social practices, the outcome of (rather than the basis for) reaching a consensus".

In conclusion, we can say that there is a variety of contributions to the study of students' participation in science classroom discourse, from different perspectives. Some works appreciate pupils' interventions primarily as followers of the teachers' demands in order to advance in their construction of scientific views. However, there is a major contribution to the understanding of students' participation in science classroom discourse when the studies take into account students' influence on classroom dynamics and on the teachers' interventions. Through the students' perspective we can better appreciate their influence in classroom knowledge, when the teachers have to change, modulate or negotiate it with the pupils.

From my point of view, there are two kinds of factors that can influence the quantity and quality of students' participations: (a) the teaching styles and strategies, the institutional context with its norms (flexibility, impositions, restrictions), the interest of the students in the activities and in the academic topic, general cultural aspects, such as the importance given to education in different contexts, and, no less important, (b) the study's analytical and theoretical approach, the proposed object of study, and the researcher's sensitivity to see and document what is not evident. However, much more work is needed in relation to the important purpose of analyzing students' participation on science classrooms discourse, in order to understand what do science means for them and to analyze learning and the construction of scientific knowledge as a social construction in educational contexts.

Note

1 The Reform consisted of free and mandatory textbooks for all children in the six primary school grades and books for teachers that contained the pedagogical programs, relevant scientific information and suggestions for working with the students.

References

Adúriz Bravo, A., & Bonan, L. (2006). Modelos y analogías en la enseñanza de la física [Models and analogies in physics education]. In M. Quintanilla & A. Adúriz-Bravo (Eds.), *Enseñar ciencias en el nuevo milenio: retos y propuestas* [Teaching science in the new millenium: challenges and proposals] (pp. 57–72). Santiago de Chile: Ediciones Universidad Católica de Chile.

Aguiar, O., Mortimer, E. F., & Scott, P. (2010). Learning from and responding to students' questions: The authoritative and dialogic tension. *Journal of Research in Science Teaching, 47*(2), 174–193.

Arruda, S. M., Villani, A., & Laburu, C. E. (2001). Perfil conceitual e/ou perfil subjetivo? In M. A. Moreira, I. M. Greca, & S. C. Costa (Eds.), *Atas do Terceiro Encontro Nacional de Pesquisa em Educação em Ciências* [Proceedings of the Third National Meeting of Research in Science Education]. Atibaia: ABRAPEC.

Bakhtin, M. (1997). *Marxismo e filosofia da linguagem* [Marxism and the philosophy of language]. São Paulo: Hucitec.

Bertalanffy, L. (1976). *Teoría general de los sistemas* [General systems theory]. México: Fondo de Cultura Económica.

Billig, M. (1987). *Arguing and Thinking: A rhetorical approach to social psychology.* Cambridge: Cambridge Unversity Press.

Billig, M. (1989). The argumentative nature of holding strong views: A case study. *European Journal of Social Psichology, 19*, 203–223.

Bruner, J. (1963). *El proceso de la educación* [The process of education]. México: UTEHA.

Bruner, J. (1984). *Acción, pensamiento y lenguaje* [Action, thought and language]. Madrid: Alianza.

Bourdieu, P., & Passeron, J. C. (1977). *La reproducción. Elementos para una teoría del sistema de enseñanza* [Reproduction. Elements for a theory of teaching]. Barcelona: Editorial Laia.

Caballero, R. M. (2001). Técnica del argumento y argumento de la técnica Heterogeneidad, intertextualidad e interdiscursividad en un texto informativo [The technique of the argument and the argument of the technique: Heterogeneity, intertextuality and interdiscursivity of an informative text]. *Discurso y Sociedad, 3*(3), 11–37.

Campos, M. A., & Gaspar, S. (2004). Análisis de la intertextualidad y la argumentación en el contexto educativo [Analysis of intertextuality and argumentation in the educational context]. *Revista Mexicana de Investigación Educativa, 9*(21), 425–449.

Candela, A. (1990). Investigación etnográfica en el aula: El razonamiento de los alumnos en una clase de Ciencias Naturales en la escuela primaria [Classroom ethnographic research: Students resoning at a science lesson in primary school]. *Investigación en la Escuela, 11*, 13–23.

Candela, A. (1991). Argumentación y conocimiento científico escolar [Argumentation and school scientific knowledge]. *Infancia y Aprendizaje, 55*, 13–28.

Candela, A. (1995). Consensus construction as a collective task in Mexican science classes. Vygotsky's theory of human development: an international perspective. *Special Issue of Anthropology and Educational Quarterly, 26*(4), 1–17.

Candela, A. (1997a). *La necesidad de entender, explicar y argumentar: Los alumnos de primaria en la actividad experimental* [The necessity of understanding, explaining and arguing: Primary education students in the experimental activity]. Ciudad de México: CINVESTAV/SEP.

Candela, A. (1997b). Demonstrations and problem-solving exercises in school science: Their transformation within the Mexican elementary school classroom. *Science Education, 81*, 497–513.

Candela, A. (1997c). The discursive construction of argumentative contexts in science education. In C. Coll & D. Edwards (Eds.), *Teaching, learning and classroom discourse: Approaches to the study of educational discourse* (pp. 89–106). Madrid: Fundación Infancia y Aprendizaje.

Candela, A. (1998). Students' power in classroom discourse. *Linguistics & Education, 10*(2), 139–164.

Candela, A. (1999). *Ciencia en el aula: Los alumnos entre la argumentación y el consenso* [Classroom science: The students between argumentation and consensus contruction]. Ciudad de Mexico, Buenos Aires, Barcelona: Paidós.

Candela, A. (2002). Evidencia y hechos: La construcción social del discurso de la ciencia en el aula. [Evidence and facts: The social construction of science discourse in the classroom]. In M. Benllock (Ed.), *La educación en ciencias: Ideas para mejorar su práctica* [*Science education: Ideas to improve its practice*] (pp. 187–215). Barcelona: Paidós.

Candela, A. (2005). Students' participation as co-authoring of school institutional practices. *Culture & Psychology, 11*(3), 321–337.

Candela, A. (2010). Time and space: Undergraduate Mexican physics in motion. *Cultural Studies of Science Education, 5*(3), 701–727.

Candela, A., & Rockwell, E. (1991). La construcción social del conocimiento escolar [The social construction of school knowledge]. *Infancia y Aprendizaje, 55*, 9–12.

Candela, A., Rockwell, E., & Coll, C. (2004). What in the world happens in classrooms? Qualitative classroom research. *European Educational Research Journal, 3*(3), 692–713.

Cazden, C. (1986). Classroom discourse. In M. C. Wittock (Ed.), *Handbook of research on teaching*. New York, NY: McMillan Publishing Company.

Cicourel, A. (1974). Some basic theoretical issues in the assessment of the child's performance in testing and classroom settings. In A. Cicourel, K. H. Jennings, R. Leiter, R. McKay, H. Mehan, & D. R. Roth (Eds.), *Lenguage use and school performance*. New York, NY: Academic Press.

De la Chaussée, M. E. (2009). Las estrategias argumentativas en la enseñanza y el aprendizaje de la química [Argumentative strategies in chemistry teaching and learning]. *Educación Química, 20*(2), 143–155.

De la Chaussée, M. E. (2010). ¡Sí, es cierto! ¡El ácido acético diluido en agua conduce la electricidad! [Yes, it is correct! Acetic acid diluted in water conduces electricity]. In R. Bernal Cuevas, & L. del C. Montes Pacheco (Orgs.), *Experiencias compartidas en educación. Quince Años en la Formación de Ingenieros* [Shared experiences in education. Fifteen years of engineer education]. Ciudad de México: UIAP, BUAP, UDLAP, UPAEP, UAT, CIFI, Lupus Magister.

De la Chaussée, M. E., & Rugarcía, A. (2007). Profesor es que eso es lo que yo no entiendo [Lo que entienden los alumnos universitarios sobre el concepto de mecanismo de reacción] [Professor, it is that this is what I don't understand (what university students understand about the concept of the mechanism of reaction)]. *Educación Química, 18*(2), 107–113.

De Longhi, A. L. (2000). El discurso del profesor y del alumno: Análisis didáctico en clases de ciencias. [The teacher's and the student's discourse: A didactic analysis in science lessons]. *Enseñanza de las Ciencias, 18*(2), 201–216.

Donaldson, M. (1978). *Children's minds*. London: Fontana.

Dorr-Bremme, D. W. (1982). *Behaving and making sense: Creating social organization in the classroom* (PhD thesis). Universidad de Harvard, Cambridge, MA.

Driver, R. (1983). *The pupil as scientist?* Milton Keynes: Open University Press.

Driver, R., Newton, P., & Osborne, J. (2000). Establishing the norm of scientific argumentation in classroom. *Science Education, 38*, 39–72.

Dumrauf, A., & Cordero, S. (2004). ¿Qué cosa es el calor?: Interacciones discursivas en una clase de Física [What is heat? Discursive interactions in a physics lesson]. *Revista Electrónica de Enseñanza de las Ciencias, 3*(2), 123–147.

Edwards, D. (1993). But what do children really think? Discourse analysis and conceptual content in children's talk. *Cognition and Instruction, 11*(3–4), 207–225.

Edwards, D., & Mercer, N. (1987). *Common knowledge: The development of understanding in the classroom*. London: Routledge.

Edwards, D., & Potter, J. (1992). *Discursive psychology*. London: Sage.

Elbers, E., & Streefland, L. (2000). "Shall we be researchers again?" Identity and social interaction in a community of inquiry. In H. Cowie & G. van der Aalsvoort (Eds.), *Social interaction in learning and instruction. The meaning of discourse for the construction of knowledge* (pp. 35–51). Amsterdam: Pergamon.

Erickson, F. (1982). Classroom discourse as improvisation: Relationship between academic task structure and social participation structure in lessons. In L. C. Wilkinson (Ed.), *Communication in the classroom* (pp. 153–181). New York, NY: Academic Press.

Erickson, F. (1986). Qualitative methods of educational research. In M. Wilkinson (Ed.), *Handbook of research on teaching* (pp. 119–161). New York, NY: McMillan Publishing Company.

Flores, F., & Gallegos, L. (1999) Construcción de conceptos físicos en estudiantes. La influencia del contexto [Physics concepts construction by students: The influence of the context] *Perfiles Educativos, XXI(85–86)*, 90–103.

Geertz, C. (1981). *La interpretación de las culturas* [The interpretation of cultures]. México: Gedisa.

Gilbert, N., & Mulkay, M. (1984). *Opening Pandora's box: A sociological analysis of scientists' discourse.* Cambridge: Cambridge University Press.

Grimshaw, A. (1990). *Conflict talk: Sociolinguistic investigations of arguments in conversations.* Cambridge: Cambridge University Press.

Gumperz, J. (1977). Sociocultural knowledge in conversational interference. In M. Seville-Troike (Ed.), *Linguistics and anthropology.* Washington, DC: Georgetown University Press.

Latour, B., & Woolgar, S. (1986). *Laboratory life: The construction of scientific facts.* Princeton, NJ: Princeton University Press.

Lave, J., & Wenger E. (1991). *Situated learning: Legitimate peripheral performance.* Cambridge: Cambridge University Press.

Leander, K. M., & Brown, D. E. (1999). You understand, but you don't believe it: Tracing stabilities and instabilities of interaction in a physics classroom through a multidimensional framework. *Cognition and Instruction, 17*, 93–135.

Lemke, J. (1990). *Talking science: Language, learning and values.* Norwood, NJ: Ablex Publishing Corp.

Lo Cascio, V. (1998). *Gramática de la Argumentación* [The gramar of argumentation]. Madrid: Alianza Editorial.

Mares, G., Guevara, Y., Rueda, E., Rivas, O., & Rocha, H. (2004). Análisis de las interacciones maestra-alumnos durante la enseñanza de las ciencias naturales en primaria [Analysis of the teacher-students interactions during science teaching in primary school]. *Revista Mexicana de Investigación Educativa, 9*(22), 721–745.

Mehan, H. (1979). *Learning lessons: Social organization in the classroom.* Cambridge, MA & London: Harvard University Press.

Mercer, N. (1995). *The guided construction of knowledge: Talk amongst teachers and learners.* Philadelphia, PA: Multilingual Matters Ltd.

Mortimer, E. F. (1998). Multivoicedness and univocality in classroom discourse: An example from theory of matter. *International Journal of Science Education, 20*(1), 67–82.

Mortimer, E. F., & Scott, P. (2003). *Meaning making in secondary science classooms.* Maidenhead, PA: Open University Press.

Naranjo, G., & Candela, A. (2006). Ciencias naturales en un grupo con un alumno ciego: Los saberes docentes en acción [Science in a group with a blind child: Teacher knowledge in action]. *Revista Mexicana de Investigación Educativa, XI*(30), 821–845.

Newman, D., Griffin, P., & Cole, M. (1989). *The construction zone: Working for cognitive change in school.* Cambridge: Cambridge University Press.

Pérez Gómez, A. (1985). *La comunicación didáctica* [The pedagogical communication]. Málaga: Universidad de Málaga.

Posner, G. J., Strike, K. A., Hewson, P. W., & Gerzog, W. A. (1982). Accomodation of a scientific conception: Towards a theory of conceptual change. *Science Education, 66*(2), 211–227.

Quintanilla, M. (Ed.). (2007). *Historia de la ciencia: Propuestas para su divulgación y enseñanza* [History of science: Proposal for its dissemination and teaching]. Santiago de Chile: Editorial Conocimiento.

Rockwell, E. (1982). Etnografía y teoría de la investigación educativa [Ethnography and educational research theory]. In *Enfoques: Cuadernos del Tercer Seminario Nacional de Investigación en Educación* (pp. 29–56). Bogotá: Centro de Investigación, Universidad Pedagógica.

Rockwell, E., & Gálvez, G. (1982). Formas de transmisión del conocimiento científico: Un análisis cualitativo [Transmission forms of scientific knowledge: A qualitative analysis]. *Educación, 42*, 97–139.

Rojas-Drummond, S., Hernandez, G., Velez, M., & Villagran, G. (1998). Cooperative learning and the acquisition of procedural knowledge in primary school children. *Learning & Instruction, 8*(1), 37–61.

Sacks, H., Schegloff, E. A., & Jefferson, G. (1974). A simplest systematic for the organization of turn taking in conversation. *Language, 50*(4), 696–735.

Saville-Troike, M. (1989). *The ethnography of communication.* Oxford: Blackwell.

Schön, D. (1987). *La formación de profesionales reflexivos. Hacia un nuevo diseño de la enseñanza y el aprendizaje de las profesiones* [Educating the reflective practitioner. Toward a new design for teaching and learning in the professions]. España: Paidós.

Searle, J. R. (1969). *Speech acts.* Cambridge: Cambridge University Press.

Sinclair, J. M. H., & Coulthard, R. M. (1975). *Towards an analysis of discourse: The English used by teachers and pupils.* London: Oxford University Press.

Tobin, K. (1990). Research on science laboratory activities: In pursuit for better questions and answers to improve learning. *School Science and Mathematics, 90*, 403–418.

Van Dick, T., & Kintsch, W. (1983). *Strategies of discourse comprehension.* New York. NY: Academic Press.

Voloshinov, V. N. (1929/1973). *Marxisim and the Philosophy of Language.* Cambridge, MA: Harvard University Press.

Vygotsky, L. S. (1984). *A formação social da mente* [The social formation of the mind]. São Paulo: Martins Fontes.

Vygotsky, L. S. (1987). *The collected works of L.S. Vygotsky.* New York, NY: Plenum.

Wells, G. (1993). Reevaluating the IRF sequence: A proposal for the articulation of theories of activity and discourse for the analysis of teaching and learning in the classroom. *Linguistics and Education, 5*, 1–37.

Willis, P. (1981). The class significance of counter school culture. In P. Woods & H. Hammersley (Eds.), *The process of schooling.* London: The Open University Press.

Wittgenstein, L. (1969). *On certainty.* Oxford: Blackwell.

Young, M. (1971). *Knowledge and control: New directions for the sociology of education.* London: Collier.

CHAPTER 11

Turning Points in Communicative Approaches to Science Classroom Discourse

Eduardo Fleury Mortimer and Phil Scott

Absctract

In this chapter we will explore examples of genuine problems that emerge when we consider that the alternation between dialogic and authoritative discourse is a norm and not an exception in science classrooms. Accordingly, the transitions between dialogic and authoritative discourse – the turning points – will be critical for planning teaching sequences. We are going to examine two teaching sequences in which these transitions take place. The first one occurred in a small town in the north of England, in the context of teaching forces to a grade 7 science class, and the second occurred in Brazil, in the context of application of a thematic project for a class of chemistry at grade 11. In analyzing these teaching sequences we will foreground some characteristics of what constitutes a turning point: (1) turning points have an entry and an exit, which can be highlighted by the teachers; (2) turning points can be phenomenon-driven or model-driven, depending on how differences are created. In the phenomenon-driven, the differences are created between students' predictions and observations, while in the model-driven differences arise as different students support different models; (3) that using turning points is a base for meaningful learning and metacognitive reflection. We conclude that major conceptual turning points can be anticipated and planned for, although most of them occurred in practice as a spontaneous way of teaching.

Keywords

turning points – communicative approaches – science classroom discourse – teaching sequences – planning

 | DOI: 10.1163/9789004409088_011

1 Introduction

The way dialogic discourse is used in science classroom has become a myth. It is as if everything could be introduced by dialogue and there were no right or wrong answers in natural sciences. However, much of the school science discourse is constituted by an authoritative discourse, with which the interlocutor cannot interfere. If we take just one example, one of the best-known laws of chemistry, Lavoisier's Law, we can explain this point better. This law says that mass is conserved in a chemical reaction, i.e., the mass of the products must equal the mass of the reactants, which is a translation of an event that occurs at the submicroscopic world, where the types and numbers of atoms in the products are the same as in the reactants, only rearranged. There is no way of sharing knowledge with chemists and be unaware of this law. And there is no way to dialogue with this law. In this case, either we assume, for the sake of school chemistry, that all reactions are to behave in this way, or we refuse the law and do not have access to the world of chemists. That is what characterizes the authoritative discourse of science.

This does not mean that dialogue is not important in science education. It is fundamental. When students are learning this law, we have an example of how important the dialogue can be for learning science. When a student is faced with reactions involving gases as products or as reactants, she tends to reject this law. After all, we burn trash to get rid of it. Every time we have a gaseous product we tend to assume that the mass is not conserved. Or when you burn steel wool and the resultant product weight more than we had initially we also have difficulties in accepting this law. So at the beginning of the learning process, we have to dialogue with different points of view taken by students if we want them to learn the Lavoisier's law. It will be through this dialogue that they will make sense of this law, and this is going to happen until the moment when the teacher can assume that everyone understands the law and starts using it univocally.

The fact of the matter is that science is an authoritative discourse which offers a structured view of the world and it is not possible to appropriate the tools of scientific reasoning without guidance and assistance. Learning science, as well as training professional scientists, inevitably involves acquiring the tools of "normal science" (Kuhn, 1962), and the canonical ways of reasoning in science (Anderson, Holland, & Palinscar, 1997). For any science teacher it is not sufficient to engage students in dialogue about their everyday views of phenomena; there is the additional and central responsibility of introducing the science perspective.

A reasonable question to ask at this point might be "why bother with the initial dialogic approaches if the teacher is bound ultimately to introduce the authoritative school science view?" The fundamental idea here is that meaningful learning involves making links between ways of thinking and talking, between everyday and scientific views of basic phenomena. The initial dialogic approaches offer the opportunity for students to express their everyday views and then later to see how these views relate to the school science perspective. In addition we would argue, based on our experience of teaching and researching in science classrooms, that the dialogic engagement is potentially motivating for students, drawing them into the problem at hand, and legitimizing their expression of whatever ways of talking and thinking they possess. In this way, the initial dialogic approaches address the teaching purposes of "opening up the problem" for the students and allowing the teacher to "explore and probe students' views" (Mortimer & Scott, 2003).

At this early stage, scientific ideas must be put into dialogue with the common ideas that people have about their world, which they express through their common language that is also a theory of experience (Halliday, 1998). There is, however, another space for dialogue in science classrooms that goes beyond this initial confrontation of scientific and everyday ideas. We can characterize this new space as another phase in the teaching sequence in which, once the students have acquired a minimum of coherence in their use of scientific discourse, they are invited to expand on employing the scientific view, so that the teacher can transfer to them the responsibility for its use. In these phases the dialogue will reappear, but at this time there will not be a dialogue mediated by the confrontation of common and scientific languages, but a dialogue (ideally at least) totally situated in the heart of the scientific language. Students will try, in possession of the tools that school science has given to them, to understand new problems and visit new phenomena. In doing so they will show inevitably different points of view, which characterize dialogic discourse. In this respect, the dialogue in this new phase is similar to the dialogue that science does in its boundaries, when scientists seek to expand science borders by working on new problems in possession of conceptual tools that are shared by the scientific community. These shared conceptual tools constitute the authoritative discourse but are used to generate different points of view that are in dispute. Consequently, more than one voice is heard in these moments.

In this chapter we will try to explore examples of genuine problems that emerge when we consider that the alternation between dialogic and authoritative discourse is a norm and not an exception in science classrooms. If this is true, then the transitions between dialogic and authoritative discourse will be critical for planning teaching sequences. Accordingly, we are going to examine two teaching sequences in which these transitions take place. The first one occurred in a small town in the north of England, in the context of teaching forces to a grade 7 science class, and the second occurred in Brazil, in the context of application of a thematic project for a class of chemistry at grade 11. The project dealt with the quality of water from a lake, which was located in Belo Horizonte, a big city in the Southeast of Brazil. In the first of these sequences we will have the opportunity to explore the transition between dialogic and authoritative discourse that happens at the beginning of a teaching sequence, in which dialogic discourse is expressed in common language and the turning point has to deal with the encounter between common language and scientific language. In the other sequence, in response to a question of how fish breathe, the students express different points of view but just in scientific language. We will seek to explore the similarities and differences between these two turning points, and reach a more general characterization of turning points. We will attempt, then, to answer two research questions: (1) which are the main characteristics of a turning point? (2) Which are the differences between turning points that happen in the beginning and the ones that occur later on in a teaching sequence? Before exploring the classroom episodes that illustrate the transitions, it is desirable to expand a little more on what authoritative and dialogic communicative approaches are.

2 Dialogic and Authoritative Communicative Approaches

The distinction between authoritative and dialogic discourse, as we put forward, is in fact based on the distinction made by Bakhtin between authoritative and internally persuasive discourse. According to Bakhtin:

> Authoritative discourse permits no play with the context framing it, no play with its borders, no gradual and flexible transitions, no spontaneously creative stylizing variants on it. It enters our verbal consciousness as a compact and indivisible mass; one must either totally affirm it, or totally reject it. (1981, p. 343)

Bakhtin exemplifies authoritative discourse in terms of religious dogma, an acknowledged scientific truth, the words of a parent, of teachers, and so on. Internally persuasive discourse, by contrast, is

> tightly interwoven with 'one's own word', it is half-ours and half-someone else's, it represents more than one voice or conceptual horizon. (Bakhtin, 1981, p. 343)

A practical way of dealing with this distinction is to frame the problem in terms of the number of points of view. We have defined dialogic discourse (Mortimer & Scott, 2003; see also Scott, Mortimer & Aguiar, 2006) as being that which is open to different points of view. At different points in a sequence of science lessons, dialogic talk inevitably takes on a different character. Thus, at the start of a lesson sequence, the science teacher might elicit students' everyday views about a particular phenomenon. Later on in the sequence, the teacher might encourage students to discuss how to apply a newly learned scientific idea in a novel context.

In the first situation the dialogic discourse involves collecting students' everyday views. A fundamentally important point here is that this kind of dialogic interaction can be played out with different levels of interanimation of ideas (Bakhtin, 1981). At one extreme the teacher might simply ask for the students' points of view and list them on the board. Here the discourse is open to different points of view, but there is no attempt to work on those views through comparing and contrasting. The teacher's approach involves a low level of interanimation of ideas. On the other hand, the teacher might adopt an approach which involves trying to establish how the ideas relate to one another (John thinks that this might be the case, but Susan seems to be suggesting something different. Nancy what do you think?). Both of these approaches are dialogic in the sense of allowing the space for different ideas to be represented, but the second approach clearly involves a higher level of interanimation of ideas. It might be the case that the teacher simply collects ideas at the start of a teaching sequence (low interanimation) and then, later in the sequence, compares and contrasts these ideas with the school science point of view (high interanimation).

In the second situation, the dialogic discourse might involve the students in working together to apply a new (to them) scientific idea to construct an explanation for a novel problem. Here we might imagine the students agreeing on some points and disagreeing on others, but working together to understand any points of difference (Oh! I see what you mean!) as they develop their explanation. The agreeing and disagreeing on points of view constitutes an ongoing, dialogic interanimation of ideas.

In general terms we can say that dialogic discourse is open to different perspectives. There is always the attempt to acknowledge the views of others, and through dialogic discourse the teacher attends to the students' points of view as well as to the school science view. Within dialogic discourse, there is the possibility of different levels of interanimation of ideas.

By way of contrast, authoritative discourse does not allow the bringing together and exploration of ideas. Here the teacher focuses attention on the school science point of view. If ideas or questions which do not contribute to the development of the school science story are raised by the students, they are likely to be reshaped or ignored by the teacher. Alternatively, if a student's idea is perceived by the teacher as being helpful to the development of the scientific story, it is likely to be seized upon and used. In these ways, authoritative discourse is closed to the points of view of others, with its direction having been set in advance by the teacher. More than one voice may be heard, through the contributions of different students, but there is no exploration of different perspectives, and no explicit interanimation of ideas, since the students' contributions are not taken into account by the teacher unless they are consistent with the developing school science account.

3 Turning Points: Methodological Issues

According to our view, the authoritative presentation of ideas alone cannot ensure meaningful learning. It is important that students have the opportunity both to make explicit their everyday ideas at the start of a teaching sequence and to apply and explore newly learned scientific ideas through talk and other action for themselves. Within the context of high school science classrooms, where dialogic discourse is universally rare, there is a tendency for it to fade out altogether as the students appropriate the school science point of view (see, e.g., Amaral & Mortimer, 2004). Thus, the paradoxical situation exists where the most fluent exponent of scientific ideas (the teacher) does all of the talking whilst the novices (the students) have little or no opportunity to speak the scientific language for themselves and to make it their own. We would argue strongly that if we expect students to engage in meaningful learning in the science classroom, they should be allowed to play with the "sharply demarcated" (Bakhtin, 1981) authoritative discourse of science in new situations, expanding its possibilities for application, making links to other areas of science, and constructing meanings that are new for them. Students need to engage in the dialogic process of exploring and working on ideas, with a high level of interanimation, within the context of the scientific point of view.

Thus, a teacher always has to alternate between dialogic and authoritative communicative approaches. If this alternation is the rule, one of the most important events in the classroom should be the *turning points,* which first and foremost are identified in terms of a change in communicative approach during the staging of a teaching intervention.

The research questions outlined above were addressed through a paired case study approach in which the teaching interactions of two teachers, one expert and the other novice, were monitored. The expert teacher works with a teaching sequence (about 5 hours) focusing on the topic of forces, with grade 7 students in a secondary school in a rural area of the North of England. The novice teacher is a student-teacher doing her teacher training, in which she had to develop a project called "Water in Focus: quality of life and citizenship". This project had the aim of training beginner teachers in the use of investigations in authentic problem solving context. The project is carried out each year, lasts for 3 months, and is organized with pairs of student-teachers working with 15/16 year old students in each of the participating public high schools. The central theme of the project is the quality of water in an urban environment focused on the Pampulha Lake, one of the top tourist attractions of the city of Belo Horizonte, being the specific focus. The lake was constructed in the 1940s and since then it has suffered a process of degradation with the increase of population around it and the various ensuing problems, such as domestic and industrial sewage. The activities of the project provide a contrast with traditional school science problems, which have a single right answer, and explore two real questions:

1. Should people be allowed to have direct contact with the lake water?
2. Should people be allowed to catch and eat fish from the lake?

We see the research reported on this chapter as constituting a first step in developing a typography of turning points and as such collected data from secondary and high schools with the aim of investigating as wide a range of approaches as possible. The lessons of both teachers were videotaped with a single camera focused on the teacher and all teacher-student interactions were audio recorded.

The data analysis involved examining the videotapes (using Videograph® and Transana®), to identify all instances of turning points. A typography of turning points was then developed based on the structure and functions of those points. A key part of this development involves identifying unambiguous classes of types of turning points and then being able to allocate examples of turning points to the appropriate classes. In carrying out the analyses of the types of turning points we took a number of steps to maximize the credibility of our interpretations and, therefore, to establish their trustworthiness (Lincoln & Guba, 1985, p. 301). First, two of us worked on the data to develop interpretations through a process of statement and review (moving back and forth between transcripts and interpretations). This was complemented by a third researcher taking on the role of 'peer debriefing' (op cit, p. 308) in probing and challenging the emergent interpretations and the bases on which these were made.

In what follows we are going to present two sequences that contain planned turning points, in the sense that they happened as a consequence of the planning of the lesson by the teacher. Both are characterized by the direction of the change in the discourse, from dialogic to authoritative. There are many differences between the context of the two teaching sequences, but the main difference that will be explored here is the language in which the dialogic discourse is played: whether everyday language or scientific language.

4 Results

4.1 *Case A: Cup on the Table Narrative*

This case focuses on teaching and learning about the normal force, more specifically, that a table exerts an upward force on a cup which is placed on it. In this case we are reporting on a short teaching sequence which has been used successfully on many occasions with children in the 11–14 age range.

The starting point here is that children of this age typically find it difficult to believe that inanimate objects such as a table can exert a force on a cup. The following transcripts (Kibble, 2006) show an interview between a researcher/teacher and two 14 year old girls.

> *Teacher:* What forces do you think are acting on the cup?
> *Girl 1:* There's *gravity*!
> *Teacher:* Is there? Tell me about that.
> *Girl 1:* Gravity pushes it down to the surface
> *Teacher:* OK and what do you think? (*turning to another girl*)
> *Girl 2:* Like ... (*comes to a stop*)
> *Teacher:* Is there any other force ... other than gravity?
> *Girl 2:* No (*shakes head*)
> *Teacher:* No? Just one force acting on the cup. OK!

When the researcher/teacher asks the first question, *'What forces do you think are acting on the cup?'*, the two girls stare unblinkingly at the cup. Words don't come easily. The forces acting on the cup? What forces? The girls are viewing this event (or what looks like a non-event to them) from an everyday point of view. As far as they can see there are probably NO forces acting on the cup. Nothing is pushing it. Nothing is pulling it. The cup is not moving. It's just sitting there doing nothing! 'Gravity' is offered as a possibility, but without great conviction.

The big question is how do we get students such as these two girls to be able to explain this event and talk about it from a scientific point of view? How might we convince the girls that the table is actually pushing up on the cup? The challenge here is to get the girls to construct a theoretical entity (the upward force of the table on the cup) which for them does not exist. How might we address this teaching and learning challenge?

4.1.1 Activity 1: The Concept Cartoon

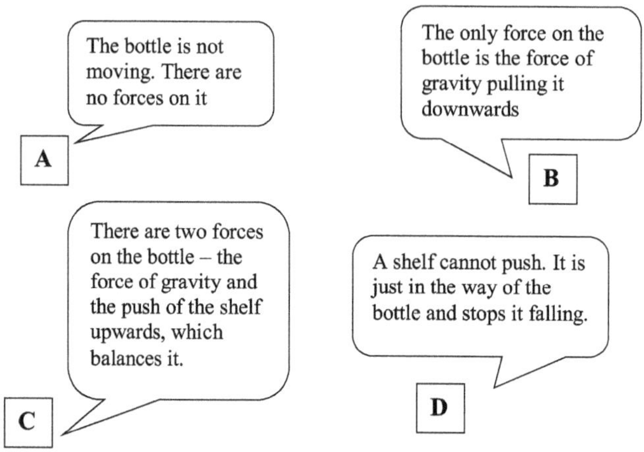

The teacher, in this case Jonathan, started the activity by organizing his class into pairs and giving each pair of students a concept cartoon showing four possibilities for the forces acting on a bottle on a shelf.

The students talk in pairs about each of these statements indicating whether they 'agree' or 'disagree' or are 'not sure' about each one. Each pair then works with another pair to compare views and to reach a consensus within the group of four. Finally Jonathan calls the class around the table at the front of the room. As we join the lesson he is talking to the whole class and the concept cartoon 'Bottle on a shelf' is projected onto the white board:

1. *Teacher:* Now I tell you what, if I was in one of your groups I'd have found that pretty confusing because of the number of different ideas .... Now I was over there with Josie and with Ryan and with Jordan and Kerry. Now they were looking at this and I tell you what ... they really didn't agree at all. There was a fundamental – that means a really important – disagreement. So I'm going to ask them if they can lead off for us and just have a look at some of the ideas they talked about ...
2. *Josie:* Well like, I don't think that a table can push. Cos gravity pulls, it's a force ... but a table can't push upwards, it's just in the way of the erm ... that's all.
3. *Teacher:* Right. Let's have a listen to what she's saying there. She's talked about the force that a lot of you have talked about, *gravity*. She's told us where she thinks that is, and what she thinks that's doing. But the disagreement between the two of them is whether the *table* can *do* anything.

> Now I think when I was listening to Ryan that he was here [*points to cartoon statement C*]. That there are two forces on the bottle, the force of gravity and the push of the shelf up which balances it, and I know 'balance' is a word that a few of you were using. And I think that Josie is here [*points to cartoon statement D*], a shelf cannot push it is just in the way of the bottle and it stops it falling. Now let's use that as a starting point …

Just as in the previous case, Jonathan has opened up a *dialogic space* (Wegerif, 2007) where students are able to express their ideas. The concept cartoon prompts the students to talk through the various possible models and it is clear that differences in points of view have been created. In fact, as Jonathan comments, the students *'really didn't agree at all'*. Josie expresses clearly the view that 'a table can't push up', which is in disagreement with the likes of Ryan, who thinks that there are two forces on the bottle. Jonathan summarizes this situation using a non-interactive/dialogic approach.

Having opened up these differences in students' thinking, the question is: what might be the next step?

In fact, what happened in practice was that the end of the lesson arrived with the issue of whether or not the table is pushing up on the bottle left unsolved. This was not entirely unplanned. As an expert teacher, Jonathan favors the use of 'cliff-hanger' lesson endings where big questions are left unsolved, to be returned to in the next lesson.

4.1.2 Activity 2: Review and the Turning Point

We return to the class for the next lesson which was taught two days later. As the class is coming into the room, some of the students are over-heard arguing about whether or not the table can push up:

> *Student 1: 'Of course it can't, it's just a table!'*
> *Student 2: 'But it must be able to, to balance the forces!'*

There is clear evidence here that the concept cartoon exercise has been effective in engaging at least some students in thinking about whether or not tables can push. Jonathan starts the lesson and refers back to the debate from the previous lesson:

1. *Teacher:* I'd like to get you to think about one of the ideas that you really *argued* about on Monday. You really *argued* about something on Monday and you did it right at the front here and you did it in little groups. What was the idea that you were arguing about? Josie was in the middle of this and Jordan was in the middle of this argument. What were you arguing about? Josie?

2. *Josie:* That a table can't push up.
3. *Teacher:* Yeah ... it was this argument about the table not being able to push up to support the gravity force that was pulling down on that heavy bottle.

At this point Jonathan has gone as far as possible in opening up the different ways of modeling the bottle on the shelf. He and the class have reached the *turning point* for this particular intervention. Jonathan now states quite clearly what is to come next:

> *Teacher:* What I want to do .... I want to leave you this morning ... with a picture of something that might help you to believe that *that* [*knocking on the table*] can push up. Now this is a very logical little argument, so you're gonna have to follow it through.

In this way, Jonathan handles the turning point with some sensitivity. It is certainly not a case of stating, 'Well as it happens Ryan was right and Josie was wrong'. Rather, we have the teacher offering a *'picture that might help you to believe'*, *'a logical argument'* which *'you're gonna have to follow'*. It is clear that the turning point has been reached and now an authoritative, logical argument is on offer. At the same time Jonathan is acknowledging that this explanation might take some believing.

4.1.3 Activity 3: The Balloon Demonstration

Jonathan now invites one of the students, Sam, to come out to the front of the class, where he hands him an inflated balloon:

1. *Teacher:* To start off with, what I'd like to do is to just get Sam to put his hands on either side of this balloon and gently squeeze it together. In fact come and stand up for me Sam [*Sam stands at the front of the room*] ... So could you just gently, like that, push that together [*Sam squeezes the balloon between his hands*]. Now what's he doing to the shape of that Sharon?
2. *Sharon:* He's making it flatter
3. *Teacher:* Going flatter. Now you're going to have to describe this, what you're doing now [*talking to Sam*]. So just let go. What are you doing with this bottom hand Sam?
4. *Sam:* Sort of like pushing it [*in quiet voice*].
5. *Teacher:* He says he's sort of like pushing it. But you *really are* pushing it, aren't you? That hand is pushing up at the same time as that one is pushing down [*teacher stands next to Sam and gestures up and down*]. In doing that he's changing the shape of the balloon isn't he? Now if you put that

balloon on the desk now. Can everybody see? Put your hand on top there now and do the same thing [*Sam pushes down on the balloon on the desk*]. Not too hard. What's he done to the shape of the balloon there Sean?

6. *Sean:* pushing it down ...
7. *Teacher:* He's pushing it down. What's he done to the shape?
8. *Sean:* Flattened it.
9. *Teacher:* Flattened it. Now, he's only got one hand on there at the moment. Where on Earth is the other force that's changing the *shape*? Where is the other force that's changing the shape? Let's hear a few people telling us.
10. *Holly:* From the table.
11. *Teacher:* Holly says the *table is* pushing. Levi what do you say?
12. *Levi:* I think the table is pushing
13. *Teacher:* The *table* is pushing. What do you say Penny?
14. *Penny:* the table is pushing
15. *Teacher:* You know this seems totally logical to me that if he changed the shape using two hands like that. If you take one hand away and push it down and get the same shape, something else must be pushing. What else do you think is pushing? Do you agree? [asking Sam]
16. *Sam:* [nods yes]

In this way Jonathan enlists the help of one of the students, Sam, and presents an argument to suggest that the table can push up, focusing attention on the forces acting on a balloon. He achieves this over turns 3–12 by taking an interactive/authoritative communicative approach, played out through I-R-E patterns of three. This pattern of interaction continues until Turn 12 when Holly provides the correct response that the other force is 'From the table'. Jonathan then conducts rapid confirmatory exchanges (Edwards and Mercer, 1987) with Levi and Penny prior to concluding the episode with an authoritative statement in Turn 17. In this way Jonathan exits the turning point by 'presenting' a logical argument centered on the analogous case of the balloon.

One key characteristic of this turning point is that here the impetus for learning comes from the differences in the students' views about the *models* presented in the concept cartoon, whereas in some cases the impetus for learning comes from observing a *phenomenon*. Here there is no phenomenon which is open to dispute. There is no arguing about whether or not bottles stand on tables. The point at issue lies with how that situation is modeled in terms of forces. In this case, therefore, the impetus for learning is generated by the students engaging in the modeling task with the concept cartoon. Here creating differences involves setting up differences in students' views about possible models.

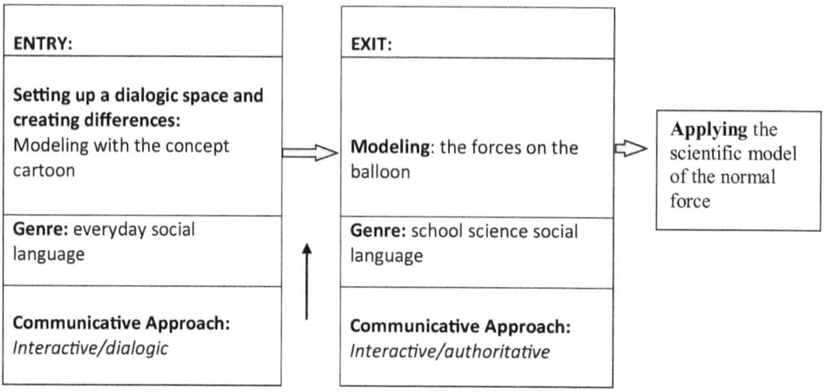

FIGURE 11.1 Turning point entry and exit on the cup on the table

4.2 *Case B: The Water in Focus Project*

This project was developed as the final part of the chemistry teaching training. The students-teachers already had a first module, which lasted for one semester, where they had opportunities to teach and to reflect on their teaching, which was videotaped. The first module was done in an experimental high school of the university. The aim of this new module was to expand the students' abilities, mainly by introducing a scientific literacy project which was inquiry-based. Thus, with the project they should develop an entirely new approach to science teaching, but without losing their already gained abilities. This means that for introducing the scientific concepts that were the basis for developing the analysis of chemical and biological parameters for the quality of water they should adopt an approach that was based on the combination of communicative approaches, teaching purposes and patterns of interactions, something they became familiar with in module 1.

We are going to exemplify a turning point through the introduction of oxygen dissolved in water. Initially, we have planned a section in which the teaching purpose was *exploring and working on students' views* (Mortimer & Scott, 2003), which necessarily involves a dialogic discourse in which students' different points of view are presented, leading to a turning point in which the matters are closing down towards the scientific point of view. This section involves the students working in groups, trying to answer four questions, the first one we are going to discuss here:

1. The fish breathe oxygen. From where do they get the oxygen they need?
   A: Do you think that this oxygen came from water molecules, which the fish break down for breathing?
   B: Or does it come from the oxygen that is dissolved in the water?

We followed Cristina, one of the student-teachers, on this part of the project and have videotaped her in carrying out this discussion with the students. Let's see what happened.

Do the fish break or not the water molecule?

1. *Cristina:* Let's have a sense of which of the options in question 1 you have chosen. Who chose that the fish break the water molecule for breathing?
2. *Around ten students* raise their hands.
3. *Cristina:* And who has chosen the other option, that the oxygen that is breathed by the fish is dissolved in the water?
4. *Around twenty students* raise their hands.
5. *Cristina:* Who are going to justify the first option, that of the fish breaking water molecules for breathing? Márcia?
6. *Márcia:* I think that there are quite a lot of fish in a river or in a lake. The only way that they get to breathing is to break the water molecule.
7. *Cristina:* And how do they break the water molecules?
8. *Márcia:* I don't know
9. *Cristina:* Who else can offer a different view on that? Carlos?
10. *Carlos:* I think water is made up from ions, H^+ and OH-. So, the fish don't need to break water molecules, they got the oxygen they need from the OH-.
11. *Cristina:* Good. Any other answer related to this option? Marcelo?
12. *Marcelo:* I think the fish can break the water molecule in O_2 and H_2, like we did once in science, electrolysis of water. But the fish have a different mechanism to do so, I'm not sure.
13. *Cristina:* Ok, anyone else? No? Let's turn to the other option. Who would like to justify why the fish get the oxygen they need from the oxygen dissolved into water? Cláudia?
14. *Cláudia:* I read this in a magazine. Although there is not a great amount of oxygen dissolved in the water, the fish have gills which they use to take the oxygen from the water.
15. *Cristina:* And how do they do that? How do they take the oxygen from water?
16. *Cláudia:* I'm not sure, I think the gills has a fine structure through which the fish can take the oxygen.
17. *Cristina:* Good. Any other answer? Marcos?
18. *Marcos:* I think we would need a lot of energy to break a molecule of water, like we had electricity when we did electrolysis. So, how is a fish going to have this energy to break water molecules if it has to obtain energy from respiration?

19. *Cristina:* Good point Marcos. Any other answer?

Up to this point, Cristina is encouraging the students to engage in interactive/dialogic interactions. According to her purpose of *exploring and working on students' views*, she has equal praise for all good answers, irrespective of their correctness (see turns 11, 17 and 19). She also offers feedback to students in order to know *how* the fish break the molecules or to know *how* the fish breathe oxygen from water. She did this for both sides in dispute: she did in turn 7 for one side and in turn 15 for the other. There is no evaluation in her discourse, and this neutral tone allowed the students to offers their contributions. Thus, we have small chains of I-R-F-R with each of the students and not a final evaluation.

The form and purposes of this episode is precisely the same as the 'Cup on the table' episode analyzed previously. Here we have the students offering their views on two possible models to account for how fish get their oxygen from the water. In the 'Cup on the Table' episode the students considered four different models for the forces acting on the cup. In both cases the consideration of models was used to create differences (Ogborn et al., 1996), with some students supporting one model and other students supporting another. These differences contribute to the creation of an impetus for learning. In the case of the 'cup on the table' the students are eager to resolve the problem in its own right. For the present case the students are not only keen to find the correct answer *per se*, but they are also keen to use this understanding in further developing the water project.

A difference between this case and the 'Cup on the table', concerning the turning points, is that the language used to express the points of view are much closer to the scientific language in the 'Water in Focus project' than in 'Cup on the table'.

In what follows, Cristina has to close down the matters, as she was not having any further different points of view on the case. In other words, Cristina has made the decision that the class has reached a turning point in the consideration of this particular conceptual problem.

26. *Cristina:* Ok folks, we are going now to consider each of your responses trying to get what really happens when fish breathe. So, let's begin with Marcelo's answer, that the fish break the water molecule in O_2 and H_2. Do you remember when you did the electrolysis of water? What do you get? Flávia?
27. *Flávia:* I think we got two gases, one in more quantity then the other.
28. *Cristina:* Which one was in great quantity?
29. *Flávia:* I'm not sure, is the oxygen?
30. *Cristina:* Which one was in great quantity, Marcelo?

31. *Marcelo:* It is the hydrogen, exactly a double in relation to oxygen.
32. *Cristina:* Good, and did you tested the properties of the two gases? Marcelo?
33. *Marcelo:* Yes, one has exploded with the flame, the hydrogen. And the other made it burn more brightly.
34. *Cristina:* Very good Marcelo. And now, can you imagine what would happen with lakes and rivers if the fish break the molecules of water? Carlos?
35. *Carlos:* They would burn?
36. *Cristina:* Yes, they would burn if only hydrogen is left, as the oxygen would be taken by the fish. So, we have clear evidence against the fish breaking down the water molecule. Besides, we have the explanation from Marcus that this break involves spending of energy and energy is also a product of respiration, so it is one of the things fish get when they breathe. It would be nonsense to have to spend energy. It is that ok for you?
37. *Several students:* Yes.
38. *Cristina:* Good. I think this left us with the other answer, that the oxygen that is breathed by the fish is dissolved into water. We should consider the argument that said that there are a lot of fish in a river or in a lake, so how can they survive with just the small amount of oxygen that is dissolved? Would anyone help me? Marcus?
39. *Marcus:* I think rivers and lakes have a huge surface, so oxygen is always dissolving there from the air. So, there is no problem with the quantity. The fish can get the small amount they need with their gills.
40. *Cristina:* Very good. So, the question 1 is settled down. Let's pass to question 2.

The turning point is clearly marked, as Cristina adopts an authoritative communicative approach. This is indicated by her intervention on turn 26 ("trying to get what really happens when fish breathe"), but is also evident from her evaluation of the answers obtained, which took more than one form. In turn 30, for example, we have a repetition of the question stated in turn 28, which means that the answer was not the desired. We also have a clear pattern of interaction from turn 30. All the turns of Cristina carry an evaluation of the last answer from the student and a new initiation. Thus, it has the form I-R-AI-R-AI …, which means that she is playing the interaction with the students through IRA triadic pattern. It is worth noticing that she did not comment the answer from Carlos, which could be considered an answer that would be difficult to analyze and refute ("I think water is made up from ions, H^+ and OH-. So, the fish cannot break the water molecule, they got the oxygen they need from the OH-"). Nevertheless, she comments all the other interventions from the students.

5 Discussion: The Key Features of Turning Points

Having considered, in some detail, these two examples of teaching and learning around turning points, we now turn to identifying their general features.

5.1 *Shift in Communicative Approach*

As stated earlier, turning points first and foremost are identified in terms of a change in communicative approach during the staging of a teaching intervention. In the examples provided so far, the switch in communicative approaches has been from dialogic to authoritative and this signals the passage from an exploration of ideas to focusing on the development of the scientific point of view.

5.2 *Turning Point Entry and Exit*

Turning points have an entry and an exit. In the cases set out above, the entry side involves activities to open up a dialogic space and to create differences,

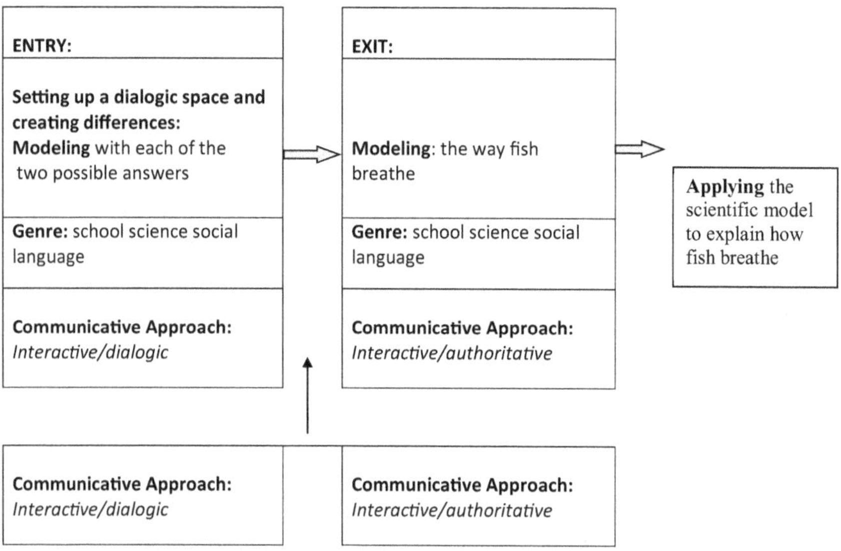

FIGURE 11.2 Turning point entry and exit on how the fish breathe

leading to an impetus for learning. The difference was created, in both cases, by using reference to alternative conceptions that are very well documented in the literature. In the case of the 'Cup on the table', it is known that students tend to reject the existence of forces acting on objects when there is no movement (Hestenes, Wells, & Swackhamer, 1992). A cup on the table has no need to be explained. Thus, the concept cartoon created the differences, raising the point that might be a force upwards acting on the cup to balance gravity. The impetus for learning emerges from the difference created and helps drive the students through the turning point so that they come to understand the scientific point of view.

5.3 *Phenomenon-Driven and Model-Driven Approaches*

Both cases set out above relate to modeling phenomena. The key difference between them is that in the first case students started from an everyday way of thinking and came to understand a scientific model while in the second students were already using scientific tools in trying to answer the question on how fish breathe. In both cases there is no immediate phenomenon being questioned and the focus of the teacher and students lies on alternative ways of modeling. We think it helpful to refer to this kind of teaching approach as being *model-driven*. We also have examples of *phenomenon-driven* teaching approaches to turning points, in which the movement through the turning point depends upon the observation of a phenomenon and the selection of a discrepant event. For example, when we asked students to observe a metal and a wood block and predict their temperature, they tend to say that the metal block is cooler, although they are in the same room. If we have a hole in each of the blocks, we can have their temperature measured and observe that both blocks have the same temperature. Here, the impetus to learning comes from the observation of a discrepant event related to a phenomenon.

In this phenomenon-driven teaching approach:

- Opening up a dialogic space raises students' and teacher's awareness of the students' reasoning about a phenomenon.
- This reasoning is challenged by a key demonstration of the phenomenon (a discrepant event).
- Differences are created between students' predictions and observations, rather than between students' views.
- The creation of differences provides an impetus for learning which leads to the introduction of the scientific point of view.
- In the model-driven teaching approaches:

- Opening up a dialogic space involves comparing different models about a particular phenomenon.
- Differences arise as different students support different models.
- This difference in students' ideas creates an impetus for learning which leads to the introduction of the scientific point of view.

5.4 *From Everyday Ideas to the Scientific Point of View versus Deciding Which Scientific Point of View Is Better to Model the Phenomenon*

A fundamental aspect in thinking about turning points is that opening up the problem on the entry side may acquire different formats, depending on the nature of the task. They can be just the first step towards developing the scientific point of view, as in the 'Cup on the table' case. In this case, the activities on the entry side of the turning point lay the ground for the development of the scientific point of view, setting up a condition of readiness for learning which is taken up on the exit side.

Introducing the scientific point of view is not a trivial matter and, as we have seen, might involve a carefully developed argument and the use of a specifically selected analogy. The development of the scientific point of view is always made in response to the students' starting points identified earlier in the dialogic exploration activities.

The second case raises a very important question to the way dialogic discourse has been used in classrooms. We have already said that within the context of high school science classrooms dialogic discourse is universally rare, which means that there is a tendency for it to fade out altogether as the students appropriate the school science point of view. Thus, there is a tendency to consider the students' different points of view only at the beginning of teaching sequences, where there are important everyday ideas that might influence the learning process. Nevertheless, as the second case presented here shows, the students can formulate alternative points of view within the scientific discourse. To put it differently, having acquired fundamental aspects of the authoritative discourse of science, students can formulate different points of view to explain a phenomenon.

The different explanations that were raised in the second case presented here were genuine attempts to articulate different aspects of the phenomenon that were in discussion. Thus, students that opt for the response that oxygen came from water molecules, which fish break down for breathing, had to articulate their answer by considering a process in which this break down of water molecules occurs. One of the students said that this process would be the same that he had done in the electrolysis of water and remembered well the electrolysis products. Another student imagined that the fish could take the oxygen from the ionized species OH-. Here also we have a correct understanding of how the water molecule is broken into ions. All these students are playing with the authoritative discourse of science to formulate a point of view regarding the question raised. This leads again to dialogic discourse, as different points of view entered the discourse. The big difference is that these different points of view are well articulate answers based on the authoritative discourse of science.

If we want that students acquire the discourse of science, one of the ways is offering new authentic problems that can be answered within this discourse. In this sense, a genuine inquiry-based science is a real opportunity for students to use their scientific discourse to make progress. But these students must have an authoritative discourse to articulate and make progress in the lessons.

5.5 *A Basis for Meaningful Learning*

To work with students on both sides of the turning point offers an opportunity for them not only to learn the correct scientific model but also to develop an awareness of how the scientific point of view relates to everyday reasoning. This point takes us back to the notion of the conceptual profile (Mortimer, 1995; Mortimer et al., 2010) and what is meant by meaningful learning. As we see it, a deep understanding of any subject matter involves being aware of the full profile of understandings and when it is appropriate to use them. For example, a teacher such as Jonathan feels secure about his students' understanding of the normal force because they both understand the scientific point of view (that surfaces can exert forces) and everyday reasoning (which suggests that they can't). The aim of an approach to teaching and learning through negotiating turning points is that students as well as the teacher should share this depth of understanding, making links between everyday and scientific reasoning and, also, between different scientific concepts (Scott et al., 2011).

5.6 *A Basis for Metacognitive Reflection*

Students' metacognitive awareness about their own thinking is raised through asking for their points of view on the entry side to the turning point. In this way the students have the opportunity to gain a sense of the path of their own learning as they negotiate the turning point.

5.7 *Turning Points and Learning Demands*

A final general point to raise from these cases concerns the nature of the turning point and the magnitude of the learning demand (Leach & Scott, 2002) involved. In the first narrative outlined here, the differences between the everyday and scientific social languages are significant: the learning demand is high. For such cases we would argue that the teaching around the turning point needs to be carefully thought through and planned, based on whatever insights to students' reasoning are available. Such turning points sit prominently in the science curriculum and experienced teachers will be aware that careful teaching is needed in these pedagogical corners even if they have never heard of the term 'turning point' (which will certainly be the case!). It is our view, as suggested earlier, that such *major conceptual turning points* can be anticipated and planned for. Teachers can gain experience and expertise in and around these areas of teaching before going on to consider others. Also, there is a literature in alternative conceptions and informal ideas in science (see, for example, Pfundt & Duit, 1994), which can be used by teachers and science educators alike to plan some of the turning points for key concepts in school science.

By way of contrast to the major conceptual turning points discussed so far, there are also situations where the teacher and students play out a shift in communicative approach in a much more spontaneous way. Examples of one such *spontaneous turning point* is something to be considered in other articles, since we have also documented it in our research.

6 Conclusions

In this chapter we explore the characteristics of turning points, which, as shown, can be planned by the teacher. This gives us a good indication that the alternation between dialogic and authoritative discourses should be a norm and not an exception in science classrooms. A well-elaborate transition from a discourse to another becomes an extremely sensitive point in the planning of the teacher. We sought to show that the transition between dialogic and authoritative discourses can occur anywhere in a teaching sequence, not only in its beginning, when students are still imbued with their everyday ideas. In advanced stages, when students have mastered the scientific tools that allow them to risk a hypothesis to solve a problem, different points of view may again appear, but most of these points of view may be well grounded in scientific discourse. This discourse is shared with the scientific community and, therefore, represents a single point of view, being an authoritative discourse.

That this dialectic between authoritative and dialogue discourses can occur in classrooms seems to be fundamental for students to progress in their understanding of science. The stronger in articulating authoritative discourse the students become, the greater the chance that they take the risk of offering different points of view in solving problems, reinstalling the dialogic discourse. For inquiry-based teaching to be successful, it is essential that the domain students have of authoritative discourse is each time bigger, as this will allow them to take the risk of formulating hypotheses and proposing models. In this case an attempt to solve practical problems only occur with a wide domain of a discourse that is shared by the scientific community and, therefore, is univocal, authoritative.

Accordingly, for a better understanding of the discursive practices that happen in real science classrooms we have to leave the easy acceptance that the discourse of science teaching has to be only dialogic and try to understand the transitions between dialogic and authoritative discourse, i.e., the turning points. This chapter only begins to develop what can be a promising research in our area.

References

Amaral, E. M. R., & Mortimer, E. F. (2004). Un perfil conceptual para entropía y espontaneidad: una carcterización de las formas de pensar y hablar en el aula de Química [A conceptual profile of entropy and spontaneity: a characterization of ways of thinking and speaking in the chemistry class]. *Educación quimica, 15*, 218–233.

Anderson, C. W., Holland, J. D., & Palinscar, A. S. (1997). Cannonical and sociocultural approaches to research and reform in science education: the story of Juan and his group. *Elementary School Journal, 97*, 359–383.

Bakhtin, M. M. (1981). *The dialogic imagination* (M. Holquist, Ed.; C. Emerson & M. Holquist, Trans.). Austin, TX: University of Texas Press.

Edwards, D., & Mercer, N. (1987). *Common knowledge: The development of understanding the classroom*. London: Routledge.

Halliday, M. A. K. (1998). Things and relation: regrammaticising experience as technical knowledge. In J. R. Martin & R. Veel (Ed.), *Reading science. Critical and functional perspectives on discourses of science*. London: Routledge.

Hestenes, D., Wells, M., & Swackhamer, G. (1992). Force concept inventory. *The Physics Teacher, 30*, 141–158.

Kuhn, T. S. (1962). *The structure of scientific revolutions*. Chicago, IL: University of Chicago Press.

Leach, J. T., & Scott, P. H. (2002). Designing and evaluating science teaching sequences: An approach drawing upon the concept of learning demand and a social constructivist perspective on learning. *Studies in Science Education, 38*, 115–142.

Lincoln, Y. S., & Guba, E. G. (1985). *Naturalistic inquiry*. Newbury Park, CA: Sage.

Mortimer, E. F. (1995) Conceptual change or conceptual profile change? *Science & Education, 4*, 267–285.

Mortimer, E. F., & Scott, P. H. (2003). *Meaning making in secondary science classrooms*. Maidenhead: Open University Press.

Mortimer, E. F., Scott, P., Amaral, E. M. R., & El-Hani, C. N. (2010) Modeling modes of thinking and speaking with conceptual profiles. In S. D. J. Pena (Ed.), *Themes in transdisciplinary research* (pp. 105–139). Belo Horizonte: Editora UFMG.

Ogborn, J., Kress, G., Martins, I., & McGillicuddy, K. (1996). *Explaining science in the classroom*. Buckingham: Open University Press.

Pfundt, H., & Duit, R. (1994). *Bibliography: Students' alternative frameworks and science education* (IPN Reports in Brief, 4th ed.). Kiel: University of Kiel.

Scott, P. H., Mortimer, E. F., & Aguiar, O. G. (2006). The tension between authoritative and dialogic discourse: A fundamental characteristic of meaning making interactions in high school science lessons. *Science Education, 90*, 605–631.

Scott, P. H., Mortimer, E. F., & Ametller, J. (2011). Pedagogical link-making: a fundamental aspect of teaching and learning scientific conceptual knowledge. *Studies in Science Education, 47*, 3–36.

Wegerif, R. (2007). *Dialogic education and technology: Expanding the space of learning*. New York, NY: Springer.

CHAPTER 12

Analyzing Discursive Interactions in the Context of Evolution Teaching with a Conceptual Profile of Adaptation

Claudia Sepulveda, Eduardo Fleury Mortimer and Charbel N. El-Hani

Abstract

Following a sociocultural perspective on learning, we report an analysis of meaning making about Darwinist explanations during discursive interactions in biology classrooms. To perform this analysis, we used a conceptual profile of adaptation in combination with a framework developed by Mortimer and Scott (2003) as a tool for analyzing classroom discourse, grounded on the dialogic theory of language of the Bakhtin circle and Vygotsky's account of the development of higher mental functions. We discuss how the results obtained through this kind of analysis make it possible to characterize in semantic, linguistic, and social terms the relationship between discursive interactions and meaning making in the science classroom, and bring contributions to the planning of teaching interventions, especially, to the management of communicative approaches and language use in the interactions between teachers and students, as a way of increasing the likelihood that students appropriate the school science perspective. An important asset in the science classroom is to explicitly approach the meaning of different ways of speaking in order to support the students in the task of becoming aware of the distinctive features of everyday and school science social languages. In the case of teaching about the theory of evolution by natural selection, it is important that the students are stimulated to master ways of speaking closer to the variational perspective zone in our adaptation profile model. This can be done, for instance, by promoting their construction of utterances in which the organisms appear as objects of the evolutionary process instead of utterances in which the organisms appear as subjects of evolutionary change; and by reconstructing narratives produced in a language characterized by personification and anthropomorfization to talk about evolution processes into narratives of chains of events without clear protagonist agents, which result in the evolution of populations of organisms.

 | DOI: 10.1163/9789004409088_012

Keywords

conceptual profile – teaching evolution – meaning making – discursive analysis

1 Introduction

Since the mid-1990s, science education research has given more attention to discourse analysis as a way of understanding meaning making in science classrooms, due to the influence of sociocultural perspectives on learning. Different studies have highlighted, from a variety of points of view, the importance of investigating classroom discourse and other rhetorical devices in science education (see, e.g., Candela, 1999; Halliday & Martin, 1993; Kelly & Brown, 2003; Kress et al., 2001; Lemke, 1990; Mortimer, 1998; Mortimer & Scott, 2003; Ogborn et al., 1996; Roth, 2005; Scott, 1998; Sutton, 1992). This new direction for science education research (Duit & Treagust, 1998) signals a move away from studies focusing on individual students' understanding of specific phenomena towards research into the ways in which understandings are developed in the social context of the science classroom. Following a Vygotskian research tradition, more emphasis has been given to the role of social mediation, through language and other socially constructed symbolic systems, in meaning making in the instructional context of the science classroom (Mortimer & Smolka, 2001; Mortimer & Scott, 2003).

The present study is included in this research tradition. Here, we report results derived from an analysis of meaning making about Darwinist explanations during discursive interactions in biology classrooms. To perform this analysis, we used a conceptual profile of adaptation in combination with a framework developed by Mortimer and Scott (2003) as a tool for analyzing classroom discourse, grounded on the dialogic theory of language of the Bakhtin circle and Vygotsky's account of the development of higher mental functions.

Conceptual profiles are models of different modes of seeing and conceptualizing the world used by individuals to signify their experience, and have been fruitfully used in a series of studies to analyze science learning in the classroom (Mortimer, 1995, 1998, 2000; Mortimer & Amaral, 1999; Amaral, 2004; Amaral & Mortimer, 2004, 2006; Coutinho, 2005; Coutinho et al., 2007; Sepulveda, 2009, 2010; Mortimer & El-Hani, 2014). Initially inspired by Bachelard's (1940) epistemological profile, they were at first developed as an alternative to Posner et al.'s (1982) conceptual change model (Mortimer, 1995, 2000). In particular, they challenged one of the central ideas in this model, namely, that students should be led to break away with their previous conceptions when learning science.

Conceptual profiles are built for a given concept and are constituted by several zones, each representing a particular mode of thinking about that concept, related to a particular way of speaking. The zones of a profile model are individuated by ontological, epistemological and axiological commitments that structure different modes of interpreting the concept at stake. Associated with these modes of thinking, there are typical ways of speaking employed in the enunciation of different perspectives on the concept. Despite the fact that each individual has her own conceptual profile, varying in the relative importance (or 'weight') of the zones, the zones in themselves are potentially shared by all individuals in a given sociocultural context. This claim is supported by sociocultural theory, which makes it possible to assume that the concepts and categories available in a given sociocultural context are held in an essentially similar form by a number of individuals inside the same culture, in a way that allows effective communication. These "collective representations" (Durkheim, 1972) have a supra-individual characteristic and are imposed upon individual cognition. When Vygotsky pointed to the social dimension of human mental processes, he was drawing from this position (Kozulin, 1990).

In the study discussed here, we departed from the assumption that a conceptual profile of adaptation is a fruitful theoretical-methodological tool to investigate the understanding of Darwinist explanatory models in the science classroom. We also proposed that the profile could be used as a tool for classroom discourse analysis, alongside with the analytical framework developed by Mortimer and Scott (2003). In particular, we see in the conceptual profile a powerful tool for the investigation of the semantic dimension of discourse, while Mortimer and Scott's analytical framework makes it possible to analyze its social and linguistic dimensions.

For most people, the phenomenon of adaptation concerns the supposedly perfect structural and functional adjustment of organic structures to the organisms' modes of living. However, when the concept of adaptation was introduced in the study of living forms, in natural theology, it was mostly used to refer to the adjustment of organic structures to one another, as an internal phenomenon of the organism, and was in turn accounted for by an appeal to the action of God (Caponi, 2006). Thus, we can see in Darwin not only a change of perspective with regard to the explanation of adaptation, cast by the English scientist in naturalistic terms, but also concerning the very nature of adaptation, which came to indicate, in the Darwinian framework, something to be explained in the relation between living beings and the circumstances of their lives. It is from this perspective that Darwin took the problem of adaptation as the main issue to be solved by a theory about the diversification of living beings.

This theory, as we all know, was first presented to the public in 1858, in a session in the Linnean Society, alongside with Wallace's work, and was exposed at book length in the following year. In *Origin of species*, as Mayr (1982) aptly argues, there is not one but several interrelated ideas composing the Darwinian theory, with a central role being played by the concept of natural selection, as the major (but, for Darwin, not single) mechanism explaining in naturalistic terms both the diversification of species and their adaptation to the conditions of living (Maynard Smith, 1969; Lewontin, 1978). After the initial controversies and even a brief eclipse, from the 1890s to the 1920s (Bowler, 2003), the theory of natural selection came to be widely and strongly accepted by the scientific communities in the 1930s and 1940s, with the construction of the evolutionary synthesis, which brought together Darwinian and Mendelian ideas (Ruse, 2002). With the large acceptance of natural selection, adaptation became, so as to say, the heart of modern evolutionary biology (Mayr, 1982; Amundson, 1996). Currently, the Darwinian concept of adaptation is at the core of one of the most fruitful and interesting debates in evolutionary biology, opposing adaptationist and anti-adaptationist thinkers since the 1970s (Ginnobili & Blanco, 2007; Sepulveda & El-Hani, 2008; Sepulveda et al., 2011). As one of the consequences of this debate, it has been pointed out that the proliferation of meanings related to the term 'adaptation' in biology is one of the main challenges to be faced by the scientific community itself. However, the polysemous nature of the concept of adaptation is not limited to science, but can be readily observed in other fields of knowledge, and, also, in everyday language, in which the term 'adaptation' is often used to designate different sorts of adjustment of living beings and human artifacts to conditions of existence, functions, and uses.

We consider, thus, that the concept of adaptation, among the diverse concepts included in evolutionary biology, is especially adequate for the construction of a conceptual profile model to be applied in the context of evolution teaching. First, the term 'adaptation' did not originate in the scientific language, but was assimilated by it from a theological discourse, which in turn brought it from everyday language. Second, the concept of adaptation plays a central role in Darwinist thinking, and, also, in evolution education. Third, it is highly polysemous, both within science and in the general culture.[1]

Based on these arguments, we decided to design an investigation in order to evaluate the heuristic potential of a conceptual profile of adaptation for analyzing evolution teaching and learning situations from a sociocultural perspective. To do so, we performed a study divided into two steps: (i) to build a conceptual profile model for adaptation; (ii) to apply this model to the analysis of discursive interactions in biology classrooms, in the context of teaching about the Darwinist theory of evolution by natural selection.

In this chapter, we will present both steps, first, explaining the conceptual profile model of adaptation, through a characterization of its zones, while briefly explaining the methodological procedure used in its construction. Then, we will show how this profile can be employed as a tool to model the semantic dimension of the discursive dynamics in the classroom, when applied together with Mortimer and Scott's analytical framework. Subsequently, we will discuss how the results obtained through this kind of analysis make it possible to characterize in semantic, linguistic, and social terms the relationship between discursive interactions and meaning making in the science classroom. Finally, we will briefly comment upon the implications of the study to the planning of teaching interventions and to teaching practice in the context of evolution education.

2 A Conceptual Profile of Adaptation

We cannot detail the methodology used to build conceptual profiles here (for more details, see Mortimer & El-Hani, 2014). We will just mention some aspects. To build a conceptual profile, it is necessary to consider a large variety of meanings ascribed to a concept and a diversity of contexts of meaning making, including at least three of the genetic domains taken into account by Vygotsky in his studies about the relationships between thought, language, and concept formation, namely, the sociocultural, ontogenetic, and microgenetic domains (Wertsch, 1985).

In order to understand the genesis of the concept of adaptation in these domains, we used the following kinds of sources: (i) information derived from secondary sources on the history of biology and epistemological treatments of the concept; (ii) information obtained in the literature on students' conceptions about adaptation and natural selection; (iii) primary empirical data gathered by means of interviews and questionnaires with high school and higher education biology students; and (iv) data gathered by means of analyses of classroom discursive interactions. We worked with these sources in a dialogical manner, that is, we avoided dealing with them in some fixed or linear sequence, but, rather, the data related to each genetic domain were all the time articulated with the data concerning other domains.

Through this methodological procedure, we built a model of the conceptual profile of adaptation composed of four zones (or modes of thinking): *intra-*

organic functionalism, *providential adjustment*, *transformational perspectives*, and *variational perspectives*. In the following sections, we will characterize these zones in terms of the epistemological and ontological commitments that ground different modes of thinking about the concept of adaptation.[2]

2.1 *Intra-Organic Functionalism*

In this zone, we find interpretations of adaptive traits that do not conceive them as demanding evolutionary explanations, i.e., explanations invoking ultimate, rather than proximate causes – if we make use of Mayr's (1982, 1988) distinction between these two kinds of causes.[3] This absence of a request for evolutionary explanation of adaptations can be associated with three modes of interpretation: (i) explanations that appeal preferentially or exclusively to proximate causes, particularly, physiological and biomechanical processes, regarded as sufficient to explain the organization of living structure and function; (ii) analyses of functional attributes of adaptive traits that focus only on their role in the maintenance of the organic system itself; or (iii) treatments of adaptations as given, i.e., as phenomena that do not require explanation.

We identified three epistemological commitments involved in the genesis and development of this mode of thinking: (i) an attachment to proximate causes; (ii) intra-organic teleology (Caponi, 2002), which favors the explanation of organic structures in terms of their causal role in the preservation of intra-organic harmony – showing, thus, some similarity with the account of adaptation in natural theology; and (iii) a conception of adaptation as a self-evident and/or self-explanatory phenomenon.

2.2 *Providential Adjustment*

We find in this zone interpretations of adaptation, in ontological terms, as a state or property of living beings, or their morphological structures, of being adjusted to the conditions of living. In causal terms, this adjustment is explained by appealing to a principle of natural economy, which assumes that nature is always in equilibrium so as to be capable of guaranteeing the perpetuation of species, and a teleological view about the organization of living forms. Accordingly, adaptation is explained as a phenomenon resulting from a necessary harmony between the organizational structure of the living system and the environmental conditions.

The mode of thinking about adaptation modeled in this zone is grounded on the following epistemological and ontological commitments: (i) the principle of the economy of nature; (ii) a teleological perspective on the explanation of living form; (iii) an ideal of order and harmony in the natural world; and (iv) an essentialist view about the nature of species.[4] As we will see, both teleology and essentialist thinking are commitments that are not exclusive of this zone, even though they also characterize it.

2.3 *Transformational Perspectives*

The main difference between this zone of the profile and the previous zone lies in the introduction of a historical, evolutionary perspective to explain the diversity of organic forms. Adaptation is interpreted in this mode of thinking as a process of transformation of the essence of a species towards a state of optimal adjustment to environmental conditions. This process takes place by means of simultaneous changes that occur with each and every individual members of the species. That is, evolutionary (phylogenetic) changes are treated as the result of an accumulation of developmental (ontogenetic) changes.

Another characteristic commitment of this zone concerns the idea that the transformations undergone by the organisms have a definite direction, i.e., all the members of the species suffer changes oriented to the same direction, and, moreover, can even follow the same sequence of steps in their evolution. Based on this linear progression, living forms reach a teleological state of optimal adjustment.[5]

Although these transformational interpretations have an evolutionary character, they still keep a kernel of essentialist thinking. Variation is not a central feature in this mode of thinking. If a person accepts the idea of evolutionary change, but is still committed to essentialist thinking, she is likely to endorse transformational perspectives.

Essentialist thinking, the focus on the individual organism, and the attachment to proximate causes are ontological and epistemological commitments of this zone that create difficulties for the development of variational perspectives on adaptation, the next zone in the conceptual profile model. However, if we compare this zone and the previous one, there is an important change in the ontological categorization of the phenomenon of adaptation that can help in learning about the variational mode of thinking. Namely, adaptation is no longer conceived as a state of being or property of an organism or morphological structure, but rather as a (historical) process of evolutionary change.

2.4 *Variational Perspectives*

In this zone we find interpretations of adaptation as a trait resulting from a process of evolutionary change, typically involving natural selection as a central agent, i.e., a process of selective spread and fixation of variants in a population under a certain selective regimen.

The proposal of the latter two zones obviously owes to Lewontin's (1983) distinction between transformational and variational explanations of evolutionary change. As we saw above, in transformational explanations, the evolution of a system takes place due to simultaneous and related changes in all the individual components of the system, while in variational perspectives – as in the Darwinian theory of evolution – the changes suffered by a system are explained as a consequence of alterations in the proportions of its (variant) components, not of individual changes of the latter (Figure 12.1).

The mode of thinking about adaptation modeled in this zone is grounded on the following ontological and epistemological commitments: (i) populational rather than essentialist thinking; (ii) the idea that organic structures and behaviors play a central role in the struggle of living beings to survive and reproduce in view of the requirements and pressures posed by ever-changing ecological circumstances; and (iii) a historical perspective on organic form (which is also a commitment of transformational perspectives). The school science discourse shows most of these commitments, which are shared, generally speaking, with the Darwinist interpretation of adaptations and their relationship with the explanation of the origins of living diversity.

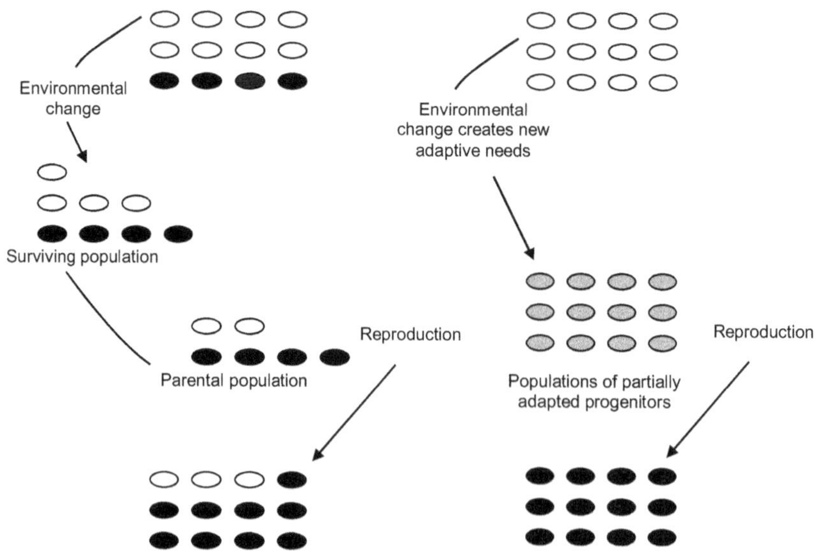

FIGURE 12.1 A schematic representation of variational (left) and transformational (right) explanations of the evolutionary change of a population as a result of environmental change. Each ellipse represents an individual in the population, and the colors, different states of a phenotypic trait

3 Methodological Procedures to Apply the Conceptual Profile to Classroom Discourse Analysis

The conceptual profile model presented in the previous section was used to guide the semantic analysis of classroom discourse. In order to do so, we integrated it with the analytical framework for discourse analysis developed by Mortimer and Scott (2003), which was used to guide the analysis of the social and linguistic dimensions of classroom discourse.

Mortimer and Scott's analytical framework is based on five interrelated aspects that focus on the teacher's role and are grouped in three dimensions: *teaching focus*, *approach*, and *actions*. The *communicative approach* is the central element in the analysis, since it is through it that we understand how the teaching focus, i.e., the *teaching purposes* and *contents*, is worked out by means of such and such actions, the *teacher interventions*, which result in certain *patterns of interaction* (see Table 12.1).

TABLE 12.1 Framework proposed by Mortimer and Scott (2003) for the analysis of interactions and meaning making in science classrooms

	Analytical aspects
i. *Teaching focus*	1. Teaching purposes 2. Content
ii. *Approach*	3. Communicative approach
iii. *Actions*	4. Patterns of Interaction 5. Teacher interventions

The teaching purposes amount to the goals guiding the selection and planning of activities that the teacher proposes to the students. They result in a script that guides the public performance of the teacher in the social plane of the classroom. The second aspect related to the teaching focus concerns the contents of the discourse. This aspect can be thought of in terms of Bakhtin's concept of social language. The contents composing the classroom discourse emerge from the contact established by everyday language and scientific language along the development of the "scientific story"[6] in the classroom. These contents are basically structured around three aspects of the social language of school science: description, explanation, and generalization (Mortimer & Scott, 2003, p. 30).

The communicative approach concerns two dimensions characterizing the teachers' exchanges with the students: (1) if there is interaction or not between teachers and students; (2) if the teachers take students' ideas into account in meaning making in the classroom. The answers to these questions define two axes from which the communicative approach can be characterized as interactive or non-interactive, and dialogic (multivocal) or authoritative (univocal).

The first axis is related to classroom interactions: when there is an alternation of speech turns in the classroom discursive interactions, the communicative approach is interactive; otherwise, it is non-interactive (Silva & Mortimer, 2005).

The second axis is organized around two extreme situations that can take place in the science classroom: on the one hand, the teacher may consider different viewpoints, i.e., voices,[7] in a dialogic approach, or, on the other, only one perspective, typically that of school science, may be taken into account, in an authoritative approach. In reality, any interaction probably contains aspects of both functions, dialogic and authoritative.

These different communicative approaches are, in turn, constructed through patterns of interaction between teachers and students. Mortimer and Scott (2003) discuss two common categories of interaction patterns: the triadic pattern I-R-E (Mehan, 1979), involving an initiation by the teacher, a response by the student, and an evaluation by the teacher, and chains of non-triadic interactions, such as I-R-P-R-P… or I-R-F-R-F…, where P means a discursive action that prompts the continuation of the student's talk and F amounts to a feedback, in which some additional information is introduced, normally by the teacher, for stimulating the student to further elaborate her talk.

Our methodological approach to investigate the discursive dynamics in the classroom using these tools was microgenetic analysis, conceived as the detailed examination of the formation of a psychological process, in which the subjects' actions and interpersonal relationships in a short time range are recorded and investigated (Wertsch & Hickman, 1987).

Our units of analysis were teaching episodes, regarded as sets of utterances that create a context for the emergence of one or more meanings related to the learning of a given concept (Amaral & Mortimer, 2006).

The teaching episodes were obtained through the following methodological steps: (1) gathering of empirical material by means of video recording of discursive interactions in the classroom; (2) organization and systematization of the data through the construction of activity maps, a strategy indicated by interactional ethnography (Gee & Green, 1998); (3) selection of episodes based on the criteria that they concerned meaning making about the concept of adaptation and/or explanatory models for adaptive changes, and they could be delimited in relation to changes in discourse content and enunciative strategies used by the teacher; (4) transcription of the utterances that composed the selected episodes.

Sound and video recording was made by using two video cameras, one fixed in the front of the class, at the right size, and another located at the middle back of the class (Figure 12.2). The latter camera was operated by a researcher in order to capture images of the teacher and some of the students as they participated in the discursive interactions.

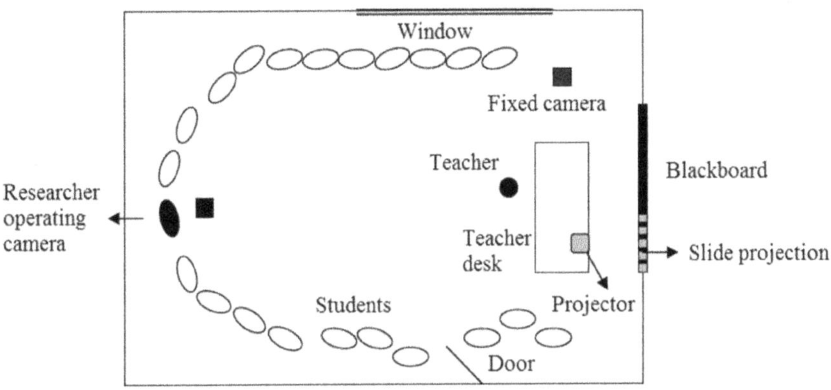

FIGURE 12.2 Spatial organization of the classroom and the position of the video cameras

As the video recordings of the classes were watched by one of the researchers (the first author of this chapter), activity maps were constructed following the methodology proposed by Amaral and Mortimer (2006). For each class, a first sketch of the map was made, in which we recorded the activities, the time in which each happened, the theme addressed, and the participants' actions. In this sketch we also indicated suggestions of segments in the classroom discursive interactions that might constitute teaching episodes. After selecting these episodes based on the criteria mentioned above, we carried a full transcription of the discursive interactions between teacher and students. Once the episodes have been selected, the activity map was reformulated so as to indicate the exact moments in the classes in which the episodes took place.

The transcriptions of the discursive interactions were made by using a coding procedure that indicates the pauses and intonations in the participants' speeches.[8] Whenever this was deemed relevant, observations were added in parenthesis pointing to gestures and other non-verbal information, such as references to illustrations and pedagogical materials that played an important role in the interaction. The episodes were organized in the form of sequences of speech turns. To maintain the confidentiality, we designated the students just by numbers, omitting their names.

4 Discourse Analysis of a Teaching Episode

The teaching episode was extracted from a class included in a teaching sequence on the theory of evolution by natural selection, which was carried out in a third-year high school classroom in a Brazilian public school. The sequence comprised seven weekly classes, each lasting 60 minutes. The episode took place in the second class, in which the teacher presented to the students and analyzed with them the case of the diversification of the Galapagos finches.

After providing information concerning the geographical distribution of the species of finches in the islands, the relationship between their beak morphology and feeding habits, and the diversity of environmental conditions in each island, the teacher asked the students to formulate explanations for both the morphological diversity of the beaks and the origins of the thirteen species of finches found in the Galapagos.

Episode: Had to adapt to survive and this led to the change of the beaks.

1. *Teacher:* Given this information/how do you explain the difference of the beaks of finches of the Galapagos Islands?
2. *Student 1:* Because of the food/they fed/they fed according to the climate/
3. *Student 2:* Depending on what they fed.
4. *Student 1:* And also because/each island had its birds/its animals/So they adapted/different individuals on each island/because of this. Because of the food they ate/also because of the climate/because each island had a climate/so each island had a kind of vegetation for them to feed on. So/ because of this variation.
5. *Teacher:* And?/Ok. Anyone else? Yes/student 3. How do you explain the difference of the beaks?
6. *Student 3:* According to the/type of food they used to survive and according to his habitat.
7. *Teacher:* Right. According to the environment in which they lived/and the food/that explains the difference of the beaks. Isn't that right? Ok. On the continent we find only one species of finch/while on the islands THIRTEEN species of this same bird genre. What explains this diversity of birds/of the group of finches on the islands? Han? What explains it/ Folks? Hello/on the continent, we have one finch species/right? And on the island we have THIRTEEN different species. What explains the diversity of this group of finches/What explains the diversity? On the continent, we have only one species and on the islands we have THIRTEEN species? What explains that? Tell me.

8. *Student 3:* The others did not adapt/to the continent. Let's say that the kind of food that the continental one eats/the others are unable to eat.
9. *Teacher:* Yes. Anybody else? What explains that on the continent we have only one and there we have thirteen?/How does this happen?
10. *Student 4:* Because his feeding habit is different from that of the other birds?
11. *Teacher:* The feeding habit is different? How could this be? We are working from the point of view of' evolution. So what happens? We have seen that/one of the points of the theory of evolution is the common ancestor. Is that true? And here is the information from the text that probably the species from the continent ((raising her voice, as student 3 tries to say something)) is the ancestor of these thirteen species/right? This helps to improve/
12. *Student 3:* Teacher/let's say that the one on the continent does not have a good evolution/yes? He does not evolve/
13. *Teacher:* On the continent, he would not have a good evolution. What would be a good evolution/Student 3?
14. *Student 3:* Does not adapt to other kinds of/
15. *Teacher:* What happened?
16. *Non-identified Student:* The capacity to adapt.
17. *Teacher:* The capacity to adapt.
18. *Non-identified student:* To the new environment.
19. *Teacher:* The capacity to adapt to the environment. If we work with the idea of a common ancestor/What is a common ancestor? It is a species that originates/that is there from others. How can we explain this? What happened? If these thirteen are originated from a common ancestor/how could this have/
20. *Student 2:* From the continent went to the islands/
21. *Teacher:* Yes.
22. *Student 2:* And he ended up adapting itself there. And there he/
23. *Teacher:* Yes. From the continent/the occupation went to the islands ((a gesture that gives the idea of migration)) and arriving there/what happens?
24. *Student 2:* He had to feed/so he/
25. *Student 1:* It is like that theory that the ancestors, there weren't/they did not have forks ...
26. *Student 2:* He goes on adapting himself/
27. *Student 1:* The teeth were like the canine tooth due to feeding/because they had to tear/and as time passed/they started to handle/cutlery and so

on/and because they did not have to feed that way/so their teeth changed over time and got to be like ours.

28. *Teacher:* Right. And turning back to the birds ((laughter)).
29. *Student 1:* So/it is that he moved from the continent/he had to learn/
30. *Teacher:* He left the continent to the island/the population went there. Arriving there/he found?
31. *Student 2:* Food/
32. *Teacher:* Different food. It is showing there/that on the islands we found there a great variety of food and environments/right? And then?
33. *Student 2:* And then he had to adapt to survive.
34. *Teacher:* They had to adapt to survive.
35. *Student 2:* And then the change of the beaks occurred.

The episode begins with an initiation by the teacher, in which she demands that the students explain the morphological diversity of the beaks of the finches. Students 1 and 2 interact, citing some factors involved in the explanation of the phenomenon. Their answers were not evaluated by the teacher, who prompted the students to continue their interaction, by encouraging other students to give their opinions (turn 5). Student 3 answered to this initiation. In the next turn, the teacher accepted his answer and made a synthesis, integrating elements found in the answers offered by the three students. Between turns 1 and 7, the communication between teacher and students take place, thus, through a chain of interaction I-R1-R2-R1-P-R3-S (where S stands for synthesis).

The way of speaking about the diversity of the beaks of the Galapagos finches used by student 3 in these speech turns show a linguistic mark characteristic of the providential adjustment mode of thinking, namely, the use of the term "according to" as a manner of establishing a relationship of a necessary adjustment between a morphological structure, the beak, and the carrying out of a vital activity for the organism, the exploration of a feeding resource.

Although student 1 also uses the term "according to" in her first speech turn, she does so in order to connect environmental conditions with the way this vital activity is carried out by the organisms. In turn 4, this student causally associates the morphological diversity of the beaks with the environmental conditions found in the islands, for instance, with the variation in the availability of feeding resources per island, due to differences in climate and vegetation. We can say, thus, that this student shows a tendency of perceiving the role played by the relationship between organisms and their ecological circumstances in the explanation of the diversification of organic forms, an important commitment for the development of a variational (Darwinist) perspective for interpreting adaptation.

In this same speech turn, student 1 makes the term "adaptation" available in the social plane of the classroom. It appears in the form of a verb, "to adapt", used in the past tense, denoting the idea of a process that already occurred. We have, thus, a first attempt to use an evolutionary perspective to interpret organic diversification.

In turn 7, the teacher makes a new initiation, proposing to the students that they explain the diversity of species of finches in the Galapagos by considering that only one related species is found in South America. Student 3 proposes, then, an explanation using the term "adaptation", previously made available by student 1 (turn 8). In this context, however, the verb "to adapt" is employed in the present tense, denoting the condition of a group of organisms being adjusted to the environmental conditions. In turn 10, student 4 presents a point of view which is similar to that of student 3, but in a hesitant manner (turn 10). The teacher negatively evaluates her answer in turn 11 and makes a new initiation.

It is important to pay attention to some linguistic aspects of the teacher's utterances. Between the speech turns 1 and 10, the most frequent questions posed by her had the following forms: "how do you explain"; "What explains". This way of formulating the questions turns the focus to proximate, mechanistic causes, or to the presentation of explanatory factors, making it less likely that the students give attention to ultimate, evolutionary causes. After turn 11, she begins to formulate questions for the students by using lexical resources that denote the ideas of succession of events or occurrence of processes. We can see this happening in questions of the form "What happened" and in the use of the terms "evolution" and "originate".

Between turns 11 and 19, the teacher interacts with student 3 and another unidentified student through a triadic pattern I-R-E. She negatively evaluates their contributions, which bring a point of view characteristic of the providential adjustment zone. These students produced utterances in which evolution and adaptation appear as properties or capacities of the organisms being adjusted to environmental conditions, using constructions such as "have a good evolution" and "[have] the capacity to adapt".

In turn 19, the teacher tries to clarify the meaning of common ancestry and offers clues concerning how it can be used to explain the origins of the diversity of Galapagos finches. Then, she insists on the question "What happened?", suggesting that the students needed to consider a chain of events, from which they could build a narrative.

Students 2 and 1 accept the teachers' proposal and begin to build an explanatory model that is closer to the one she expects, by means of the construction of narratives. The first narrative is put together between the speech turns 19 and 26, through interactions between the teacher and students 2 and 1, framed as a chain I-R2-P-R2-E/I-R2-R1-R2.

As we can observe in Figure 12.3, the role of the narrator is shared by student 2 and the teacher. Student 2 proposes, between turns 20 and 22, that the ancestral species went to the islands and "adapted there". The teacher makes an intervention, then, that changes the order in which the event of adaptation takes place. The narrative begins, then, to show the following sequence of events: the ancestral bird went from the continent to the islands, and, when it arrived there, it had to feed, and, then, it began to adapt itself. This was an important operation to move meaning making about the concept of adaptation towards an evolutionary perspective: the term "adaptation", previously used to designate a self-evident phenomenon, begins to refer to a phenomenon resulting from some other event or process.

The meaning ascribed to the term "adaptation" – a process resulting from a necessity felt by the organism – and the type of agency in the narrative – the

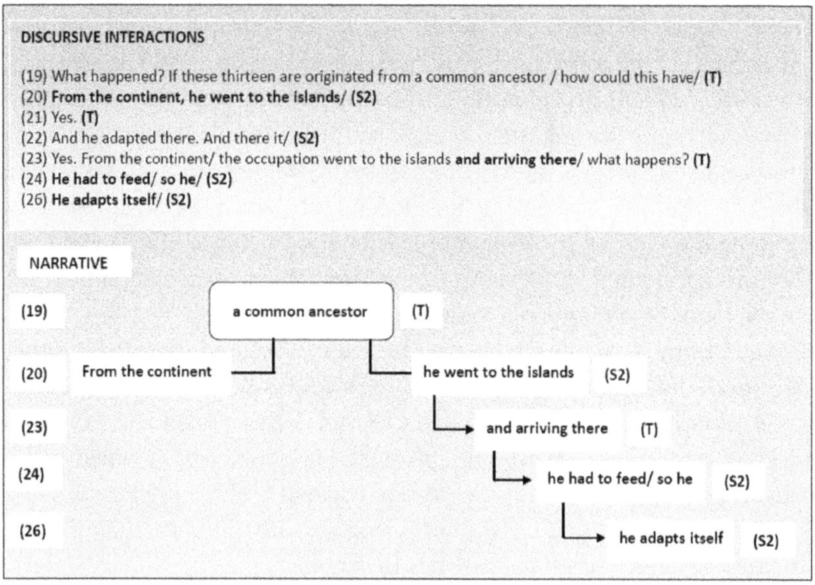

FIGURE 12.3 Construction of narrative by student 2 (S2) through interactions with the teacher (T), between speech turns 19 and 26. Top, we show the speech turns in which the narrative was constructed. The numbers between parentheses at the left side indicate the speech turns, while those at the right side indicate who produced the utterance. The events composing the narrative are highlighted in black. Bottom, we show a reconstruction of the narrative that was produced by means of this discursive interaction. The arrows indicate the order in which events occur, and the agent of the narrative is shown by a circle.

organism as the protagonist of the adaptive process – produce a way of speaking that reveals a negotiation towards the construction of a transformational perspective.

In turns 25 and 27, student 1 builds a narrative with the purpose of creating an analogy between the situation interpreted by student 2, the diversification of the beaks of the finches, and the changes in the morphology of the human dental arch throughout the evolution of our species. The narrative is composed by the following sequence of events: our ancestors began to handle cutlery and, consequently, abandoned their previous way of eating, by tearing up meat, and, then, the teeth changed until reaching their current morphology. Despite the problems found in this narrative, what is most important, from a semantic point of view, is that organisms are not the agents of adaptive change anymore, but rather a morphological structure of the organisms **suffers** the change: "*their teeth changed over time*".

In turn 28, the teacher asks the students to come back to the development of the narrative related to the problem that was originally proposed and, between turns 29 and 35, a third narrative is constructed by students 1 and 2 and by the teacher. This narrative is built by means of an I-R-E triadic pattern.

The participation of students 1 and 2 and the teacher in the construction of this narrative is shown in Figure 12.4. Student 1 established the agency and the initial event. In speech turn 30, the teacher, when evaluating the answer offered by student 1, seeks to change the agent of the narrative proposed by the student, from "he" to "the population" of ancestral birds. However, in the subsequent initiation, she uses again the pronoun "he", which is adopted by student 2 in speech turn 33. In turn 34, the teacher makes another attempt to change the agent of the narrative, when she rephrases the student's answer by using the verb "to have" in the third person of the plural. At the end of the episode, student 2 produces an utterance in which the agent of the change of the beaks is not clear – "*and then the change of the beaks occurred*" – something that can be treated as an advance with regard to the previous utterance in which the organism was the subject of the evolutionary change.

This new narrative shows some changes in relation to the previous narrative about the same theme. The connection between the events "to arrive at the islands" and "having to feed" or "having to adapt" (see Figure 12.3) was better qualified, when it was introduced the event that an ancestral bird found a diversity of food resources in the archipelago (see Figure 12.4). This additional element was made possible by the teacher's evaluation at turn 32. In this second narrative, differently from the first, the end point is the beak

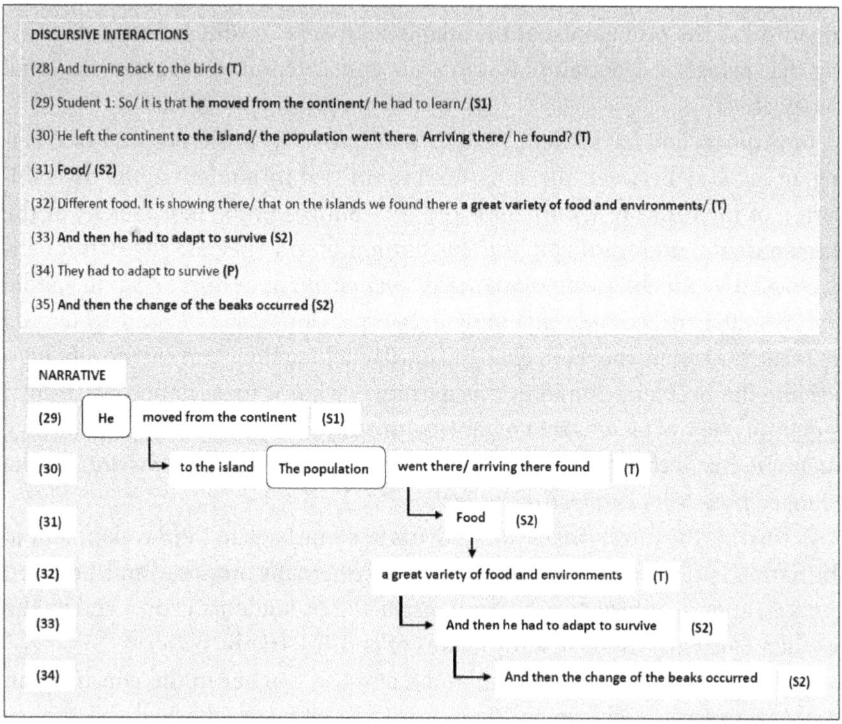

FIGURE 12.4 Construction of a transformational explanatory narrative for the diversification of the beaks of Galapagos finches by means of a discursive interaction between the teacher (T), student 1 (S1) and student 2 (S2), between speech turns 29 and 35. The figure uses the same structure explained in the caption of Figure 12.3.

diversification, not the adaptation of the birds to the island. Another important feature is that the birds appear as protagonists of the actions of migrating, finding different food resources and adapting, but not of the change of the beaks, which now appears as a consequence of the adaptive process, "*the change of the beaks occurred*".

The narrative resulting from this interaction shows the following sequence of events: the ancestral bird migrates from the continent to the islands (S1), finds a diversity of environments and food resources (T; S2), adapts to survive (due to a necessity) and the change of the beaks occurs (S2). Thus, the discursive interactions are indeed shifting alongside the episode towards the scientific story.

In this episode, the teacher has two intentions, to explore the students' ideas and to introduce and develop the scientific story. The interactions between teacher and students are related to a theoretical explanation of the diversification of the beaks of the Galapagos finches. Even though the teacher encouraged the students, at the beginning of the episode, to put forward their explanatory models, making no evaluations, she began at turn 11 to establish the directions in which the students were to develop their explanations. Therefore, an authoritative/interactive communicative approach prevailed in this episode. Regarding meaning making about the concept of adaptation, there was a negotiation around the commitments underlying the way of thinking corresponding to the providential adjustment zone and those of the transformational perspectives. Table 12.2 presents a summary of modes of thinking and speaking negotiated in this episode, as well as the discursive aspects involved in this process.

5 Relationships between Discursive Interactions and Meaning Making in Evolution Teaching

By applying this kind of analysis to a set of thirteen episodes, it was possible to identify stages of appropriation of the school science perspective by students in the classroom and to relate them to enunciative strategies used by the teacher. We were also able to address other aspects of the discursive context, such as the emphasis given to certain contents.

In the beginning of the teaching sequence, organic diversity was described by the teacher in theoretical terms, based on the notion of biodiversity, pointing to the diversity of forms and sizes shown by living beings, to unicellular and multicellular organization, and to the different taxonomic groups in which organisms are classified. Questioned about the possible explanations to the phenomenon of diversity, the students reacted by taking organic diversity to be a given phenomenon, which does not demand – as other natural phenomena – any explanation.

After a theoretical and empirical description of the specific case of the diversification of thirteen species of finches in the Galapagos Islands was presented, the students came to recognize the origins of organic diversity as a legitimate problem. Initially, they tried to interpret it under the light of the principle of the economy of nature or evoking the agency of some force or process that would maintain the organisms harmonically adjusted to the environmental conditions. These are epistemological commitments that ground the mode of thinking represented by the zone providential adjustment in the adaptation profile model described above.

At this point, the term "adaptation" was brought to the social plane of the science classroom by the students, who used it with the meaning stabilized in

TABLE 12.2 Discursive aspects that interact in meaning making about the concept of adaptation in the analyzed teaching episode

Teaching purposes	Exploring students' ideas: Explanations for the diversity of the beaks of finches in the Galapagos Islands Introducing and developing the scientific story: introducing the notion of common ancestry and an evolutionary perspective
Discourse content	Theoretical explanation
Communicative approach	Authoritative/interactive
Patterns of interaction	(1–7): I-R1-R2-R1-P-R3-S (7–19): I-R-E triads (19–28): I-R2-P-R2-E/I-R2-R1-R2-R1-E (28–35): I-R-E triads
Modes of thinking	(1–19) Adaptation as a property or state of being of the organism (19–35) Adaptation as the process of evolutionary change
Ways of speaking	(1–19) Affirmation of the relation of necessary adjustment between morphological structure and vital activity of the organism. Recurring terms and expressions: *According to, capacity to adapt.* (19–35) Narrative in which organisms or groups of organisms are protagonists of a transformation towards the adjustment to the needs of survival. Recurring expressions: *it goes on adapting itself, had to adapt*
Perspectives on meaning making about the concept	Negotiation around the commitments of the zone of providential adjustment towards the development of commitments of the transformational zone

the everyday social language. The verb "to adapt" was used as a sort of principle that could explain a harmonic and necessary adjustment of the organism to the environment, which, in turn, would give origin to diversity.

After the second class, the students began to appropriate elements of the school science language, more specifically, related to the notion of common ancestry. By doing it, they adopted the construction of narratives as a form of explanation. At that point, there was a change in the way of thinking and speaking about adaptation in the social plane of the classroom, concerning the ontological status of the concept: it was not longer used to designate a capacity of adjusting or a state of being adjusted, but was rather employed to name a process of change. This new ontological commitment led to the development of an evolutionary perspective to interpret organic diversity, an epistemological commitment shared by the transformational and variational zones in our adaptation profile model.

We identify a series of enunciative strategies articulated by the teacher that were important in managing the classroom discourse so that this advance could be promoted: (1) the introduction of the concept of common ancestry and the teaching action of marking it as a key idea, by using an authoritative communicate approach; (2) the use of phraseological resources suggesting the occurrence of a sequence of events, such as, for example, the interrogation "What happened?", which she used recurrently, stimulating the students to build narratives; (3) the support given to the students for elaborating narratives to explain the origins of the thirteen species of finches and putting forward predictions about what would happen with a hypothetical bird population in the face of a situation of food shortage.

Between the second and the fourth class, the students gradually appropriated some additional notions of school science, such as the idea that both environmental and genetic factors play a causal role in the origins and diversification of organic form, and the idea that phenotypic changes take place in the populations alongside several to many generations. As a consequence of this appropriation, the students built more and more sophisticated explanations, while still keeping the commitment to a transformational perspective.

After the fourth class, the students began to more and more frequently use terms of the school science social language to communicate in the classroom, including "evolution", "generation", "species", "inheritance", "inherit", "descendant", but still in a hesitant and uncertain way. In this class, the notion that diversity originates from a gradual evolutionary process reached a high level of univocality among the students. In linguistic terms, an evidence for this finding lies in the frequent use of the gerund in the description of adaptive change events, which gave origin to the different species of Galapagos finches: "*they were evolving*", "*they were altering*", "*he was changing*", "*they were adapting*".

Regarding the appropriation of the school science social language, the following teacher's discursive movements were important: (1) to replace empirical referents with theoretical referents in the students' utterances[9]; (2) to reconstruct utterances elaborated by the students in a teleological and anthropomorphic language in a way of speaking closer to the school science social language.

The fifth class was a significant moment with regard to the construction of univocality around the school science perspective. Student 3, who was until that moment committed to a mode of thinking characteristic of the providential adjustment zone, proposed and put into negotiation a transformational perspective for explaining phenotypic change in a bird population simulated in a game carried out in the classroom (*clipbirds*[10]). A group of four students, which was also developing transformational explanations, reacted negatively to that student's proposal. Two of them (students 1 and 2) counter-argued by constructing a narrative that described a change in the proportion of phenotypic variants in a population, namely, large- and small-sized beaks, as a consequence of differential survival and reproduction. Even though this narrative was constructed using everyday language, it showed the distinctive feature of a variational explanation: the change of a system was explained by the change in the proportion of its components (see Figure 12.1).

At the end of the teaching sequence, in the seventh class, four students (2, 5, 6, and 7) interpreted a hypothetical case of emergence of insecticide resistance in an agricultural pest based on a Narrative explanations constructed in terms of a variational perspective.

According to our analysis, the following strategies contributed to the development of a variational perspective in the social plane of the science classroom: (1) the introduction of the concept of intrapopulational variation by means of the teacher's authoritative communicative approach and the effort to make it intelligible and plausible to the students alongside the sequence; (2) the insistence on the need of interpreting adaptive change from a population perspective; (3) the emphasis of phenotypic variation in populations, the comparison between the efficiency of these variations in carrying out the same task, and the treatment of demographic phenomena; (4) the explanation that both internal forces, related to genetic variability, and external forces, concerning environmental factors, participate in the causal processes involved in the evolution of organic form.

6 Concluding Remarks

The application of the adaptation profile model to the analysis of teaching episodes made it clear that not only epistemological and ontological commitments structuring ways of thinking but also ways of speaking about adaptation are negotiated in the process of understanding Darwinist explanations. This conclusion is supported, for example, by the analysis of the role played by transformational narratives in the construction of a variational evolutionary perspective in a class of students departing from non evolutionary interpretations of adaptation. The first narrative built by the students for explaining organic diversity showed organisms as the agents of their own adaptive change. The distinctive hallmarks of this way of speaking about adaptation was the recurrent usage of the expression "had to adapt", teleological formulations, and them appeal to personification and anthropomorphization – characteristic features of everyday language.

These narratives were a "thinking device" (Lotman, 1988; cited in Wertsch, 1991) used by some students to approach the school science perspective. They exhibited two heuristic aspects that operated as conceptual seeds for the development of a variational explanation for organic diversification: the historical approach to explaining adaptation and the idea that events of environmental change play an important role in the evolutionary process. However, the negotiation of some other commitments was needed for developing the Darwinist account of adaptation from these conceptual seeds. For example, it was necessary to distinguish the causal role of the environment in evolutionary change in a Darwinist perspective from the idea of direct action of the environment on the organisms. In this manner, the students were able to understand that the environment played the role of a selective agent of variations in populations of organisms.

This process of meaning negotiation should not be understood in isolation from the students' appropriation of ways of speaking. One of the key discursive strategies for the development of the Darwinist understanding of adaptation in the social plane of the classroom consisted, for example, in the change of the type of agency in the narratives on the origins of adaptations. By proposing challenges and providing scaffolds for the students in the elaboration of their narratives, the teacher managed to shift this role from individual organisms to populations of organisms, and, then, to selective pressures imposed by the environment. In these latter narratives, organisms ceased to play the role of agents and become, rather, objects of evolutionary forces, both internal and external to them. The linguistic hallmarks of this new way of speaking are found in the recurrent usage of expressions such as "*underwent changes*", "*was favored*", "*was selected*".

Regarding the implications to teaching planning and classroom practice, the results of this study point to the importance of two concerns: (1) with the promotion of an understanding of the epistemological and ontological commitments that structure modes of thinking about the concepts we intend to teach; (2) with the students' capacity of mastering both the social language of school science and the scientific speech genres. If we succeed in these two aspects, we will be able to support the development of the students' toolkit for the mediation of verbal thinking and its heterogeneity, as proposed by Wertsch (1991). According to this standpoint, new ways of speaking can be made available by means of discursive interactions in the classroom and, when appropriated by the students, constitute mediating tools that help them negotiate new meanings generated by the contact between different cultural perspectives and, thus, develop new modes of thinking.

In the case of teaching about the theory of evolution by natural selection, it is important that the students are stimulated to master ways of speaking closer to the variational perspective zone in our adaptation profile model. This can be done, for instance, by promoting their construction of utterances in which the organisms appear as objects of the evolutionary process, as in the statement "*resistant bacteria were favored*", instead of utterances in which the organisms appear as subjects of evolutionary change, as in "*the bacteria evolved resistance*". Narratives produced by using a language characterized by personification and anthropomorfization can be reconstructed along the discursive interactions between teachers and students, so that they leave room to narratives of chains of events without clear protagonist agents, which result in evolutionary outcomes in populations of organisms. The themes of the utterances can be shifted from a focus on phenomena related just to the internal constitution and functioning of organisms to relationships between organisms and environments. The analysis of how a single structure effectively serves a function can give room to a comparative analysis of the differential efficacy of phenotypic variants in fulfilling that function, in a specific population embedded in specific environmental conditions.

Another important asset in the science classroom is to explicitly approach the meaning of different ways of speaking in order to support the students in the task of becoming aware of the distinctive features of everyday and school science social languages.

Finally, we consider that the results of the research reported here point to the heuristic value of classroom discourse analysis combining conceptual profiles to investigate the semantic dimension of discourse and analytical frameworks such as Mortimer and Scott's (2003) to examine its social and linguistic dimensions. This analytical work can bring contributions to the management of communicative approaches and language use in the interactions between teachers and students, as a way of increasing the likelihood that students appropriate the school science perspective and master the school science language in the social plane of the classroom.

Notes

1 Mortimer and El-Hani (2014) discuss these criteria for selecting concepts that can be fruitfully modeled by means of conceptual profiles.
2 In this work, we did not deal with axiological commitments as a separate category. However, it will become clear in the next sections that such commitments are interwoven with some epistemological and ontological commitments which we examined.
3 Ernst Mayr proposes that every biological phenomenon is the result of distinct kinds of causes, ultimate and proximate causes. The first kind of cause is involved in physiological, developmental, behavioral processes, typically answering questions of the form "How?" Proximate causes concern how organisms function. The second kind of cause is related to processes leading to evolutionary changes and typically answers questions of the form "Why?" Ultimate causes explain why organisms show the organic structures and functions that we observe today.
4 Shtulman (2006, p. 171) interprets essentialist thinking in terms of the idea that the external appearance and behavior of a species are determined by some superior causal power or "essence", which is a (not necessarily divine) innate potential of the members of a species developing the same characters.
5 We do not intend to claim that teleology is not valid at all in biology, but, rather, that teleological explanations of the evolutionary process are not acceptable under the light of the current state of scientific knowledge. This does not mean that, say, physiological processes or behaviors cannot admit teleological explanation. For more details, see Caponi (2002), Nunes-Neto and El-Hani (2009, 2011), El-Hani and Nunes-Neto (2009).
6 Mortimer and Scott (2003, p. 18) use the expression "scientific story" to designate the way the scientific perspective is narrated to the students in the social plane of the classroom, so as to make it accessible to them. They depart from Ogborn and colleagues' (1996) claim that science teaching puts forth an approach to natural phenomena that is expressed in ideas and conventions characteristic of the language of school science, so as to compose a kind of script, which is similar to a story.
7 We are employing here the notion of "voice" in accordance with the Bakthin's circle, i.e., to refer to the perspective of the speaking subject, to its conceptual horizon, intention, and worldview (Wertsch, 1991, p. 51).
8 The discursive interactions were freely translated into English by the authors. We tried to keep them as faithful to the original statements as possible, but this has

been sometimes hard, given their colloquial nature and the consequent usage of a number of expressions and contractions which are not so easily translatable into another language. Furthermore, we did not try to reproduce occasional grammar and writing mistakes in the English translations. The original transcriptions in Portuguese are available on request from the authors. The only punctuation marks used were questions marks – to indicate interrogative intonations – and periods or full stops, because they were the only ones that could be inferred with enough confidence. The slash (/) was used to indicate short pauses in the middle of utterances and abrupt truncations between speech turns. Longer pauses were represented by the symbol +, with a proportional relationship between the number of symbols and the duration of the pause. Uppercase font was used to represent intonations indicating emphasis.

9 We follow here Mortimer and Scott's (2003, p. 131) understanding of empirical referents as those constituting elements or properties of a system or object that are directly observable, and theoretical referents as constituents or properties that are not directly observable in a system or objects, but are rather entities created by scientific theoretical discourse. There are, certainly, a number of epistemological issues to discuss regarding this differentiation between empirical and theoretical referents, but this is not the place to expand on the issue. For the sake of our arguments, we consider that the distinction as proposed by Mortimer and Scott is currently sufficient.

10 The game *clipbirds* was developed by Al Janulaw and Judy Scotchmoor, and is available at http://www.ucmp.berkeley.edu/education/lessons/clipbirds/, retrieved Feb. 10th 2012. For more details on our adaptation of the game, see Vargens and El-Hani (2011).

References

Amaral, E. M. R. (2004). *Perfil conceitual para a segunda lei da termodinâmica aplicada às transformações físicas e químicas e dinâmica discursiva em uma sala de aula de química do ensino médio* [Conceptual profile for the second law of thermodynamics applied to physical and chemical transformations and discursive dynamics in a high school chemistry classroom] (Thesis, Doctorate in Education). Faculdade de Educação, Universidade Federal de Minas Gerais, Belo Horizonte, MG.

Amaral, E. M. R., & Mortimer, E. F. (2004). Un perfil conceptual para entropía y espontaneidad: una carcterización de las formas de pensar y hablar en el aula de Química. *Educación quimica, 15*, 218–233.

Amaral, E. M. R., & Mortimer, E. F. (2006). Uma metodologia para estudar a dinâmica entre as zonas de um perfil conceitual no discurso da sala de aula [A methodology to study the dynamics among zones of a conceptual profile in classroom discourse]. In F. M. T. Santos & I. M. Greca (Eds.), *A pesquisa em ensino de ciências no Brasil e suas metodologias* [Science education research in Brazil and its methodologies] (pp. 239–296). Ijuí, RS: Editora Unijuí.

Amundson, R. (1996). Historical development of the concept of adaptation. In M. R. Rose & G. V. Lauder (Eds.), *Adaptation* (pp. 11–53). San Diego, CA: Academic Press.

Bachelard, G. (1940). *La philosophie du non.* Paris: PUF.

Bowler, P. J. (2003). *Evolution: The history of an idea* (3rd ed.). Berkeley, CA: University of California Press.

Candela, A. (1999). *Ciencia en el aula: Los alumnos entre la argumentación y el consenso* [Classroom science: The students between argumentation and consensus contruction]. Ciudad de Mexico, Buenos Aires, Barcelona: Paidós.

Caponi, G. (2002). Explicación selecional y explicación funcional: La teleología en la biología contemporánea [Selectional explanation and functional explanation: the teleology in current biology]. *Episteme, 14*, 57–88.

Caponi, G. (2006). El vivente y su medio: Antes y después de Darwin [The living being and its environment: before and after Darwin]. *Scientiae Studia, 4*, 9–43.

Coutinho, F. A. (2005). *Construção de um perfil conceitual de vida* [Construction of a conceptual profile of life] (Thesis, Doctorate in Education). Faculdade de Educação, Universidade Federal de Minas Gerais, Belo Horizonte, MG.

Coutinho, F. A., Mortimer, E. F., & El-Hani, C. N. (2007). Construção de um perfil para o conceito biológico de vida [Construction of a profile for the biological concept of life]. *Investigações em Ensino de Ciências, 12*(1), 115–137.

Duit, R., & Treagust, D. (1998). Learning science: From behaviorism towards social constructivism and beyond. In B. J. Fraser & K. G. Tobin (Eds.), *International handbook of science education* (pp. 3–25). Dordrecht: Kluwer.

Durkheim, E. (1972). *Selected writings.* Cambridge: Cambridge University Press.

El-Hani, C. N., & Nunes-Neto, N. F. (2009). Function in biology: Etiological and organizational perspectives. *Acta Biologica Colombiana, 14S*, 111–132.

Gee, J. P., & Green, J. L. (1998). Discourse analysis, learning and social practice: A methodological study. *Review of Research in Education, 23*, 119–169.

Ginobili, S., & Blanco, D. (2007). Gould y Lewontin contra el programa adaptacionista: Elucidación de críticas [Gould and Lewontin against the adaptationist program: elucidation of criticisms]. *Scientiae Studia, 5*(1), 35–48.

Halliday, M. A. K., & Martin, J. R. (1993). *Writing science.* London: Falmer Press.

Kelly, G. J., & Brown, C. (2003). Communicative demands of learning science through technological design: Third grade students' construction of solar energy devices. *Linguistics and Education, 13*, 483–532.

Kress, G., Jewitt, C., Ogborn, J., & Tsatsarelis, C. (2001). *Multimodal teaching and learning: The rhetorics of the science classroom.* London: Continuum.

Kozulin, A. (1990). *Vygotsky's psychology: A biography of ideas*. New York, NY: Harvester Wheatsheaf.

Lemke, J. L. (1990). *Talking science: Language, learning and values*. Norwood, NJ: Ablex Publishing Co.

Lewontin, R. (1978). Adaptation. *Scientific American, 249*, 212–222.

Lewontin, R. (1983). The organism as the subject and object of evolution. *Scientia, 118*, 63–82.

Maynard Smith, J. (1969). The status of neo-Darwinism. In C. H. Waddington (Ed.), *Towards a theoretical biology, vol. 2: Sketches* (pp. 82–89). Edinburgh: Edinburgh University Press.

Mayr, E. (1982). *The growth of biological thought*. Cambridge, MA: Harvard University Press.

Mayr, E. (1998). *Toward a new philosophy of biology*. Cambridge, MA: Harvard University Press.

Mehan, H. (1979). *Learning lessons: Social organization in the classroom*. Cambridge, MA: Harvard University Press.

Mortimer, E. F. (1995) Conceptual change or conceptual profile change? *Science & Education, 4*, 267–285.

Mortimer, E. F. (1998). Sobre chamas e cristais: A linguagem cotidiana, a linguagem científica e o ensino de ciências [On flames and crystals: Everyday language, scientific language, and science teaching]. In A. Chassot & R. J. Oliveira (Eds.), *Ciência, ética e cultura na educação* [Science, ethics, and culture in education] (pp. 99–118). São Leopoldo, RS: UNISINOS.

Mortimer, E. F. (2000). *Linguagem e formação de conceitos no ensino de ciências* [Language and concept formation in science teaching]. Belo Horizonte, MG: Editora UFMG.

Mortimer, E. F., & Amaral, L. O. F. (1999). A conceptual profile for molecule and molecular structure. In N. Psarros & K. Gavroglu (Eds.), *Ars mutandi: Issues in philosophy and history of chemistry* (pp. 89–101). Leipzig: Leipziger Universitäts Verlag.

Mortimer, E. F., & El-Hani, C. N. (Eds.). (2014). *Conceptual profile: A theory of teaching and learning scientific concepts*. Dordrecht: Springer.

Mortimer, E. F., & Scott, P. (2003). *Meaning making in secondary science classrooms*. Maidenhead: Open University Press.

Mortimer, E. F., & Smolka, A. L. B. (2001). Linguagem, cultura e cognição: Um olhar sobre o ensino e a sala de aula [Language, culture, and cognition: A look at teaching and the classroom]. In E. F. Mortimer & A. L. B. Smolka (Eds.), *Linguagem, cultura e cognição: Reflexões para o ensino e a sala de aula* [Language, culture, and cognition: Reflections for teaching and the classroom] (pp. 9–20). Belo Horizonte, MG: Autêntica.

Nunes-Neto, N. F., & El-Hani, C. N. (2009). O que é função? Debates na filosofia da biologia contemporânea [What is function? Debates in current philosophy of biology]. *Scientiae Studia, 7*(3), 353–401.

Nunes-Neto, N. F., & El-Hani. C. N. (2011). Functional explanations in biology, ecology, and Earth system science: Contributions from philosophy of biology. In D. Krause & A. A. P. Videira (Eds.), *Brazilian studies in history and philosophy of science, Boston studies in philosophy of science* (Vol. 290, pp. 185–200). Dordrecht: Springer Verlag.

Ogborn, J., Kress, G., Martins, I., & McGillicuddy, K. (1996). *Explaining science in the classroom*. Buckingham: Open University Press.

Posner, G. J., Strike, K. A., Hewson, P. W., & Gertzog, W. A. (1982). Accommodation of a scientific conception: Toward a theory of conceptual change. *Science Education, 66*, 211–227.

Ruse, M. (2002). Evolutionary biology and teleological thinking. In A. Ariew, R. Cummins, & M. Perlman (Eds.), *Functions: New essays in the philosophy of psychology and biology* (pp. 33–60). Oxford: Oxford University Press.

Roth, W.-M. (2005). *Talking science: language and learning in science classrooms*. Lanham, MD: Rowman and Littlefield.

Scott, P. H. (1998). Teacher talk and meaning making in science classrooms: A Vygotskian analysis and review. *Studies in Science Education, 32*, 45–80.

Sepulveda, C. (2009). Aplicação de um perfil conceitual para adaptação à análise de interações discursivas no ensino de evolução [Application of a conceptual profile for adaptation to the analysis of discursive interactions in evolution teaching]. In E. F. Mortimer (Ed.), *Anais do VI Encontro Nacional de Pesquisa em Educação em Ciências* [Proceedings of the VI National Meeting of Research in Science Education]. Belo Horizonte, MG: ABRAPEC.

Sepulveda, C. (2010). *Perfil conceitual de adaptação: Uma ferramenta para a análise de discurso de salas de aula de Biologia em contextos de ensino de evolução* [Conceptual profile of adaptation: A tool to discourse analysis in biology classrooms in the context of evolution teaching] (Thesis, Doctorate in History, Philosophy, and Science Teaching). Programa de Pós-Graduação em Ensino, Filosofia e História das Ciências, Universidade Federal da Bahia/Universidade Estadual de Feira de Santana, Salvador, BA.

Sepulveda, C., & El-Hani, C. N. (2008). Adaptacionismo versus exaptacionismo: O que este debate tem a dizer ao ensino de evolução [Adaptationism vs. exaptationism: What does this debate have to tell to evolution teaching?]. *Ciência e Ambiente, 36*, 93–124.

Sepulveda, C., Meyer, D., & El-Hani, C. N. (2011). Adaptacionismo [Adaptationism]. In P. Abrantes (Ed.), *Filosofia da biologia* [Philosophy of biology] (pp. 162–192). Porto Alegre, RS: ARTMED.

Silva, A. C. T., & Mortimer, E. F. (2005). Aspectos teórico-metodológicos da análise das dinâmicas discursivas das salas de aula de ciências [Theoretical-methodological aspects of the analysis of discursive dynamics in the science classroom]. In R. Nardi & O. Borges (Eds.), *Anais do V Encontro Nacional de Pesquisa em Educação*

Científica [Proceedings of the V National Meeting of Research in Science Education]. Bauru, SP: ABRAPEC.

Shtulman, A. (2006). Qualitative differences between naïve and scientific theories of evolution. *Cognitive Psychology, 52*, 170–194.

Sutton, C. (1992). *Words, science and learning*. Buckingham: Open University Press.

Vargens, M. M. F., & El-Hani, C. N. (2011). Análise dos efeitos do jogo clipsitacídeos (clipbirds) sobre a aprendizagem de estudantes do ensino médio acerca da evolução [Analysis of the effects of the game Clipbirds on high school students' learning about evolution]. *Revista Brasileira de Pesquisa em Educação em Ciências, 11*(1), 143–168.

Wertsch, J. V. (1985). *Vygotsky and the social formation of mind*. Cambridge, MA: Harvard University Press.

Wertsch, J. V. (1991). *Voices of the mind: A sociocultural approach to mediated action*. London: Harvester Wheatsheaf.

Wertsch, J. V. & Hickmann, M. (1987). Problem solving in social interaction: A microgenetic analysis. In: Hickmann, M. (Ed.) *Social and functional approaches to language and thought* (pp. 251–266). New York, NY: Academic Press.

CHAPTER 13

Argumentation from Toulmin's Perspective in Teaching Science

Marta A. Pesa, Stella M. Islas, Silvia del Valle Bravo and Celia Medina

Abstract

Contributions from philosophy, sociology of science and cognitive psychology of science, and cognitive psychology are presented to justify the urge to develop reasoning and argumentation abilities for science learning. Some indicators are suggested to guide the teaching and learning processes in the different fields of scientific knowledge.

Keywords

argumentation – science teaching – scientific discourse – Toulmin

1 Introduction

Over the last decade, key groups of science education researchers have more and more focused their interest on the issue of scientific argumentation, both in its theoretical aspects and its central role in the teaching and learning of science. Evidence of the increasing importance of this field may be found in the plethora of recent publications, the growing number of papers on this topic in conferences, and the rising volume of research being undertaken at an international level.

Without ignoring the importance of conceptual content in the process of the construction of scientific knowledge, more attention is drawn nowadays to the study of the instruments, criteria, models, and rules students use to research, justify, evaluate, and value scientific knowledge (Duschl, 1998; Osborne et al., 2001). From this perspective, scientific discourse gains special importance, not only as an expression of principles, laws, theories and concepts, but also as a discourse structuring the rational operations used to relate, on the one hand,

 | DOI: 10.1163/9789004409088_013

data, explanations, and conclusions, and, on the other, the epistemological principles that guide inquiries.

If we look at the reality of our classrooms, scientific education continues to emphasize facts and the presentation of closed and finished theories, without paying attention to the processes associated with the development of critical and reflective reasoning or the formulation of arguments, both of which are fundamental to making science learning a new way to perceive and know the world (Simon, Erduran, & Osborne, 2006; Driver, Leach, & Millar, 1996). The latter requires the development of competence in understanding and formulating scientific arguments, analyzing how evidence is used to construct explanations and justifications, and examining the data which form the bases of scientific ideas and theories (Simon 2006; Jimenez Aleixandre & Diaz de Bustamante, 2003).

The richness of argumentation lies at different levels and is related to many different aspects of teaching and learning. To begin with, it is related to the construction of knowledge in which phenomena can be interpreted through models – and through the choice and justification of the most appropriate model by which to explain and explore the available evidence. It also relates to communication, to discussions among peers and experts in articles and reports in which results are presented, as well as in their participation in controversies in which two or more conflicting ideas differ as to how to tackle a problem. Finally, argumentation is also an important opportunity for the development of scientific competences within the framework of the scientific culture (Jimenez Alexaindre, 2000, 2010; Islas, Sgro, & Pesa, 2009).

Contemporary philosophy and history of science emphasize the idea that scientific practices (Kitcher, 1993; Rousse, 2002, 2007; Galison, 1987) are not simply the accumulation of evidence about what the world is, but they imply, in addition to the formulation of explanatory and predictive theories open to discussion and refutation, the construction of models, along with the development of experimental, argumentative, and discursive skills and abilities. These fields of inquiry seek to understand not only the products of science (what we know), but the complex processes of its intellectual achievements (how we know it) (Dumrauf & Cordero, 2005), in which controversies, debate, dissent, and the need for justified selections between competing perspectives predominate (Duschl, 1998).

Some authors consider argumentation as a crucial part of classroom dialogues which strengthen learning (for example, Kubli, 2005). Others focus their attention on the development of teachers' competences in educating critical citizens (for example, Tenreiro-Vieira, 2004). In any case, guiding ideals include improved scientific comprehension, better reading and writing skills in the field, and the development of the ability to identify the differences between everyday language and scientific language (Sardá & Sanmartí, 2000). All approaches explicitly acknowledge the significance of developing argumentation abilities, due to their epistemic and discursive importance regarding the construction of knowledge, the development of heuristic strategies for learning to reason, and for making the justification of statements about the world explicit (Henao & Stipcich, 2008).

Within this framework, our research team deals with discursive argumentation and its implications in teaching science. Our research has interdisciplinary characteristics, since it integrates contributions from both the philosophy and sociology of science, from cognitive psychology, and from our own experiences as science teachers and researchers in science education. We will try to provide some contributions here regarding the following topics:

- What do we understand via argumentation? The limitations of deductive logic in the analysis of scientific argumentation and the use of Toulmin's theory as a framework will be analyzed.
- Debates between physics researchers will be examined, including examples such as those regarding light-matter interaction and the concept of a negative refraction index.
- We will investigate the complementarities between the epistemological approach to argumentation and cognitive psychology, especially those aspects referring to the progressive development of discursive and reasoning skills in the science learning process.

2 Logic and Argumentation

2.1 *Limits of Deductive Logic*

Since its beginning, philosophy has dealt with the analysis of our methods for reasoning and for justifying statements, i.e., the ways in which we argue. Since Aristotle, this analysis has been conducted systematically and formally, resulting in, among other things, logic as a discipline which mainly studies the validity and correction of arguments and reasoning. However, the question of validity is not the only aspect of arguments. The theory of argumentation focuses on their persuasive efficacy, as well as on their usefulness in justifying our beliefs and statements in different situations, including the epistemic contexts in which models and scientific explanations are developed and evaluated.

Until the mid-20th century, metascience had paid little attention to the theory of argumentation. This approach was predominantly logical and strategic:

An epistemology whose principal tool was mathematical logic was favored in order to avoid psychological analyses, as the latter were considered unreliable.

The question of argumentation was linked not only to the validity of scientific inferences, but also, and more deeply, to the problematic circumstance that it is in fact reason and methods of argumentation that guarantee and explain the success of science. That is, it was linked to the question of scientific rationality and, for a long time, to decisions about that which should be considered as **the** (unified, singular, correct) scientific method.

Discussions on whether science uses an inductive or deductive method are almost endless. Even by simply starting with the name of each method, the question of precisely which arguments should predominate in its scientific activity becomes evident. What is implicit in such debates are two different ways of conceiving knowledge: whether it is based on sensory experience, or if, on the contrary, it is more related to our creations and constructions.

The classic empiricist tradition identified the scientific method with inductive inferences which, arising from "neutral" observations of facts, would lead to general laws.[1] In turn, the rationalist tradition, having geometry as an ideal, thought natural science could be constructed with deductions from an initial truth.

Regardless of how scientists arrive at the formulation of their hypotheses and laws, the debate has continued during the 20th century. The problem is that, even though non-deductive reasoning, such as induction, allows for the introduction of new information, it does not provide a firm basis for conclusions, since it is ampliative. By contrast, deductions, which do provide a firm basis, do not allow for the generation of new information.[2]

Philosophers of science in the 20th century, including the neopositivists (comprising logical positivism and logical empiricism), were aware that the reasoning and inference methods scientists use to formulate general laws and hypotheses are diverse. Furthermore, according to the "standard view" (neopositivism, see Popper, 1959), some of these methods cannot be analyzed in terms of formal deductive logic. Thus, their analysis of the philosophy of science was limited to the context of justification, leaving out what they called the context of discovery (Reichenbach, 1938; Popper, 1959) – and thereby acknowledging that inductions, analogies, or intuitions could be part of the possible mental procedures used to arrive at a hypothesis.

Once the analysis of the context of discovery was abandoned, it seemed that the philosophy of science could be free from the problem of induction. However, not everything was as it seemed to be. No matter how thoroughly the hypothetical deductive method is applied, and the participation of inductive inferences denied, just as Popper intended, induction reappears in the positive test.

According to Popper (1959), and with the support of deductive logic, we can be sure when a hypothesis is false, but we can never be sure if it is true. In principle, if the logical rules are followed, in order to empirically confront a hypothesis, observational consequences have to deductively derive from the hypothesis; in other words, basic or particular statements which describe a possible observation are derived therefrom. If, as a result of an empirical or observational test, such statements contradict the hypothesis, then (if the auxiliary hypotheses and the initial conditions remain the same), the hypothesis is false and must be abandoned. Up to this point, the logical rule of *modus tollens* works perfectly and the deductive process is impeccable.

On the other hand, if the observable consequences do not contradict the hypothesis, then, according to Popper, we cannot say that the hypothesis is true, or approximately true, or probably true, etc. It is only possible to say that it has been corroborated, i.e., for the time being, we cannot assert that it is false. Popper does not speak of confirmation because, according to deductive logic, we cannot derive the truth of a universal statement from a particular statement in which it is implied, just as we cannot derive the truth of a cause from a consequence. Then, one may wonder, why would we consider the initial hypothesis to be corroborated? After all, being corroborated means that a particular statement, a basic one, does not deny the universal statement but affirms it, and this is an inductive jump. Besides, why would we prefer to keep the corroborated hypothesis if we cannot inductively project it into the future? Facing these objections, Popper was forced to accept there was a certain inductivist blow in this process; however, he continued in his subsequent work to deny the participation of induction (Popper, in Schlipp, 1974).

To sum up, in every positive test there is an inductive jump from the data obtained in the experiment to the affirmation of a general hypothesis. This jump is comparable to the examples Toulmin (1995) provides in reference to the substantial arguments which will be analyzed in the following section.

Towards the end of the 1960s, the standard philosophical view lost much of its influence. Simultaneously, the historicist turn in the philosophy of science showed that deductive logic could not continue to be the main tool of analysis for understanding scientific developments and that an analysis restricted to the finished scientific product – i.e. well established theories – was insufficient. Moreover, philosophers never found **the** scientific method; what they have found is that there is no unique method. Further, they acknowledged the need to analyze not only the context of justification, but also the contexts of discovery, application, and science education. However, the influence of logical deductive analysis has been so tenacious that, even nowadays, the fact that the scientists can and do use varied reasoning methods (deductive, inductive, abductive, analogic, etc.) is still questioned.

As a consequence of these shifts, along with the contributions of the historicist turn, contemporary philosophy of science counts on input not only from the history of science but also from psychology, sociology, and theory of argumentation. Thus, an analysis of reasoning which departs from a strictly deductive point of view, one which stems instead from a pragmatic and discursive perspective, has emerged. From such an approach, non-deductive reasoning can be accepted as legitimate without question, as simply an extension of our daily argumentation.

2.2 *Stephen Toulmin's Theory of Argumentation*

Stephen Toulmin's influential book, *The Uses of Argument* (first edition 1958/last edition 1995), provides different bases for the analysis of scientific arguments or reasoning. Some consider Toulmin's work to fall within what is referred to as informal logic; others consider it part of what is referred to as theory of argumentation. The truth is, it is an analysis of the natural language we use not only in scientific reports, but in different environments, such as police and judicial investigations, scientific controversies, and political debates. In his analysis, Toulmin uses what is known in philosophy as "modal logic", a kind of logic which analyzes those arguments that contain modal concepts such as possibility, necessity, contingency, and impossibility.

His interest does not center on the validity of arguments according to their syntactic and formal structure, but on their efficacy in legitimizing a statement according to its material or semantic content. The author departs from what he calls the "*legal analogy*", which fundamentally consists of comparing "*rational evaluation procedures with the legal ones*" (Toulmin, 1995). For Toulmin, the validity of an argument does not lie so much in its empty formal structure as in the procedures that take place as it is argued.

It is possible to say that the distinction Toulmin establishes between analytical or theoretical arguments and substantial ones is comparable to the distinction between deductive and inductive arguments. In fact, analytical or theoretical arguments correspond to deductive ones; similarly, validity is not related to the context, or to the procedure to justify our statements, but to the syntactic structure. The substantial argument, in turn, includes, among others, inductive arguments.

Toulmin holds that deductive arguments are not substantial, i.e., they are not weighty, because, in the first place, even though they guarantee conserving the truth, if we depart from a truth in science, we cannot always be sure that our premises are true. Secondly, since in deductions nothing can be concluded which is not already implicit in the premises, they do not add anything new; therefore, they turn out to be scarcely fruitful, not only for the production of new knowledge, but also for those cases in which we might expand knowledge from one field to another.

It is worth pointing out that the pertinence of statements with the pretension of being knowledge, as well as those statements (which make up the argument) used to justify them, should be considered in their historical context and in the context of the argumentative field in which they are made.

The general procedure Toulmin finds in arguments from different argumentative fields (ethics, justice, aesthetics, science, etc.), consists of departing from a statement with pretensions of knowledge or truth, a statement which in the reconstruction of the arguments will take the place of the conclusion (C) and which is legitimized by a set of data (D).

Toulmin then notes that when someone objects to our statement we will find ourselves forced to produce warrants (W) of the relations we established between the data and our statement or conclusion. These warrants are general hypothetical statements that authorize or legitimize the transition from the data to the conclusion. They are the reasons presented in order to justify the connection between the data and the statements of knowledge.

Mere experimental results, combined only with mathematics and certain formal criteria, are not sufficient to produce scientific knowledge (Lang da Silveira & Ostermann, 2002). The scientific community always orients its experiments and legitimizes its conclusions guided by theoretical knowledge, models, laws, and theories.

If resistance is still found in our objector, we can resort to backing up the warrants (Backing, B) according to the data previously established.

To quote an example from Toulmin (1995, p. 92):

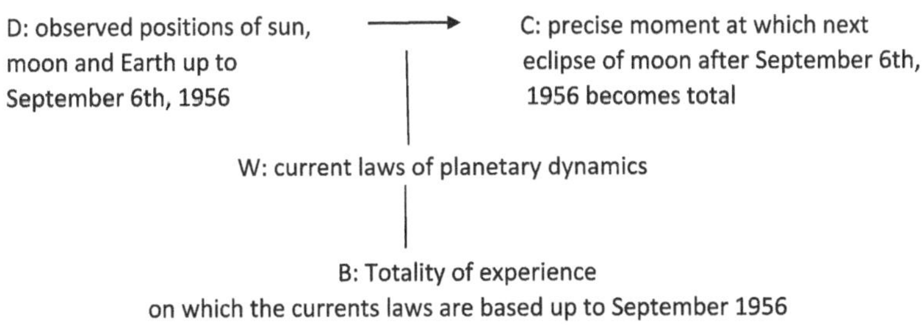

In addition to these previous elements, Toulmin acknowledges other elements destined to establish the limits and nuances of our statements, which are related to the degree of certainty of the statements, which is, in turn, dependent upon certain conditions and is subject to exceptions. Thus, warrants certify the soundness of an argument, but they can confer different degrees of force to the conclusions. Qualifiers (Q), such as "necessarily", "probably", "presumably", etc., describe this force. In the cases where there clearly are exceptions, the conditions of rebuttal or refutation (R) are formulated as indicators of the circumstances in which the authority of warrant would have to be set aside, i.e., restrictions or exceptions that apply to the conclusion or to the limits to its validity.

The strength of Toulmin's model lies in its ability to evaluate arguments. Data, statements, backing, refutation, and qualifiers are invariable characteristics of arguments and are discipline-independent. What counts as warrant, data, or backing characteristically depends on the discipline concerned with a given argument.

For Toulmin, sound and robust arguments are substantial arguments that can resist criticism and objections. This is different from deductive arguments, in which it is only possible to conclude that which is already contained in the premises.

To sum up, Toulmin's analysis of the validity of arguments is not an atemporal and purely formal matter as found in deductive logic, but offers questions regarding the procedures used to base our statements upon good reasons.

For this author, theories of knowledge and logic cannot remain separate. Furthermore, logic cannot be restricted to deductive logic alone. According to Toulmin, the problem is that logic has followed the mathematical model as a guide – and therefore demands that all kinds of argumentation do the same. For him, the fact that all substantial arguments are not conclusive (as the analytical ones are) does not make them any less legitimate or valid. In his opinion, there is no "inductive jump" that needs to be justified; in fact, there is no logic gap unless we consider arguments solely from an analytical point of view.

2.3 *Open Aspects of Science Teaching Research*

The aspects mentioned above show the importance of focusing our attention on the development of argumentation abilities in the classroom as a strategy for integrated science learning (Cudmani, Pesa, & Salinas, 2000). In order to design didactic proposals, it is relevant to consider the epistemic traits of knowledge construction in the classroom, asking ourselves questions such as:

- What do students consider as data?
- What do they understand by conclusion or statement of knowledge? How do they justify or support them?
- When do they consider an explanation satisfactory? Do they value justifications?
- Which work environments favor argumentative discourse in the classroom?

This spectrum of questions will expand as we advance in our knowledge about the foundations of argumentation processes.

3 Argumentative Debates among Physics Researchers, Using as Examples Light-Matter Interaction and the Concept of a Negative Refraction Index

The mid-20th-century contributions of the sociologist Robert Merton brought about a trend toward research which focuses on the characteristics of science as an institution and tradition (Vessuri, 1991), i.e., the study of the sociology of science. The sociological study of science has dealt with many different subjects, including the cognitive strategies used by researchers (Nersessian, 1992), the ways in which the flow of information is regulated (Campanario, 2004), the kind of inferences involved in reasoning at scientific events (Dunbar, 2000), the way systems of distributed cognition are formed (Giere, 2002), and cognitive-cultural systems (Nersessian, 2006). The broad range of studies extant demonstrates the complexity of scientific practices.

The communicative processes within scientific communities, and those between these communities and society, are among the diverse dimensions of scientific practices.

> Communication is essential to the nature and practice of science. The fact that the number of publications and articles has been used as an indicator of the growth of scientific knowledge establishes the production of publications of the research process as a valid measure of its activity. (Russell, 2001)

Scientific communication has special characteristics, since each community establishes norms and standards around such elements as length of papers, themes, valid methodology, number of authors, grammatical and writing styles, the quantity and quality of bibliographic references, and what constitutes important data, in order to lend value and validity to a publication. There are also tacit criteria, such as ways of arguing, i.e., the use of verbal strategies (or written ones) to defend one's own ideas and/or to argue against someone else's, with the intention of generating debate, questioning results, or coming to an agreement (Van Eemerem & Grootendorst, 2002).

When argumentation in the scientific community is the object of study, recording empirical data is an arduous task. The synchronic argumentative exchanges that occur, for example, in scientific events, or in sessions in a conference, usually remain hidden (or accessible only through ethnographic research or the written record of resulting conclusions). Nonetheless, it is possible to have access to other, diachronic exchanges such as those in scientific journals and/or publications specifically devoted to the discussion of methodology, design, and results in research.

Along this line, we will present research (Islas et al., 2009) on communications within a community of physicists. We studied electronic articles entitled *Comment* and *Reply* in the debates regarding light-matter interaction and negative refraction indexes (Islas, 2010).

We examined 26 specialized journals available online, not with the purpose of drawing general conclusions regarding the way scientific discussion was conducted, but in order to obtain empirical data which would exemplify the reasoning present in certain scientific debates. In particular, we identify those cases in which the validation of a conclusive statement is centered more on the semantics of the conclusions than on the syntax of the inferences.

The debates used (Islas, 2010) focus on the following topics:

- The limits of validity established for a principle/law/result.
- The pertinence of the analogies used.
- The type of model associated with light.
- The margins of uncertainty delimiting the results.
- The different theoretical views with which the same experimental result is justified and interpreted.

Moreover, within the limitations of this research, some characteristics of these debates and the reasoning utilized therein can also be identified (Islas et al., 2009), to wit:

- The use of analogical inferences is frequent; even though they do not allow scientists to draw conclusions with logical validity, they offer guidelines for the interpretation of results.
- Most reasoning can be described as abductions.
- Controversies arise from disagreements regarding the attributes each scientist considers to be essential in the physics situations under analysis.
- Semantic differences in the argumentations expressed as different interpretations of the same phenomenon are frequent.

These results show the complexity of the construction of scientific knowledge and contribute indicators to guide classroom work when structuring instructional activities. Just as Duschl (1998) points out, science teaching should reflect the characteristic traits of knowledge construction through debates, seeking consensus, and making aspects that are open or in question explicit, along with those that are in conflict. All of these are characteristics of the process of transforming data into explanations, providing opportunities for students to develop, evaluate, and revise alternative solutions.

However, teachers' epistemological conceptions (Harres, 1999) do not coincide with the actual work of the community of researchers. Teachers tend to focus on closed theoretical outlines, leaving aside the process of the production of knowledge and minimizing social traits, including the internal controversies and debates regarding which of all the possible interpretations of a phenomenon is the most appropriate one.

In addition to the study of current debates, it is possible to discern the importance of argumentation in the models of theoretical change contributed by philosophy of science. From the second half of the 20th century on, philosophers of science have pointed out the importance of debates in the development of the history of science (Kuhn, Feyerabend, Laudan, etc.). Within this frame is Nudler's view (2009), which offers controversial debates as a model of change. In other words, debates in which two or more conceptions differ regarding a given problem make up the engine of development in a discipline. These debates are possible because, according to Nudler, there is no incommensurability; thus, such controversies develop on common ground, wherein scientists share metaphysical and methodological suppositions, procedural rules, and values.

To conclude, controversies are an engine for scientific progress because – in addition to entailing the refinement and modification of theories – they make explicit and clarify assumptions, thereby fostering the development of thought.

A teacher committed to these understandings could help students to meaningfully comprehend not only scientific information, but also the social aspects of science. In this way, we will be able to educate scientifically literate citizens capable of independent thought, as well as respectful and rational debate.

4 Complementarity between the Epistemological Approach to Argumentation and Cognitive Psychology

According to Gérard Fourez (1997), specialized technical knowledge does not suffice to ensure participation and rational decision making vis-à-vis current scientific-technological issues. The ability to build rational frameworks regarding concrete problems is also necessary.

This approach to concrete problems reveals the fact that approximation to the world from a particular discipline is partial and generally too limited; therefore, an interdisciplinary approach to problems is necessary. Fourez (1997) therefore proposes an interdisciplinary perspective, the purpose of which is not to create a new discourse beyond particular disciplines, but to construct a new, specific "practice" by which to approach the problems of daily life. This is an essentially "political" practice, since it involves a "negotiation" among different points of view in order to *decide* on an adequate representation *with regard to an action*.

Other authors (Jiménez Alexaindre & Díaz de Bustamante, 2003) have also pointed out that the development of critical thinking is part of our social competence as citizens, understanding critical thinking as "the ability to develop an independent opinion, acquiring the faculty of reflecting on reality as well as participating in it" (p. 362). At the same time, critical thinking is closely related to the development of reasoning abilities, i.e., the ability to relate a statement, explanation, or option to the evidence that supports it.

Jiménez Alexaindre and Díaz de Bustamante's notion has several points in common with Fourez's notion of rationality frameworks and Toulmin's rationality, in the sense that all three involve an individual's rational activity in facing a given situation and assessing the reliability of statements based on the available evidence. Indeed, this can be considered a social process, based on a dynamic and dialogical negotiation of meaning by which participants must give reasons, receive objections, make rebuttals, and perhaps modify or refine initial claims (Toulmin, 1955). This perspective on developing critical thinking and the capacity for substantiated decisions in order to ensure full participation as citizens in today's world necessarily leads to a conception of science teaching that is based on the processes of argumentation. Regarding this idea, Duschl (1998) points out that the teaching of science should emphasize the development of scientific language as a discourse that critically examines and evaluates the process of the transformation of evidence into explanations.

These proposals are consistent with contributions from many cognitive psychologists' theories of learning (Ausubel, Vygotsky, Vergnaud, Gowin, Johnson-Laird, etc.), which clearly state the importance of language and the negotiation of meaning in the learning process.

Vygotsky (1977) states that "language is, above all, a means of social communication, a means of expression and comprehension … language combines the communicative function with the thinking function". Similarly, Ausubel (1968) asserts that language makes the development of complex ideas and concepts acquired through meaningful learning possible.

At the same time, Vergnaud (1990, 1994) attributes great importance to language in the process of conceptualization. In Vergnaud's theory of conceptual fields, previous knowledge acts as a precursor to new knowledge, playing a fundamental role in the development of progressive control over an area of knowledge. In this scenario, learner activity in a given situation is led by schemes that generate a sequence of actions which depend on the parameters of the situation. Vergnaud (1990, 1996) defines a scheme as the "organized totality which allows[individuals] to generate a different kind of behavior according to the particular characteristics of each situation".

Vergnaud's theory of learning holds that a concept gradually becomes meaningful to individuals through situations and problems, since from them they abstract the properties they consider relevant; these will then constitute their concepts-in-action and theorems-in-action. As long as these invariable operations can be expressed explicitly, they have the potential to become concepts and theorems. In this process, linguistic expressions, symbols, and symbolic representations are considered as the path through which cognitive complexity is achieved. The mediators in this process of explicitation are words and argumentative discourse, which allow the individual to make conscious use of the operative invariants.

In Vergnaud's theory of learning, the concept of scheme involves the individual's use of rational processes – such as relating data, drawing conclusions from data, and assessing the validity of a theoretical model – in a given situation, in order to explain the behavior in a particular experimental system. These procedures are understood as the path through which concepts become meaningful.

It is clear from these different theoretical frameworks, each of which attempt to explain how meaningful learning can be achieved, that language and human interaction are essential factors in the achievement of such learning. These varied frameworks all point to a close relationship between communicative competencies and the learning of scientific models. Language plays a fundamental role in the processes of teaching and learning, as a structural element of the concepts and reasoning used to solve problematic situations.

5 Conclusions

At the beginning of this chapter, we pointed out that argumentation – as framed here by Toulmin (1995) – has recently become an important research focus, due to its theoretical aspects as well as its vital role in science education. In this vein, Jiménez Alexaindre (2003) states that argumentation is not only part of scientific competence but also contributes to the development of basic educational competences such as:

- Competences related to the learning process (linguistic communication, explicitation of thought processes, etc.).

- The formation of responsible citizens, capable of critical thought, decision making, and concrete actions.
- Development of the scientific culture necessary for understanding and assessing scientific practices.

In this chapter, contributions from philosophy, the sociology of science, and cognitive psychology have been presented that justify the urge to develop reasoning and argumentation abilities for science learning. Some indicators have been suggested to guide the teaching and learning processes in the different fields of scientific knowledge.

One of the aspects that deserves further analysis is the role of the teacher in the development of these reasoning abilities. Although the appropriation of a culture by any individual depends on their own activity in the construction and reconstruction of concepts, it also depends on the help provided by their environment, and, thus, on the quality of the mediations occurring during the learning process (Vergnaud, 2013).

So, from a standpoint that blends the contributions presented in this chapter, the teacher takes up the role of a mediator. Teachers choose and present adequate situations, make clear the objectives of activities, contribute to the organization of those activities, and facilitate inferences. In this respect, teachers must conduct discussions with the aim of helping students to express their thoughts and overcome biases in doing so, all the while guiding the process towards the development of sound argumentation. Teachers face a significant challenge at this point in providing the necessary help to students in order to ensure that they are able to avoid fallacies in their (or others') reasoning.

In fact, research on human reasoning in different contexts (González Labra, 1998) has identified a variety of reasoning biases that could explain learning difficulties. By biases we mean systematic tendencies to consider irrelevant factors in the inference process. These are the usual result of a particular way of processing information: People usually use heuristics based on their automatic, involuntary, and generally emotional processing of situations. In this sense, biases are approximative rules that simplify problems, allowing intuitive resolutions that result from a compromise between speed and efficiency (Tversky & Kahneman, 1984). Though pragmatically useful in daily life, these rules produce conclusions that are far from formally correct (or scientifically valid).

The improvement of education and the development of competences within the field of argumentation greatly depend on the professional development and improvement of teachers. Teachers should, besides having deep knowledge of disciplines, be ready to:

- help students state their thoughts explicitly;
- help students justify their statements with appropriate scientific evidence;

- develop and establish criteria for the construction and evaluation of arguments during discursive interactions in the classroom; and
- help students understand scientific information in the context of the social web of science.

Thus, a good didactic performance in the teaching of science is necessarily based on an awareness of the difficulties that arise in the cognitive tasks involved in the process of the construction of scientific knowledge, and also of the obstacles that usually have to be faced.

Notes

1 Inductive inferences do not only go from the particular to the general. There can also be inductions that go from general to general, etc. What identifies them is their ampliative quality, i.e., the conclusion has more information than the premises from which it comes (Vickers, 2010).

2 It is worth pointing out that for logic the only formally valid inferences are the deductive ones, since in them the conclusion demonstrably follows from the premises, i.e., given true premises, if we reason correctly, the truth will necessary follow. In inductive inferences, however, the conclusion does not necessarily follow, but the premises provide a greater or lesser degree of support. The problem with induction is that there is no way of justifying it from a logical point of view, since its ampliative character prevents inductive reasonings from being logically valid inferences (Vickers, 2010).

References

Ausubel, D. (1963). *The psychology of meaningful verbal learning*. New York, NY: Grune and Stratton Ed.

Campanario, J. (2004). Algunas posibilidades del artículo de investigación como recurso didáctico orientado a cuestionar ideas inadecuadas sobre la ciencia. *Enseñanza de las Ciencias, 22*(3), 365–378.

Cudmani, L. C. de, Pesa, M., & Salinas, J. (2000). Hacia un modelo integrador de enseñanza de las ciencias. *Enseñanza de las Ciencias, 18*(1), 3–14.

Driver, R., Leach, J., Millar, R., & Scott, P. (1996). *Young people's images of science*. Buckingham, England: Open University Press.

Dumrauf, A., & Cordero, S. (2005). La enseñanza de la termodinámica a través de procesos argumentativos. *Enseñanza de las Ciencias, Numero extra.*

Dunbar, K. (2000). How scientists think in the real world: Implications for science education. *Journal of Applied Developmental Psychology, 21*(1), 49–58.

Duschl, R. (1998). La valoración de argumentos y explicaciones: promover estrategias de retroalimentación. *Enseñanza de las Ciencias, 16*(1), 3–20.

Fourez, G. (1997). *Alfabetización científica y tecnológica*. Argentina: Ed. Colihue.

Galison, P. (1987). *How experiments end*. Chicago, IL: University of Chicago Press.

Giere, R. (2002). Distributed cognition in epistemic cultures. *Philosophy of Science, 69*(2), 632–644.

González Labra, M. J. (1998). *Introducción a la psicología del pensamiento*. Madrid, Spain: Ed. Trotta.

Harres, J. B. S. (1999). *Concepções de professors sobre a natureza da ciencia* (Doctoral dissertation). Pontificia Universidade Católica do Rio Grande do Sul, Porto Alegre, Brazil.

Henao, B., & Stipcich, S. (2008). Educación en ciencias y argumentación: La perspectiva de Toulmin como posible respuesta a las demandas y desafíos contemporáneos para la enseñanza de las Ciencias Experimentales. *Revista Electrónica de Enseñanza de las Ciencias, 7*(1), 47–62. Retrieved from http://www.saum.uvigo.es/reec/index.htm

Islas, S. (2006). *Los debates en clases de física. Memorias del Primer Congreso Nacional "Leer, escribir y hablar hoy la ciencia, la literatura, la prensa"* [CD-Rom]. Universidad Nacional del Centro de la Provincia de Buenos Aires.

Islas, S. (2010). *La formación epistemológica de los profesores de física respecto de los debates producidos al interior de la comunidad de investigadores en la disciplina* (Doctoral dissertation). Facultad de Filosofía y Humanidades, Universidad Nacional de Córdoba.

Islas, S., Sgro, M., & Pesa, M. (2009). La argumentación en la comunidad científica y en la formación de profesores de física. *Ciência y Educação, 15*(2), 291–304.

Jiménez Alexaindre, M. P. (2010). *Competencias en argumentación y uso de pruebas*. Barcelona: Ed. Grao.

Jiménez Alexaindre, M. P., Bungallo Rodriguez, A., & Duschl, R. (2000). "Doing the lesson" or "doing science": Argument in high school genetics. *Science Education, 84*, 757–792.

Jiménez Alexaindre, M. P., & Diaz de Bustamante, J. (2003). Discurso de aula y argumentación en la clase de ciencias: cuestiones teóricas y metodológicas. *Enseñanza de las Ciencias, 21*(3), 359–370.

Kitcher P. (2001). *El avance de la ciencia*. México: UNAM.

Kollar, I., Fischer, F., & Slotta, J. (2005). Internal and external collaboration scripts in web-based science learning at schools. In T. Koschmann, D. Suthers, & T. W. Chan (Eds.), *Computer supported collaborative learning: The next ten years* (pp. 331–340). Mahwah, NJ: Lawrence Erlbaum Associates.

Kubli, F. (2005). Science teaching as a dialogue: Bakhtin, Vygotsky and some applications in the classroom. *Science & Education, 14*(6), 501–534.

Lang da Silveira, F., & Ostermann, F. (2002). A insustentabilidade da proposta indutivista de "descubrir a lei a partir de resultados experimentais". *Caderno Brasileiro de Ensino de Física,* (Suppl.), 7–27.

Leitão, S. (2000). The potential of argument in knowledge building. *Human development, 43*, 332–360.

Longino, H. (2006). The social dimensions of scientific knowledge. In E. N. Zalta (Ed.), *The Stanford encyclopedia of philosopy*. Retrieved from https://plato.stanford.edu/entries/scientific-knowledge-social/

Mulhall, P., & Gunstone, R. (2008). Views about physics held by physics teachers with differing approaches to teaching physics. *Research in Science Education, 38*, 435–462.

Nersessian N. (1992). How do scientists think? Capturing the dynamic of conceptual change in science. In R. Giere (Ed.), *Cognitive models of science* (pp. 3–45). Minneapolis, MN: University of Minnesota Press.

Nersessian, N. (2006). The cognitive-cultural systems of the research laboratory. *Organization Studies, 27*(1), 125–145.

Nubiola, J. (2001). La abducción o lógica de la sorpresa. *Razón y palabra, 21*. Retrieved from http://www.razonypalabra.org.mx/anteriores/n21/index.html

Nudler, O. (2009). *Espacios controversiales: Hacia un modelo de cambio filosófico y científico*. Argentina: Editores Miño y Dávila.

Osborne, J., Collins, S., Ratcliffe, M., Millar, R., & Duschl, R. (2001, November). *What ideas about science should be taught in school science? A Delphi study of the expert community* (pp. 1–34). Sixth International History, Philosophy, and Science Teaching Conference, Denver, CO.

Popper, K. (1959). *The logic of scientific discovery*. London: Hutchinson.

Reinchebach, H. (1938). *Experience and prediction: An analysis of the foundations and structure of knowledge*. Chicago, IL: University of Chicago Press.

Rouse, J. (2002). *How scientific practices matter: Reclaiming philosophical naturalism*. Chicago, IL: University of Chicago Press.

Rouse, J. (2007). Naturalism and scientific practices: A concluding scientific postscript. In M. M. Chienkuo & R.-L. Chen (Eds.), *Naturalized epistemology and philosophy of science*. Holland: Rodopi Bv Editions.

Russell, J. (2001, June). La comunicación científica a comienzos del siglo XXI. *Revista Internacional de Ciencias Sociales, 168*, 1–15.

Sardá, J. A., & Sanmartí, N. (2000). Enseñar a argumentar científicamente: Un reto de las clases de ciencias. *Enseñanza de las Ciencias, 18*(3), 405–422.

Schlipp, P. A. (Ed.). (1974). *The philosophy of Karl Popper* (Vols. 1 & 2). La Salle, IL: Open Court.

Simon, S., Erduran, S., & Osborne, J. (2006). Learning to teach argumentation: Research and development in the science classroom. *International Journal of Science Education, 28*(2–3), 235–260.

Tenreiro-Vieira, C. (2004). Formação em pensamento crítico de professores de ciências: Impacte nas práticas de sala de aula e no nível de pensamento crítico dos alunos. *Revista Electrónica de Enseñanza de las Ciencias, 3*(3). Retrieved from http://www.saum.uvigo.es/reec/inodex.htm

Toulmin, S. (1995). *The uses of argument*. New York, NY: Cambridge University Press.

Tversky, A., & Kahneman, D. (1984). Juicios en situaciones de incertidumbre, heurísticos y sesgos. In M. Carretero & A. G. García Madruga (Eds.), *Lecturas de psicología del pensamiento*. Madrid: Alianza.

Van Eemeren, F., & Grootendorst, R. (2002). *Argumentación, comunicación y falacias*. Chile: Ediciones Universidad Católica de Chile.

Vessuri, H. (1991). Perspectivas recientes en el estudio social de la ciencia. *Interciencia, 16*(2), 60–68.

Vergnaud, G. (1990). La théorie des champs conceptuels. *Recherches en Didactique des Mathématiques, 10*(23), 133–170.

Vergnaud, G. (1994). Multiplicative conceptual field: What and why? In H. Ghershon & J. Confrey (Eds.), *The development of multiplicative reasoning in the learning of mathematics*. Albany, NY: State University of New York Press.

Vergnaud, G. (1996). A trama dos campos conceituais na construção dos conhecimentos. *Revista do GEMPA, 4*, 9–19.

Vergnaud, G. (2013). Pourquoi la thèorie des champsconceptuels? *Infancia y Aprendizaje, 36*(2), 131–161.

Vickers, J. (2010). The problem of induction. In E. N. Zalta (Ed.), *The Stanford encyclopedia of philosophy*. Retrieved from http://plato.stanford.edu/archives/fall2010/entries/induction-problem/

Vygotsky, L. (1977). *Pensamiento y lenguaje*. Buenos Aires: La Pléyade.

CHAPTER 14

Science Textbooks: A Discursive Perspective

Isabel Martins

Abstract

In this chapter we present an overview of research that has been conducted in Brazil about Science textbooks for the last 25 years under critical discourse perspectives. The text starts by presenting the bases of critical discourse analysis in order to provide a theoretical basis to frame the science textbook both as a sociocultural artifact as well as a genre where different semiotic modes are articulated in meaning making interactions that take place in classrooms. Ways through which different kinds of knowledge and discourses (e.g. scientific, pedagogical, journalistic, amongst others) are recontextualized in the textbook are discussed alongside with compositional and rhetorical aspects. Examples of analyses include textual aspects which have been recently emphasized, (re)constructed and ignored in Brazilian Physics textbooks. Relationships are established between such discursive moves and both epistemological and contextual features of science education.

Keywords

science textbooks – critical discourse analysis – meaning making – Physics textbooks

1 The Argument

Science textbooks have often been targeted as objects of research by the science education community (Clément et al., 2007; Bensaude-Vincent & Lunders, 2000; Koulaidis & Tsatsaroni, 1996). In Brazil, textbooks are often referred to with mixed feelings (Cassab & Martins, 2008). If, on the one hand, they are considered as a powerful didactic resource that plays a very important role in structuring classroom activities, they can also, on the other hand, be seen as an instrument that controls teaching practices and undermines teachers' pedagogical autonomy. Nonetheless, they are widely available and frequently

 | DOI: 10.1163/9789004409088_014

used by teachers in general, and these aspects justify efforts toward exposing and criticizing conceptual and methodological inconsistences in textbooks (Santos, Joaquim, & El-Hani, 2012; Bizzo, 2000), as well as investigations which focus on other facets relevant to understanding the discursive, historical, political, ideological, and economic dimensions involved in textbook production, distribution, reception, and uses in classrooms (Cassab & Martins, 2003; Fracalanza, 1992; Freitag, Motta, & Costa, 1997).

In recent years, the scope of Brazilian research concerned with science education and textbooks has broadened (Fracalanza, 1992; Megid Neto, 2006), especially with respect to the need for problematizing language issues in textbooks (Martins, Gouvêa, & Vilanova, 2012). Examples include research on students' textbook reading practices (Silva & Almeida, 1998; Martins, Gouvêa, & Piccinini, 2005); science teachers' criteria for choosing textbooks (Cassab & Martins, 2008); historical and cultural influences on textbook representations (Selles & Ferreira, 2004); analyses of images and illustrations (Piccinini & Martins, 2004; Otero & Greca, 2004; Freitag et al., 1997; teachers' and curricular representations of the textbook (Megid Neto, 2006; Fracalanza, 2003); as well as analysis of discursive genres (Braga, 2003) and of underlying rhetorical aspects of textbook presentation (Ribeiro & Martins, 2007; Nascimento & Martins, 2003). Such investigations have been conducted under a number of perspectives, notably those related to the field of discourse analysis.

In this chapter, I will present and discuss examples of Brazilian research into science textbooks that was carried out under theoretical and methodological perspectives associated with discourse studies, especially those of critical discourse analysis. I will argue that the dialogue between science education and discourse studies allows for sound and novel insights not only into the nature and use of textbooks but also around learning and teaching science. The chapter discusses both the theoretical bases and the results of investigations conducted by our research group and presents an exercise of analysis which highlights some of these features.

2 Framing Textbooks through Critical Discourse Lenses

Many studies regarding aspects of the language used in Brazilian textbooks share the assumption that language cannot be conceived of outside of its social and historical dimensions, a seminal proposition put forth by Mikhail Bakhtin (2000). Another idea that underlies discursive perspectives is Michel Foucault's notion that discourse involves articulations between knowledge and power that lead to the institutionalization of "truths". Discursive approaches also consider texts as symbolic objects that contain traces of their social processes of production. From this perspective, texts are seen not as mirror images of social practices but as elements that constitute such practices (Fairclough, 1992). This is the case for the textbook, further understood here as a locus where new meanings arise as a result of the interweaving of the textbook's many constitutive discourses, namely, scientific, pedagogical, and everyday discourses, among others.

In contrast with traditional linguistic perspectives, which have language structure as a starting point, critical discourse perspectives are motivated by questions and problems of a social nature. From this perspective, texts and the social practices in which they are inscribed are seen as inseparable dimensions of discursive events. These two dimensions are mediated by discursive practices, that is, processes of textual production, circulation, and reception (Fairclough, 1992). Social practices, discursive practices, and texts are articulated so as to structure the basis for what Lemke (1996) calls a social theory of discourse, that is, a system that establishes connections between unique events, realized as texts on the micro-social scale, and as global structures identified with ideological, discursive, and political practices at the macro-social level. Texts are therefore inseparable from social situations. Bakhtin's idea of dialogism also accounts for textual heterogeneity in the sense that texts actualize previous discourses and anticipate new ones in the contexts of multiple relationships between social voices. These dynamic processes are textually marked by citations, paraphrases, presuppositions, and discursive representations. They constitute intertextual relations, that is, dialogue between texts and discourses. Similar processes happen with the ways through which genres and styles constitute textbooks so as to reveal interdiscursive relationships.

By emphasising the inextricable relationship between social practices and discourses (Fairclough, 1992), critical discourse analysis (CDA) portrays discursive change as a reflection, as well as a promoting factor, of social change. In other words, CDA is interested both in social effects on texts and the social effects of texts. Thus, texts are understood as an instrument that may exert control upon discursive interactions in the classroom, as well as a semiotic artifact to which participants in classroom interactions respond in an active, transformative, and interested manner (Martins, 2007).

3 Insights into Textbooks

3.1 *The Textbook as a Sociocultural Artifact*

In an earlier paper, I argued that

> the textbook is a sociocultural artifact, that is, its social conditions of production, circulation and reception are set with reference to social practices established in the society. As such, it has a history that is not detached from the very history of school education. (Martins, 2006)

This means that discursive analyses of textbooks must be carried out against a background of considerations about public policies for education in Brazil that include, for example, the increase in both availability and length of compulsory education (Brazil, 1996); the creation of government-funded programs of textbook evaluation and distribution (Brazil, 2016); and school and curricular reforms that foreground education for citizenship, contextualization, and interdisciplinarity in the curriculum (Brazil, 2013). No less important are the earlier influences of post-Sputnik science education programs – such as the Physical Science Study Committee (PSSC), Biological Sciences Curriculum Study (BSCS), and The Chemical Education Material Study (ChemStudy – in Brazilian secondary school education, as well as in teacher education and in the structuring of the Brazilian science education (research) community, as revealed by the increasing numbers of graduate studies programs, academic journals, and conferences (Nardi, 2014).

Science textbooks are not just "carriers" of scientific knowledge but materializations of school science discourse. As such, they recontextualize a number of science-related discourses (e.g. scientific, pedagogical, everyday, and media discourses). Recontextualization occurs through a process whereby hierarchical relationships from one field of knowledge are transformed and reconfigured so as to constitute a new field of knowledge. Textbooks are a privileged locus of recontextualizations, revealing plural, complex, intricate relationships between different discourses on/of science education in contemporary context. Analyses of the editorial processes for Brazilian textbooks show different ways through which the textbook is changed in response to demands posed by pedagogy, curriculum recommendations, and other educational policies. For instance, we have found allusions, paraphrases, and interpretations incorporating discourses related to (i) curriculum recommendations (e.g., references to daily life contexts and decision making); (ii) official evaluation programs (e.g., eliciting previous knowledge, development of communication skills); and (iii) science education research (e.g., inquiry and experimentation, relationships between science, technology, and society) (Pinhão, 2012; Moreira, 2013).

With respect to the last point, textbook authorship provides insights into the possible sources of the knowledge recontextualized in textbooks. It is worth noting the increasing number of science textbooks published in

Brazil between 2000 and 2010 that were authored by active, leading science education researchers. An analysis of one such group of textbooks has shown explicit consideration of the knowledge derived from science education research in books intended for both teachers and students, namely, references to the history of science, problematizations regarding the nature of science, and reflections about the language of science (Moreira & Martins, 2015). Textbooks also express aspects of pedagogic knowledge, for instance by structuring contents in ways that entail recapitulations and cross referencing between subject matter presented in different chapters, or by including problems and exam questions. The organization of contents into units that take into account specific time-space features of educational environments (e.g. 50 minute lessons, access to computers, or the availability of science laboratories in schools) is another example of how the characteristics and objectives of the classroom lead to reconfigurations of scientific knowledge as taught in schools. Relationships between science and current affairs, contemporary scientific discoveries and controversies, all find their way into textbooks through the inclusion of reproductions of popular science texts or media reports.

Thus, we characterize textbooks as strongly sensitive to demands at both the macro and micro levels. Textbook production is organized in such a way as to respond to changes both in the broad educational scenario and in the classroom. It is in this sense that textbooks reflect, create, and modify relationships between culture, in general, and scientific culture, in particular, as mediated by the social agenda of the educational system. As such, textbooks are fundamental elements in the enactment of curriculum policy (Abreu, Gomes, & Lopes, 2005), a process that entails a discursive dimension and occurs in multiple settings (Bowe, Ball, & Gold, 1992).

3.2 *The Textbook as a Semiotic Hybrid*

Following more general trends of communication in society, but at the same time keeping a correspondence with inherent features of scientific texts, textbooks are constituted by a diversity of languages, including verbal (text writing), mathematics (equations, graphs, notations), and imagery (drawings, photographs, maps, diagrams), and may thus be considered a semiotic hybrid (Lemke, 1988). Each of these different semiotic modes can be considered as more or less suited to dealing with concrete communicative demands (show, describe, explain) and the relationships between them can involve subordination, complementation, opposition, or preparation (Kress et al., 2001; Piccinini & Martins, 2004).

This view on the textbook suggests the importance of studying reading practices in a way that problematizes the relationships between different semiotic modes. Our studies involving semi-structured interviews with a group of 14-year-old science students about their textbook reading practices revealed that the text is read and interpreted with respect to students' repertoire of previous readings of both scientific and everyday texts and images, as well as recollections of classroom situations related to the thematic content in question (Martins & Gouvêa, 2014). This is consistent with the idea that making meaning is a historical process that is realized through establishing relationships between discourses. One example involved students' readings of a well-known image of a metallic spoon inside a transparent glass filled halfway with water, presented in connection with a discussion of refraction in their textbook. While all students identified everyday objects (glass, spoon), only a few mentioned the optical effect. The accompanying abstract schematic diagram of refraction as represented by light rays traversing different media was also described in terms of analogies with everyday objects (e.g., sticks). Our analyses also revealed that readings are not dependent on the graphic aspects of the image, as in the case of the schematic diagram of refraction, where the simplicity of the arrows contrasted with sophisticated adjacent theorization about the nature of light. In other words, some of the intended readings of such diagrams are only possible in the context of scientific discourse which, in this case, involved essential considerations about the role of modeling in the construction of scientific knowledge.

Our analysis of science textbooks demonstrated how images are recruited for a number of epistemological goals, for instance, to provide conceptual definitions (e.g., magnetic field lines or the DNA molecule) or to organize patterns through a display of cases (e.g., the periodic table), as well as for pedagogical goals, such as motivation and exemplification (Martins, 2001). Moreover, curriculum recommendations for contextualized approaches to scientific content (Martins et al., 2003) are often realized though images, which include references to daily life situations that recur in different texts (e.g., labels containing the nutritional components of industrialized foodstuffs or images of waste in landfills). If, on the one hand, this phenomenon represents an attempt to establish links between science and everyday life, it may, on the other hand, bear the risk of a pervasive standardization that could jeopardize the exploration of the diversity of relationships between science, culture, and aspects of the daily life of different social groups.

Advances in graphic production and the broadening of digital technologies available for communication and information also contributed to changes in the composition and layout presentation of textbooks. The inclusion of digital contents, made available through electronic media, brings the potential for reconsiderations of traditional hegemonic pedagogical trends, not to mention offering support for innovations in terms of methodological approaches.

3.3 *The Textbook as a Genre*

The text of a textbook is not a simple adaptation of scientific text, exclusively through didactic transpositions of content reference, for the purpose of school education. In earlier studies (Martins; Cassab; ROCHA, 2001; Martins; Damasceno, 2002), we have shown how science textbooks recontextualize a variety of discourses, namely the scientific, the media, the everyday, and the pedagogical, among others, transforming and incorporating texts related to these (originating from scientists, popular science texts, newspaper articles, comics, curriculum recommendations, etc.) in the construction of a discourse which has its own identity and that reflects relationships between school culture and scientific culture, in the context of the social commitments and goals of educational practice. The recognition of this discursive hybridity in textbooks has proven particularly productive in the characterization of the textbook as a genre in its own right (Braga, 2002). As such, its textual features are directly related to the discursive practice of schooling, that is, a collective enterprise that involves practices of production, communication, and evaluation, and that is linked to particular institutional organizations. From this perspective, the textbook materializes school-science knowledge, as opposed to science knowledge, and is constructed and legitimized with reference to the goals and epistemological commitments which are typical of the social sphere of the school. In this sense, it is possible to consider the textbook as a new genre, that is, as a new discursive construction that takes place in the context of schooling and that, as such, incorporates – and takes part in – interactions with other discourses, such as the scientific, pedagogical, and everyday. In other words, school knowledge does not correspond to a simplified version of scientific knowledge, a view which is consistent with analyses carried out in the field of curriculum studies (Ferreira & Selles, 2004) and that emphasizes the role of historical changes, conflicts, and struggles in the emergence of school disciplines (Goodson, 1983; Gomes; Selles; Lopes, 2013).

In its discursive aspects, such as composition, layout, thematic choices, and rhetorics, the textbook bears markers of the social activity of teaching science in schools. This is why the textbook can be characterized by a lexical stability (e.g., specific vocabulary) as well as by the recurrence of both grammatical (e.g., general statements, questions) and editorial organizational structures (e.g., exercises, additional readings, comprehension questions, etc.). There are also overt relationships between some presentation formats, including the double-page spread and the organization of activities within the spatial and temporal constraints of the classroom, that reinforce the idea that a genre is "a way of acting in a socio-discursive culture and not just a way of writing" (Marcuschi, 2008, p. 17).

Genres express, therefore, an inextricable relationship between textual forms and social practices. For instance, a job interview is a specific type of discursive interaction, whose textual features mark asymmetries and power relations between participants. The importance of genres goes well beyond the identification of typical stabilized discourse forms insofar as they essentially in fact characterize ways through which people are positioned within social practices. This is why, considering the contemporary demand for the incorporation of educational goals regarding the provision of tools for citizenship and decision making, genre analysis seems to be an appropriate means by which to discuss the extent to which textbooks represent students' possibilities for social participation and relationships to scientific knowledge. Pinhão's genre analyses (2010) have helped us understand how structural choices present in texts collaborate to challenge or support not only the dissemination of a number of discourses related to health and environment in science education, but also other aspects such as power relations and possibilities for social participation. For instance, individual – as opposed to collective – decision making was a more commonly found pattern in the texts analyzed. Furthermore, most texts did not find ways to portray significant dimensions of health and environment as situated in a wider network that relates social, political, and conceptual relationships. Also, it seems that cognitive and behavioral perspectives were more influential than socio-historical ones. In other words, although there is some degree of hybridity in textbook discourse, some views are construed as hegemonic and, as a consequence, less emphasis is given to the potentially transformative critical dimension involved in people's discourses about and attitudes toward health and environment. Another aspect highlighted by Pinhão's analyses is that decision making appears to be directly linked to technocratic perspectives, since specialized knowledge is often portrayed as the main basis for sound decisions.

4 Stability and Change in Textbooks

Considerations about the discursive hybridism of textbooks brought by discursive perspectives can be further exemplified through an analysis of five Brazilian high school physics textbooks, conducted with the purpose of identifying meanings related to both stability and change in the formats used for the presentation of scientific concepts (Martins, 2007). Based upon critical discourse analysis (CDA), we discuss the extent to which such features in the textbook's language can be linked to broader factors, namely, educational, cultural, societal, and the like. Given that theoretical framework, our inquiry considers the historical and socially situated character of texts, along with their relationships with discursive practices in society. Our main research question was: which discourses have been constructed, ignored, stabilized, and transformed in Brazilian textbooks over the last 40 years?

Our analysis consisted of the identification of textual realizations of intertextuality and its subcategories. In other words, inspections of the texts were carried out with the goal of identifying traces of the incorporation and transformation of different texts in the form of citations, paraphrases, discursive representations, presuppositions, negations, reformulations, etc.

The corpus consisted of set of Brazilian physics textbooks published in regular editions since 1970. They were selected because of their high degree of acceptance and recognition in both school and academic communities in Brazil. The books present their methodological approaches to the teaching of physics clearly and do not contain conceptual errors. Some of them had been written from lectures in university physics departments, others by experienced school teachers. There were examples of texts that had been authored by groups involving both academic and school teachers, some of which were connected to teaching projects aimed at improving physics teaching in Brazil. All selected textbooks were published by leading Brazilian publishing houses, and some of them had several revised editions published between 1970 and 2010.

Our CDA involved an analysis of conjuncture in order to discuss relevant aspects of the socio-historical and political contexts in which these textbooks were published. Amongst those, we highlight the emergence and consolidation of a community of physics educators that centers around the Teaching Secretary for the Brazilian Physical Society, an institution that has organized 18 national symposia for physics teachers and 12 national research conferences, and which publishes two academic journals related to teaching physics. No less important are some of the public policies implemented during these 40 years, including changes in the weekly hours of physics lessons; recommendations for interdisciplinarity and contextualization in physics curricula; national programs of appraisal; the free distribution of textbooks; the implementation of a national exam for high school certification and admission in federal universities; increased availability of professional development programs for teachers, including those at the graduate studies level; and science education gaining status as an area of knowledge in the Ministry of Education Higher Degrees Agency's evaluation directory.

A further level of analysis involves relating specific texts and discourses in the textbooks to more general macrosocial features. To that end, textual analyses are performed so as to identify aspects, such as lexical and grammatical choices, as well as intertextual and interdiscursive strategies, that allow for the construction of meanings for features such as recurrence and ruptures in text presentation.

In the investigation reported here, we have defined our corpus of analysis as the subset of texts related to the topic of classical mechanics, not only because of its epistemological and didactic importance in the study of physics but also for its structuring role in curriculum development.

5 Stabilized Discourses

Overall, it was possible to observe that the books were fairly stable with respect both to the lexical and grammatical structures used, which reflected a strong commitment to introducing scientific vocabulary and specialized language, as well as definitions and examples. Similar problems, open-ended questions, suggestions for experiments, and supplementary readings are examples of elements that help transform science knowledge into school science knowledge. These elements were constant features in all editions examined. Likewise, free fall equations, inclined planes, and Newton's cannonball are among the canonical examples of conceptual entities that allow for the immediate identification and recognition of the texts as linked to physics.

An interesting and recurrent example, present in nearly all books analyzed, refers to the presentation of Newton's second law of motion. All books, with one exception, chose to introduce the topic via Euler's statement, which specifies that there is a direct proportional relation between the net force acting on a given body and the variation of its velocity in time, defining the proportionality constant as the body's inertial mass. In fact, this differs from Newton's original statement, which establishes a relationship between force and the variation of linear momentum. The recurrence and stability of the $F=ma$ formulation reinforces a didactical approach which prioritizes the study of physical situations where forces act continuously upon a body.

If, on the one hand, it has the advantage of dealing with examples that do not demand complex mathematical treatment and relate to a large set of everyday phenomena, on the other hand, the presentation of ideas as static closed statements disguises the discursive character of the construction of scientific ideas. For instance, textbooks that follow the historical evolution of ideas and that explore aspects of the discursive character of the construction of scientific ideas and texts by discussing alternative approaches to the same idea can help students to perceive the actual nature of the scientific enterprise. Such an approach to scientific ideas – found in just one of the books analyzed – also allows for discussion of the history of their development, of the socio-political context in which they are proposed, and of the reasons why some tend to be more widely disseminated and accepted, along with the phenomenological domain they best describe. Investigations of scientific discourse have explored how these aspects are discursively constructed and provided the essential elements for a wider perception of the nature of science and of scientists' activities. Bazerman's (1999) analysis illustrates the discursive dimension of scientific knowledge by investigating how Thomas Edison constructed different textual representations for light and electricity in numerous texts, including letters, newspaper interviews, correspondence exchanged with patent boards and with investors who granted financial resources to fund his work, leaflets, papers, etc. Such representations allowed the incorporation of new meanings to current systems of knowledge and, simultaneously, influenced people to understand and to participate in the "electrification" of the different domains of daily life (Bazerman, 1999). This analysis also demonstrates the strong dependence between the formats and styles of the texts produced by scientists and characteristics of their target audiences, an idea that is not at all present in most textbooks analyzed.

Images were also found to be important allies in the achievement of stable presentation, conferring identity to physics textbooks as such. For instance, diagrams – such as those of the Atwood machine or strobe photographs of bodies in accelerated motion – were commonly and frequently used in all of the mechanics textbooks analyzed, conferring on them both an identity and an immediate association with the didactic genre. Pictures of blocks of different sizes (representing bodies with different masses) being pulled by forces of differing intensities (represented by differently sized/-oriented arrows) were frequently associated with the introduction of the study of motion through one-dimensional movements, as well as with the giving of priority to scalar over vectorial treatment. Other examples found were machine diagrams (e.g., pulleys and levers) and parabolic trajectories of obliquely thrown projectiles.

6 Deconstructed Discourses

Although geared to discursive stability, textbooks sometimes transgress. Interestingly, the use of some images also featured as indicators of changes in textbooks. Improvement in desktop publishing facilities, which became cheaper and more widespread over the decades in question, had a significant impact on the graphic presentation of textbooks. Over time, high quality photographs, comic strips, images from newspapers and Internet sites, as well as innovative layout formats, became increasingly typical in the books analyzed.

Insofar as textual presentation is concerned there was very little variation of the traditional conceptual hierarchy between kinematics and dynamics that has been firmly established as a privileged didactic sequence, and that can be typically described as: definitions of space and time, followed by the study of velocity and acceleration, to be applied in free fall and ballistic problems. The 'causes' of motion are only described after the study of a 'description of motion' is finished and involve the presentation of Newton's laws and the principles of conservation of linear momentum and energy, all in the context of typical problems such as motion in inclined planes, pulleys, and collisions. The study of circular motion and rigid bodies is not always part of curricula. In the name of simplicity, formal descriptions of ideal movements are favored over the discussion of movements which are encountered in daily life situations and which involve, for instance, rotations and impulsive forces. This same hierarchy has been present in books published in different parts of the country, written by authors with different professional experience (university lecturers, researchers, or practicing teachers) and aimed at different target audiences, for the past 40 years.

Although it was possible to detect changes with respect to this pattern, most of the books analyzed did not contain structural changes. Ruptures with this mainstream approach were seen in only one of the five textbooks analyzed. Oriented towards a discussion of 'everyday physics', this book takes daily life contexts and technological objects as motivations for introducing physical principles. This approach, which develops the study of physics as a context for understanding and participating in society, is clearly inspired by Paulo Freire's ideas of education as an emancipatory and liberating praxis (Freire, 1970). In coherence with Freirian perspectives, scientific knowledge is quite prominent in the text; it is, however, always articulated with relevant daily life contexts. For instance, the mechanics text starts with the vectorial treatment of rotation and translation movements, with reference to contexts such as reading maps and finding your way around a new city. It then goes on to explore the idea of conservation in the study of car crashes, in order to introduce the concept of linear motion – which will later be used to present Newton's second law. The bases for the choices made by the authors can be understood with reference to the conditions of production of the book. An inspection of the book's preface presents it as the result of a project involving university lecturers and science education researchers who have worked with Paulo Freire himself, science teachers who have worked with these researchers in graduate studies programs, and practicing teachers who used the material in their own classrooms and gave feed-back to the authors.

A significant example of change can be seen in connection with intertextual references to results from science education research in the form of allusions to students' low motivation and interest in physics; the tension between universalism and multiculturalism in science education; indirect references to students' insufficient performance in assessment programs; or to the need to relate scientific and everyday knowledge.

An interesting case of intertextuality refers to the didactic reconstruction of another set of physics education research results. Albeit timidly, textbooks started stimulating discussions about well-known alternative conceptions, such as notions around bodies needing a force to keep in motion or force and velocity necessarily having the same direction and orientation. The texts also started to address the reader directly, proposing debates or activities that establish cognitive conflict and confront prediction and evidence. In three of the books analyzed, we found references to students' ideas, proposals for discussions about their validity and explanatory power, and experimental activities leading to results designed to challenge those ideas.

Apart from that, other changes observed were related to the inclusion of photographs of everyday situations, excerpts from various media (newspapers, popular science), and use of informal language. Some of these changes can be understood as bearing intertextual relations with current curriculum recommendations and national programs for textbook evaluation. Brazilian Curriculum Parameters for high school education (launched in 1999 and 2002) emphasize the role of interdisciplinarity and contextualization in curriculum development, and stress the need to explore the different 'languages' of science and to foster the critical reading of texts associated with science. Similarly, the current Brazilian Textbook Evaluation Program reinforces the need for discussing the nature of science, the role of imagination in the development of scientific ideas, the tentative and provisional character of scientific explanations, and the relationships between science, technology, society and the environment, as well as for avoiding mechanical and decontextualized learning experiences. Both programs link science learning and citizenship, and value the ability to use scientific knowledge to inform responsible decisions in relevant everyday situations.

The influence of both sets of recommendations is especially noticeable in the increasing number of images now found in the textbooks. As plentiful and high quality as they are, the new images are, in most cases, recruited to serve the purpose of providing a quick response to these recommendations; because of this, many function merely as add-ons that are not coherently articulated with the written text. In some cases, the new images contrast quite sharply with the traditional, authorized didactic images of science, such as drawings of electric circuits or magnetic field lines; here again there is no counterpart in the verbal text to provide cohesion to the combined images and text as a whole. In other words, the observed changes did not always correspond to substantial transformations in the textbooks, since both the selection of topics and conceptual hierarchy are maintained.

Overall, it is possible to say that curriculum development guidelines as well as official appraisal scheme demands have influenced transformations in contemporary physics textbooks. There is, however, the danger that the need to conform to evaluation criteria may lead to excessive homogeneity, to a future in which textbooks will look very much like one another. Having said that, curriculum recommendations and criteria for textbook appraisals set out by public policies have had, by and large, a positive impact on textbooks, since textbook writers and editors are now strongly discouraged from methodological approaches that lead to rote learning. Writers are also much more aware of the need for respecting cultural diversity and for eliminating allusions to ethnic and gender prejudices. Another positive influence, though quite modestly realized, is the inclusion of topics of modern and contemporary physics in physics textbooks, which have typically been outdated as compared to biology textbooks, which have generally been much more agile in incorporating contemporary science knowledge and debates.

Two significant absences were also noted throughout the materials analyzed. The first one relates to the lack of incorporation or inadequate incorporation of the sociopolitical dimension of scientific knowledge. For example, in all texts examined, projectile motion was illustrated by instances of packages falling from airplanes, while no reference was made to the fact that ballistic motion equations allow for the precise launching of contemporary mass destruction missiles. When present at all, the political impacts of science and technology were generally confined to examples from the past, as in the case of the Nagasaki and Hiroshima atomic bombs. Another significant absence was the almost complete lack of references to the affective and emotional aspects of scientific activity (Lemke, 2006).

7 Conclusions and Implications

In this chapter, we sought to explore and demonstrate the potential for discursive approaches to the analysis of textbooks through our examination of a set of Brazilian textbooks. Our analyses aimed to reveal traces of the social processes involved in the construction of meaning in physics textbooks, and showed that, even with the increased inclusion of non-scientific texts, the physics textbook format has not changed in its essential conceptual aspects. Furthermore, we suggest that a recognition of the role of intertextuality in the constitution of texts might lead to a greater awareness of the historically and socially situated character of textbooks in both teaching and learning and allow for the construction of different perceptions and attitudes towards knowledge within educational practices.

References

Abreu, R. G. de, Gomes, M. M., & Lopes, A. C. (2005). Contextualização e tecnologias em livros didáticos de biologia e química. *Investigações em Ensino de Ciências, 10*(3), 405–417.

Bakhtin, M. M. (1981). Discourse in the novel. In *The dialogic imagination* (C. Emerson & M. Holquist, Trans.; M. Holquist, Ed.). Austin, TX: University of Texas Press.

Bakhtin, M. M. (1986). *Speech genres and other late essays* (V. W. McGee, Trans.; C. Emerson & M. Holquist, Eds.). Austin, TX: University of Texas Press.

Bakhtin, M. M. (2000). *Marxism and the philosophy of language.* Cambridge, MA: Harvard University Press.

Bazerman, C. (1999). *The languages of Edison's light.* Cambridge, MA: MIT Press.

Bensaude-Vincent, B., & Lunders, A. (2000). *Communicating chemistry: Textbooks and their audiences, 1789–1939.* London: Neale Watson.

Bizzo, N. M. V. (2000). Falhas no ensino de ciências. *Ciência Hoje, 159*, 26–31.

Bowe, R., Ball, S. J., & Gold, A. (1992). *Reforming education and changing schools.* London: Routledge.

Braga, S. A. M. (2003). *O texto do livro didático de ciências: Um gênero discursivo* (Doctoral dissertation). UFMG, Faculdade de Educação. Belo Horizonte.

Braga, S. A. M., & Mortimer, E. F. (2003). Os gêneros de discurso do texto de biologia dos livros didáticos de ciências. *Revista Brasileira de Pesquisa em Educação em Ciências,* 3(3).

Brazil. (1996). Lei 9.394, de 20 de dezembro de 1996. Estabelece as Diretrizes e Bases da Educação Nacional. *Diário Oficial da União.* Brasília, *DF, 134*(248), 27833–27841. Retrieved from http://portal.mec.gov.br/seesp/arquivos/pdf/lei9394_ldbn1.pdf

Brazil. (2005). Ministério da Educação e Cultura. *Plano nacional do livro didático, 2005.* Retrieved from http://www.mec.gov.br

Brazil, FNDE. (2016). *Programa nacional do livro didático, 2016.* Retrieved from http://www.fnde.gov.br/programas/livro-didatico

Brazil, MEC. (2013). *Diretrizes Curriculares Nacionais Gerais da Educação Básica.* Brasília: MEC; SEB; DICEI. Retrieved from http://portal.mec.gov.br/index.php?option=com_docman&view=download&alias=15548-d-c-n-educacao-basica-nova-pdf&category_slug=abril-2014-pdf&Itemid=30192

Brazil, PNDL. (2011). *2012 Guia de livros didáticos: Física.* Brasília: Ministério da Educação, Secretaria de Educação Básica.

Carneiro, M. (1997). As imagens no livro didático. In *Anais do I encontro nacional de pesquisa em educação em ciências* (pp. 366–373). Águas de Lindóia, SP.

Cassab, M. (2003). *Significando o livro didático: Com a palavra, os professores de ciências* (Master's thesis). UFRJ, Núcleo de Tecnologia Educacional para a Saúde, Rio de Janeiro.

Cassab, M., & Martins, I. (2003a). Um balanço dos estudos recentes conduzidos com o livro didático de ciências. In *Anais do encontro regional de ensino de biologia, 2.* Niterói, RJ.

Cassab, M., & Martins, I. (2003b). A escolha do livro didático em questão. In *Anais do IV encontro nacional de pesquisa em educação em Ciências.* Bauru, SP: CD-ROM.

Cassab, M., & Martins, I. (2008) Significações de professores de ciências a respeito do livro didático. *Ensaio, 10*(1), 7.

Clément, P., Bernard, S., Quessada, M.-P., Rogers, C., & Bruguière, C. (2005). Different theoretical backgrounds for different didactical analyses of biology school text-books. In *European science education research association conference, 5.* Barcelona, Spain: CD-ROM.

Fairclough, N. (1992). *Discourse and social change.* Cambridge: Polity Press.

Ferreira, M. S., & Selles, S. E. (2004). Análise de livros didáticos em ciências: Entre as ciências de referência e as finalidades sociais da escolarização. *Educação em Foco, 8*(I & II), 63–78.

Fracalanza, H. (1992). *O que sabemos sobre os livros didáticos para o ensino de ciências no Brasil?* (Doctoral dissertation). Unicamp, Faculdade de Educação.

Fracalanza, H., & Megid Neto, J. (Eds.). (2006). *O livro didático de ciências no Brasil.* Campinas, Brazil: Editora Komedi.

Freire, P. (1970). *Pedagogy of the oppressed.* New York, NY: Herder and Herder.

Freitag, B., Motta, V. R., & Costa, W. F. (1997). *O livro didático em questão.* São Paulo, Brazil: Cortez.

Halliday, M. A. K. (1985). *An introduction to functional grammar.* London: Edward Arnold.

Koulaidis, V., & Tsatsaroni, A. (1996). A pedagogical analysis of science textbooks: How can we proceed? *Research in Science Education, 26*, 55–71.

Lemke, J. (1996). *Textual politics.* London: The Falmer Press.

Lemke, J. (2006). Research for the future of science education: New ways of learning, new ways of living. *Ensenãnza de las Ciencias, 24*(1), 5–12.

Martins, I. (2006). Analisando livros didáticos na perspectiva dos estudos do discurso: Compartilhando reflexões e sugerindo uma agenda para a pesquisa. *Pró-posições, 17*, n. 1(49), 117–136.

Martins, I. (2007). *Stability and change in the physics textbook discourse.* Paper presented at the ESERA Conference, Malmö. Retrieved from http://195.178.227.107/esera/Files/302.doc

Martins, I., Gouvêa, G., & Piccinini, C. L. (2005). Aprendendo com imagens. *Revista Ciência e Cultura, 57*(4), 38–40. Retrieved from http://cienciaecultura.bvs.br/scielo

Martins, I., Gouvêa, G., & Vilanova, R. (2012). *O livro didático de ciências: Contextos de exigência, critérios de seleção, práticas de leitura e uso em sala de aula.* Rio de Janeiro. Retrieved from http://www.nutes.ufrj.br/arquivos/O_livro_didatico_de_Ciencias.pdf

Megid Neto, J., & Fracalanza, H. (2003). O livro didático de ciências: problemas e soluções. *Ciência & Educação, 9*(2), 147–157.

Moreira, M. C., & Martins, I. (2015). A recontextualização de discursos da pesquisa em educação em ciências em livros didáticos de ciências: um estudo de caso sobre como aspectos de pesquisas vinculadas a referenciais construtivistas constituem um texto didático autorado por pesquisadores. *Revista Brasileira de Pesquisa em Educação em Ciências, 15*(2), 238–257.

Nardi, R. (2014). Memórias do Ensino de Ciências no Brasil: a constituição da área segundo pesquisadores brasileiros, origens e avanços da pós-graduação. *Revista do IMEA-UNILA, 2*, 13–46.

Nascimento, T. G. (2003). *O texto de genética no livro didático de ciências: Uma análise retórica* (Master's thesis). UFRJ, Núcleo de Tecnologia Educacional para a Saúde.

Nascimento, T. G., & Martins, I. (2005). O texto de genética no livro didático de ciências: Uma análise na perspectiva da retórica crítica. *Investigações em Ensino de Ciências, 10*(2). Retrieved from http://www.if.ufrgs.br/ienci/

Orlandi, E. P. (1999a). *Análise de discurso: Princípios e procedimentos.* São Paulo: Pontes.

Orlandi, E. P. (1999b). *Discurso e leitura* (4th ed.). São Paulo: Cortez.

Otero, M. R., & Greca, I. M. (2004). Las imágenes en los textos de física: entre el optimismo y la prudencia. *Caderno Brasileiro de Ensino de Física, 21*(1).

Piccinini, C. L., & Martins, I. (2004). Comunicação multimodal na sala de aula de ciências: Construindo sentidos com palavras e gestos. *Ensaio Pesquisa em Educação em Ciências, 6*(1). Retrieved from http://www.fae.ufmg.br/ensaio/

Quesado, M. (2003). *A natureza da ciência e os livros didáticos de ciências para o ensino fundamental: uma análise textual* (Master's thesis). UFRJ, Núcleo de Tecnologia Educacional para a Saúde.

Ribeiro, R. M. L., & Martins, I. (2007). O potencial das narrativas como recurso para o ensino de ciências: uma análise em livros didáticos de física. *Ciência e Educação, (UNESP), 13*, 293–309.

Santos, V. C., Joaquim, L. M., & El-Hani, C. (2012). Hybrid deterministic views about genes in biology textbooks: A key problem in genetics teaching. *Science & Education, 21*, 543–578.

Selles, S. E., & Ferreira, M. S. (2004). Influências histórico-culturais nas representações sobre estações do ano em livros didáticos de ciências. *Ciência e Educação, 10*(1), 101–110.

Silva, H C., & Almeida, M. J. P. M. (1998). Condições de produção da leitura em aulas de física no ensino médio: Um estudo de caso. In M. Almeida & H. Silva (Eds.), *Linguagens, leituras e ensino de ciências.* Campinas, SP.

PART 5

The History, Philosophy, and Sociology of Science in Science Teaching

CHAPTER 15

The History, Philosophy, and Sociology of Science in Science Teaching

André Ferrer P. Martins

Abstract

In this chapter, we assume that the importance of the History, Philosophy and Sociology of Science (HPSS) for Science Education is well known. In the areas of Science Education and Science Education research, HPSS has found an unmistakable niche in recent decades. HPSS has potential in science education as both content in itself and didactic strategy. Using this (apparent) consensus as background, we will discuss three issues: (1) Difficulties and obstacles to the effective use of historical-philosophical approaches in basic (primary and secondary) and higher education; (2) The risks of an epistemologically equivalent treatment of everyday and scientific knowledge; and (3) The likely harm caused to high quality science education by an extreme relativist view of science. In our discussion, we point out that reasons for teaching science must go beyond the perspective of understanding the natural and technological world, also including the way in which this understanding is achieved. Science consists of special knowledge, with its own language and epistemology, which differ from other forms of knowledge (everyday knowledge, in particular). Learning about science also involves learning about these differences. This is important to avoid an extreme relativist view of science. HPSS plays a crucial role in this process. Although learning about science is not provided exclusively by a historical-philosophical approach, the latter has proved to be an excellent path to achieve this goal. HPSS, along with other theoretical-methodological tendencies and approaches, continues to be a fertile field providing elements that lay the foundation for epistemologically-oriented scientific education.

Keywords

history of science – philosophy of science – sociology of science – nature of science

 | DOI: 10.1163/9789004409088_015

1 Introduction

Texts devoted to the history and philosophy of science in science teaching usually begin by pointing to the relevance and benefits of using a historical-philosophical approach in the science teaching and learning process. Thus, in order to avoid disheartening readers who expect this type of approach based on the title of this chapter, I will reserve a few paragraphs for this topic in the limited space of this introduction. However, an extended discussion is unnecessary since the importance of the history and philosophy of science for science education is well enough known. This is the first aspect to be underscored.

A few observations must be made: first, I would like to refer to the history, philosophy *and sociology* of science (HPSS), instead of simply to the history and philosophy of science (HPS), given the importance of the sociology of science both in general and in the arguments that follow. Second, I would like to highlight the difference between academic research fields in HPSS and in HPSS *in science teaching*. There are, nonetheless, researchers working in both fields, which can be considered distinct but interconnected areas. The focus of this study is on HPSS in *science teaching*.

In recent decades, HPSS has found an unmistakable niche in the areas of *science education* and *science education research*. There is considerable evidence to support this: a significant number of articles about HPSS in specialized academic journals; the inclusion of HPSS in conferences on science teaching; and the incorporation of recommendations about the didactic use of HPSS in different national science curricula, in Brazil and abroad (AAAS, 1990; Brazil, 2000, 2006; NC, 2007); among other aspects.

HPSS has potential for *science education* as both *content in itself* and *didactic strategy*. As content to be taught, this area has gained intrinsic value and has been playing an important role in higher education. This is mainly true for education programs for science teachers, although it is also present in elementary, middle, and high school textbooks. As a didactic strategy, it could facilitate the understanding of concepts, models, and theories,[1] in addition to providing a deeper awareness of the nature of scientific activity. In both cases – which are in fact overlapping – the presence of HPSS at different teaching levels has been considered relevant in that it could contribute to: (i) humanizing science and its teaching, by considering political, economic, social, and ethical aspects, among others, that have influenced and continue to influence the practice of science; (ii) problematizing the common view of science and the scientist; (iii) facilitating understanding of the content to be taught; (iv) stimulating reading; (v) motivating students with learning difficulties and/or those not attracted to the more formal aspects of science (algorithms, equations, etc.); (vi) promoting critical thinking in general; and (vii) improving the planning and execution of classroom activities, taking into account students' alternative conceptions and the historical construction of scientific concepts.[2] This list – which is based on studies conducted by a number of authors, including Zanetic (1989), Gil Pérez (1993), Matthews (1992, 1994a), Campanario (1998), Peduzzi (2001), El-Hani (2006), and Martins (2006) – is by no means exhaustive.

In the area of *science education research*, HPSS has made many contributions, including the following: providing the *theoretical foundations* of teaching and learning models (e.g., the conceptual change model (CCM) by Posner et al., 1982, based on the work of Thomas Kuhn, Imre Lakatos, and Stephen Toulmin, among others); establishing *critiques* of the perspectives put forth by these models (e.g., Villani's [1992] critique of CCM, based on Laudan's work); and questioning the theoretical bases of constructivism (e.g., Matthews, 1994b; Osborne, 1996; Ogborn, 1997; Geelan, 1997; Marín Martínez et al., 1999; Laburu & Carvalho, 2005).

HPSS is therefore important for science education from both the *theoretical* and *applied* points of view. As a result, it has become essential to incorporate historical, philosophical, and sociological dimensions into science teacher education programs (Carvalho & Gil Pérez, 1998; Marandino, 2003; Colombo de Cudmani & Salinas de Sandoval, 2004; Duarte, 2004). Thus, HPSS emerges as a *key element of teacher education.*

The scenario outlined so far seems to indicate that the inclusion of HPSS in science education is practically a matter of consensus and that there are no major problems. Using this (apparent) consensus as background, I will in the rest of this chapter discuss a number of questions that, in my opinion, are relevant to current science education and related to HPSS, as concerning with the follows:

1. Difficulties in and obstacles to the effective use of historical-philosophical approaches in primary, secondary, and higher education;
2. The risks of an epistemologically equivalent treatment of everyday and scientific knowledge;
3. The likely harm caused to high quality science education by an extreme relativist view of science.

2 Why Aren't We Moving Forward?

The widespread acknowledgement of the importance of HPSS for pedagogic purposes contrasts strongly with daily science classroom practices. In Brazil, official curriculum documents related to the scientific areas of primary and secondary education reinforce the relevance of historical-philosophical approaches (Brazil, 2000, 2006). These documents support the sociocultural contextualization of science and the development of competencies that enable the student to

> Recognize the historical meaning of science and technology, perceiving their role in human life at different periods and the human capacity to transform the environment.
>
> Understand the sciences as human constructs and how they developed by the accumulation, continuity or rupture of paradigms, relating scientific development to social transformation. (Brazil, 2000, p. 13)

Nevertheless, few primary or secondary students have any contact with the history, philosophy, or sociology of science during their school years.[3] This contributes to the perpetuation of naïve conceptions of science and its development.

Most of the reasons for this deficiency are well known. Part of the blame lies with the small amount and low quality of available didactic materials. Indeed, most textbooks on the market make little or no mention of HPSS. Moreover, the history in these manuals is, as a rule, simplistically portrayed and distorted – which brings us back to criticisms made in the 1970s about the use of HPSS for didactic purposes (pseudohistory, quasi-history,[4] etc.). These criticisms persist, given that pseudohistory continues to be a matter of concern (Allchin, 2004).

In Brazil, this is only part of the problem regarding educational resources. Secondary school teachers generally feel trapped by the content required for college entrance examinations and – not coincidentally – traditionally present in high school textbooks. Any change in this status quo may not be well received – and requires efforts that teachers are not always willing to make. If these factors were not enough, many teachers also consider HPSS as *just additional content* to be taught (for which there is no available time) rather than as a strategy/approach that may reorient content selection (Martins, 2012).

We now arrive at a key aspect which accounts for another large portion of the blame for the relatively low number of successful student experiences with regard to HPSS: teacher education. Here, the issue becomes more complex, since, even though teachers in training generally study a discipline or disciplines with historical-philosophical content, this does not mean that they are able to plan and execute HPSS-related activities in high school classrooms or other contexts. Even after having graduated, teachers often do not feel confident enough or have the necessary didactic skills to work with elements of HPSS. Although many teachers share an interest in and awareness of the value of this approach, there is a lack of "know-how" (to a certain extent, this is merely the "tip of the iceberg" as regards what Shulman [1986] called *pedagogical content knowledge*).

In a study we conducted with current and future high school physics teachers at our university (UFRN), it was clear that, in addition to the aforementioned problems, there is a naïve view of the role played by HPSS in didactic purposes. Teachers who had already worked with this perspective viewed HPSS as a type of "illustration", something peripheral to the curriculum, to be added as an *introduction to 'regular' topics and themes.* In effect, the use of HPSS was limited to a motivational strategy, aimed at awaking students' interest in other issues. It seemed that teachers did not take the prospect of actually *learning physics* with HPSS very seriously (Martins, 2007).

The difficulties facing teachers are understandable: how can we expect teachers to be suitably prepared to work with HPSS in primary and secondary education if most of their training as teachers arises out of extremely fragmented curricular structures that barely touch on this topic? How can we require teachers to master the didactic skills needed to work with HPSS if they did not encounter such skills in their own educations, especially when, in most cases, teachers' entire educational experiences were grounded in a more directive teaching model that does not value constructive criticism?

In Brazil, teacher training courses in the natural sciences traditionally involve only a single historical-philosophical discipline, although there are exceptions. Thus, future teachers end up having limited contact with HPSS. This would not be a problem if the other disciplines were somewhat concerned with using a historical-philosophical approach (I am obviously not proposing that all the disciplines focus on the historical-philosophical approach …). Teacher trainers were, for the most part, not made aware of the importance of HPSS, nor did they receive specific preparation in this area. As a result, they often end up with naïve conceptions of science and its development (studies about the "nature of science" illustrate this fact, as I discuss below). It will be the responsibility of any discipline containing historical-philosophical content to *deconstruct* views of science that were constructed by future teachers throughout their schooling.[5]

To a certain extent, this absence shows the insufficient achievements of HPSS in science teaching: we have not 'won over our peers', that is, we have not sensitized most teacher trainers involved in teacher education programs to incorporate HPSS more consistently into their curricula. Even if it is not the structural 'backbone' of the courses offered, HPSS may help, from an epistemological viewpoint, in the choices and actions involved in both the planning and implementation of the disciplines that make up teacher training curricula. The challenges lie in the collective construction of pedagogical projects for courses and in investing in the education of teacher trainers.

Lastly, I would like to point out another question that I consider pressing: although HPSS has found a niche, it has perhaps not yet adequately brought together historians/philosophers/sociologists and educators. Efforts are being made in this direction, such as the presentation of education-related papers at conferences specifically devoted to the history and philosophy of science. Challenges in this field involve the *mutual recognition of the specificities* in each area and *opening a fruitful dialogue* between these areas. Each area – whether history of science, etc., or science teaching – has its own specificities (theoretical and methodological) that need to be known and respected. Thus, the use of pseudohistory as well as the use of a "*pseudo-teaching*" is problematic ... (I am referring to naïve conceptions about the process of science teaching and learning). Only open dialogue between the areas, respecting the difficulties of each perspective, can contribute to the increasingly well-founded and fruitful use of HPSS in science teaching.

3 How Can We Best Connect Scientific and Everyday Knowledge

For a long time, high school level science teaching in Brazil was propaedeutic, that is, aimed at preparing students for university. The resulting teaching model, which was largely based on textbooks devoted to entrance examination preparation, was marked by *decontextualization*. Science models and theories were studied as an end in themselves, with no relationship to students' daily life experience. Consequently, most of the subjects studied made no sense to the students, given the abstraction of certain scientific models and the fragmented manner in which the topics were presented.

In physics, for example, what is the sense of calculating the resulting force between two electric point charges at infinity and placed one meter apart? Nearly all mechanics work in this context deals with problems that disregard friction, air resistance, and so on, in stark contrast to what may be observed in daily life.

Several perspectives are vying to change this scenario. On the one hand, such developments as the production of low cost materials for school laboratories (which is especially relevant for developing countries), increased inclusion of modern and contemporary physics topics, and discussions related to environmental education and STSE (science, technology, society, and environment), are each in their own way focused on making science teaching more significant to students' own lives and worlds.

On the other hand, changes to and restructuring of national curricula are aimed at reorienting science teaching toward content *contextualization*. The previously mentioned Brazilian reference documents for science curricula are emphatic regarding this need. Since we intend to head toward universalization at the high school level, we must develop scientific education curricula and practices that make sense to all, including those individuals who will not pursue scientific careers.

Problems begin to arise when we attempt to identify what is meant by *contextualization*. Although the term 'contextualize' means to relate to the context, it has been generally understood only in terms of *everyday* context (the context of 'applying' knowledge). This is not the only possibility, since we can refer, for example, to *historical* context (the context of the 'origin' of a particular aspect of knowledge). However, the prevailing interpretation of the guiding principle of contextualization has led teachers and teacher trainers to seek to give more meaning to science learning by relating its topics and curricular contents to students' daily life experience.

This perspective has been fruitful. After all, studying electricity and magnetism is more interesting when electric shower heads, light bulbs, and magnetic cards are used instead of point charges at infinity; the same is true of understanding thermodynamics using refrigerators rather than thermometric scales and dilatation coefficients; or identifying and differentiating chemical substances at a supermarket instead of doing it based only on theoretical definitions, etc. Without the proper measures, however, this approach could lead to epistemological misunderstandings.

The crucial question here is: would scientific concepts, laws, and theories actually be identifiable from our immediate reality? Are we, as science teachers, providing this view to our students by approaching science contents using technological objects and daily life experience as references?

The history and philosophy of science help us to unravel this knot. The knowledge that science produces about reality is based on *models*, which are free creations of the human mind and not copies of that reality. The history of science shows how certain models (such as the geocentric) were supplanted by others, and how certain ideas currently accepted by science (the existence of atoms, for example) were not always embraced. The philosophy of science has long since overcome conceptions intended to attribute definitive, true, and verifiable knowledge to science. For this reason as well, our tentative, conjectural, and changeable knowledge of reality cannot be a copy of reality.

The notion of the *model* as a *representation of reality* is of fundamental importance to science teaching. Yet the map must not be mistaken for the actual territory (Korzybski, 1931). When teachers interpret the electric current that passes through a shower head or heater in terms of concepts such as electric field, electric force, free electrons, etc., they must be aware that to a certain extent these things 'are not there'. Historically, models that essentially treated electricity as a fluid were dominant and remain quite plausible for interpreting many of the phenomena studied at the high school level. Similarly, much science teaching utilizes two-dimensional tools for more complicated concepts. For instance, when drawing a pendulum on the blackboard, the teacher must keep in mind that it is not a real pendulum and that the wire of the pendulum drawn has no mass. There is no friction at the support point. Its oscillation frequency is constant for small angles. Mass is concentrated at one point. It oscillates on a fixed plane …

A large number of other examples could be cited: the manner in which species are classified in biology or the double helix model of DNA, not to mention tectonic plates in the geosciences and the big bang theory in cosmology, just to name a very few.

From an epistemological viewpoint, it is important to characterize science objects as representations. Ogborn (1997) refers to the *theoretical objects of science,* taking into account the abstractions involved in their constructions. The French philosopher Gaston Bachelard (1996) frames representations as *secondary objects* (in contrast to the *primary objects* of immediate perception), which are re-presented to our awareness as a function of abstract thinking. In this sense, scientific knowledge emerges in the *rupture* with immediate reality and not in the continuity of everyday knowledge. According to Bachelard, first/immediate experience is actually an epistemological obstacle to scientific knowledge. It is the first error to be overcome (Bachelard, 1996).

At this point it is worth exemplifying our perspective by quoting a brief well-known excerpt from the introduction to the book *The Nature of the Physical World,* by Sir Arthur Eddington, where the author speaks of the existence of "two tables":

> One of them has been familiar to me from earliest years. It is a commonplace object of that environment which I call the world. How shall I describe it? It has extension; it is comparatively permanent; it is coloured; above all it is substantial. […]
>
> Table No. 2 is my scientific table. It is a more recent acquaintance and I do not feel so familiar with it. […]
>
> My scientific table is mostly emptiness. Sparsely scattered in that emptiness are numerous electric charges rushing about with great speed; but their combined bulk amounts to less than a billionth of the bulk of the table itself. Notwithstanding its strange construction it turns out to be an

> entirely efficient table. It supports my writing paper as satisfactorily as Table No. 1; for when I lay the paper on it the little electric particles with their headlong speed keep on hitting the underside, so that the paper is maintained in shuttlecock fashion at a nearly steady level. If I lean upon this table I shall not go through; or, to be strictly accurate, the chance of my scientific elbow going through my scientific table is so excessively small that it can be neglected in practical life. […]
>
> I need not tell you that modern physics has by delicate test and remorseless logic assured me that my second scientific table is the only one which is really there – wherever "there" may be. On the other hand I need not tell you that modern physics will never succeed in exorcising that first table – strange compound of external nature, mental imagery, and inherited prejudice – which lies visible to my eyes and tangible to my grasp. (Eddington, 1928, pp. ix–xii)

I believe that at this point my stance is already quite clear: I consider it essential to differentiate between scientific and everyday knowledge. There are fundamental epistemological differences between these two forms of knowledge, regarding proposals, methodologies, degrees of coherence, and generality, among others. In light of these differences, the search for the contextualization of knowledge – typically interpreted as the attempt to attribute greater meaning to the content to be taught by drawing connections to students' daily lives – must take this matter into account. Building scientific knowledge in the classroom using everyday elements and student experiences as references will only be consistent if both the continuities and the *ruptures* between the everyday and the scientific are taken into consideration. I fear that the majority of teachers engaged in the search for contextualization do not consider these epistemological differences and end up transmitting the false impression that scientific knowledge is simply 'there', as a copy of reality, to their students. By doing this, they promote an erroneous conception of science. HPSS can help to guide us through this maze.

To conclude this section, two warnings are warranted: first, the importance of contextualization or 'daily science' approaches must not be denied. We should remember the propaedeutic curriculum and the meaningless, context-less content that we intend to supplant .... Science education has been improving and gaining more significance for students in that curricular topics and content are being designed with their daily life experience in mind.

Second, everyday knowledge must not be underestimated.[6] Multicultural studies have consistently demonstrated the relevance that different forms of knowledge have in diverse cultures, as guides to the practices and actions of individuals in their social environment (e.g., Hodson, 1993; Cobern, 1996; Cobern & Loving, 2001).[7] Everyday knowledge, even in a culture dominated by science, guarantees our identity and the possibility of social communication. Although other fields, such as ethnoastronomy and ethnomathematics, provide different explanations from the scientific ones, they do offer satisfactory explanations from social perspectives. We must avoid, however, the pitfall of treating scientific and everyday knowledge in an epistemologically equivalent manner.

4 Relativism Is Good, But Not That Good

The previous section pointed out some relevant pitfalls through which HPSS can help to guide science teaching. Underlying what was discussed are concerns about building the proper image of science, its concepts, models, theories, and methodologies. Let us pause for a moment: the word 'proper' may cause shivers in many people .... Is there such a thing as a 'proper image of science' to be transmitted by teaching? This brings us to the third question concerning HPSS and science teaching.

In recent decades, the 'nature of science' (NOS) has been widely studied from a number of viewpoints.[8] A relative consensus has been reached in terms of the many mistaken conceptions regarding science with which science education should be concerned. Based on Fernández et al. (2002), a non-exhaustive list of such conceptions would include the following:

- Empirical-inductivist and non-theoretical conceptions, which conceive of observation and experience as 'neutral', that is, absent of hypotheses and a priori ideas;
- Rigid conceptions (algorithmic, exact, infallible), based on the existence of a single 'scientific method' which is conceived of as a sequence of steps to be mechanically followed;
- Non-problematic and non-historical conceptions, which omit the problems that give rise to knowledge and present them as already settled without showing their evolution, errors, and historical missteps;
- Exclusively analytical conceptions, which highlight the fragmentation of fields of study without paying attention to the significant unification processes that the historical development of science has undergone;
- Merely accumulative conceptions of scientific development (linear growth), which ignore the crises and reformulations that science has experienced;
- Individualist and elitist conceptions, which attribute scientific knowledge to the action of isolated geniuses, ignoring collective contributions and the existence of cooperative groups;

- Decontextualized conceptions, which attribute social neutrality to scientific activity, and which do not acknowledge the complex relations between science, technology, and society, portraying the scientist as a being 'above good and evil'.

Taken together, the above assertions comprise, to a large extent, what could be characterized as the *commonsense conception of science and its development* which is widely present in our culture, disseminated by the media (newspapers, internet, magazines, TV, etc.) in general, and even promulgated by science textbooks. We know that both students and teachers at various levels of learning and teaching hold such mistaken views (Lederman, 1992, 2007).

When we consider what would be 'proper' in terms of the nature of science, the consensus is not as widespread. After all, there are several theoretical frameworks characterizing scientific endeavor. Science is a complex activity that resists attempts at definition. Nevertheless, it is also possible to produce lists of aspects to be considered in efforts toward scientific education which is aimed at constructing proper concepts for doing science. McComas et al. (1998, p. 513), for example, propose the following items, taken from eight documents regarding international science education standards:

- Scientific knowledge, while durable, has a tentative character;
- Scientific knowledge relies heavily, but not entirely, on observation, experimental evidence, rational arguments, and skepticism;
- There is no one way to do science (therefore, there is no universal step-by-step scientific method);
- Science is an attempt to explain natural phenomena;
- Laws and theories serve different roles in science, therefore students should note that theories do not become laws even with additional evidence;
- People from all cultures contribute to science;
- New knowledge must be reported clearly and openly;
- Scientists require accurate record keeping, peer review and replicability;
- Observations are theory-laden;
- Scientists are creative;
- The history of science reveals both an evolutionary and revolutionary character;
- Science is part of social and cultural traditions;
- Science and technology impact each other;
- Scientific ideas are affected by their social and historical milieu.

Lederman (2007, pp. 833–835), underscoring the non-absolute character of these types of lists, indicates several key points of knowledge for primary and secondary students. Briefly:

- The crucial distinction between observation and inference;
- The distinction between scientific laws and theories;

- Scientific knowledge as involving human imagination and creativity;
- Scientific knowledge as subjective and/or theory-laden;
- Science as a human enterprise practiced in the context of a larger culture, wherein practitioners (scientists) are the product of that culture;
- Scientific knowledge as never absolute or certain, but rather tentative and subject to change;
- The distinction between NOS and science processes or scientific inquiry.

We could also add: "the knowledge that science works with models, and not with copies of reality", although this may be understood from other items of the list(s).

When we look carefully at these last two lists, we can identify many aspects that originated in the historical, philosophical, and sociological analyses of science performed by thinkers such as Popper, Kuhn, Lakatos, Feyerabend, Toulmin, Bachelard, Fleck, Laudan, Sarton, Hessen, Koyré, Duhem, Merton, and Bourdieu, among others. The apparent consensus that derives from these lists leads to a stance of *moderate relativism*, by pointing to the existence of arguments used in the construction of science that are not strictly rational, the roles of imagination and creativity in science, the relationships between scientific knowledge and society, the non-existence of a rigid method, etc. The sociology of science, in particular, played an important role in constructing this view and in criticizing conceptions of science and scientists marked by positivism.

Science education would benefit from the deconstruction of the common-sense view of science, represented by the items presented above in the first list, as well as from the construction of another view, based on the characteristics of the last two lists. To be clear: although there may be disagreement with respect to the particular conception of science to be presented, this discussion refers to the disposition of the general population and not to that of HPSS experts.

At this point, the last of the problems that I would like to point out in this chapter emerges: could the – necessary – deconstruction of naïve views of science lead to the construction of anti-scientific positions and, for other reasons, to naïve conceptions of doing science as well?[9] To what extent can the *presentation* of moderate relativism result in the *adoption* of an *exacerbated relativism* that does not point out the epistemological peculiarities of science or differentiate it from other forms of knowledge?

This concern may seem exaggerated or irrelevant. I would like, however, to defend the opposite, based on several concrete examples regarding, first, the social debate around scientific questions and, then, the initial impact of discussions about NOS among university students.

Affirming that science is part of our social and cultural traditions does not mean, for example, that science and religion establish equivalent forms of dialogue with reality, or that they have similar concerns. Saying that science is not guided by a single, rigid method composed of fixed steps that begin with the observation of phenomena does not mean that no methodologies are adopted in the daily practice of scientists, or that it is not guided by any defined and collectively agreed upon procedural rules. Affirming that a scientific theory cannot be definitively confirmed does not also mean that there is no evidence in its favor or good reasons for it to be adopted. Confusions of this nature seem, to a moderate relativist, to stem simply from a lack of knowledge. However, in extreme relativist readings, they may be defensible positions.

Following 2009's "Year of Darwin" celebrations, we witnessed renewed and vigorous debates involving the evolutionism-creationism question. In Brazil, as in the rest of the world, this commemorative year was marked by publications in national newspapers and periodicals, TV debates, exhibits, etc. Although this was undoubtedly a great opportunity for the dissemination of the theory of evolution among the general population, it in fact also incurred a significant polarization between science and religion. This latter was intensified by publications such as *The God delusion* (Dawkins, 2006) and, in retrospect, inevitable in the country with the largest number of Catholics in the world.[10] What interests us most here is the way in which an exacerbated relativist view was used to undermine the debate and to disseminate a mistaken concept of doing science. From a historical-philosophical viewpoint, the position that science is only *one* form of knowledge – and one which is in no sense in conflict with religious thinking – has been widespread even among minimally knowledgeable individuals. Furthermore, since scientific thinking cannot be proven, and given that there is no rigid method to guarantee the acquisition of truth, *then* everything falls within the field of speculation and hypothesis ...! Well, the *existence* of different forms of knowledge is obviously not at stake, nor is the capacity of each of them to make us *happy* and fulfilled from a human standpoint. The problem, however, lies in taking these forms of knowledge as epistemologically equivalent.

Brazil has also recently witnessed debates about the use of embryonic stem cells in scientific research; as a result, permission was required from its Supreme Court for such studies to legally proceed. Once again there was a polarization between science and religion. The fact that science has no definitive answers to questions such as "What is life?" or "When does life start?" (after all, scientific knowledge is tentative!) has been used as an argument to weaken the scientific viewpoint: *If* science does not have the answers, *then* we must trust the religious authorities. Indeed, we were left with the impression that the law could only have been approved *because* there is a widely disseminated – misunderstood – view of science that attributes excessive authority to scientific knowledge! If extreme relativism was widespread among the general population, who knows if the result would not have been unfavorable for the scientific perspective?

An examination of the initial impact of discussions on NOS among university students complements the previous examples. Many students do not enter university agreeing with absolutely *all* the aspects associated with a mistaken view of science. This is partially because of the relativization of scientific knowledge in society, which results from social criticisms of science, and partially owing to the importance attributed to traditional knowledge. Students manifest views that mix, for example, elements of an empirical-inductivist conception with aspects compatible with more current views of doing science (such as the mutability of scientific knowledge). What impact do NOS studies (guided by HPSS) have on this *background*?

I have observed a very rapid change from moderate to extreme relativism, mainly among Humanities students (both undergraduate and graduate students), due to their initial contact with NOS-related discussions. The deconstruction of a commonsense view of science, associated with elements contained in the humanities frameworks themselves (such as the importance attributed to culture, the non-existence of absolute truths, methodological plurality, among others), may lead to hasty analyses of the field of natural sciences. In the absence of the historical elements and specific content of the natural sciences that might or might not sustain their assertions, many students tend toward an exacerbated relativism: *if* science works *only* with models that are *merely* representations of reality, generating mutable knowledge which cannot be definitively proved, *then* ... anything goes! This is a kind of philosophical 'short circuit' that leads to yet another mistaken and naïve conception of science.

What matters to me is the following: is this what we want for science teaching? Can the extreme relativism that collaborates with efforts toward the proper characterization of science guide citizens in making science-related decisions? Could the deconstruction of mistaken conceptions of doing science and of its development contribute to the dissemination of inadequate and anti-scientific views? Do we not run the risk of "throwing the baby out with the bath water"?

5 History, Philosophy, Sociology, and Science Education

At the start of every semester, all science teachers should ask themselves: *why do I teach science?* The search to understand natural phenomena and the world around us usually come to mind at this time, but these are certainly not the only reasons for teaching science. Many argue that science is fundamental to understanding the technology that permeates modern existence, since scientific theories are the functional bases of the devices and equipment used in our daily lives (Menezes, 2005). In this line of reasoning, ordinary citizens *depend* on scientific knowledge to *cope with* the technological world around them. To a certain extent, this argument is a fallacy.

Although science is undoubtedly necessary to explain the functioning of technological devices in modern industrial society, it is not up to scientific education to guarantee all students a refined and detailed understanding of current technological paraphernalia in all their complexity. In addition, it is increasingly evident that we are much more *users* of these products than anything else. Ordinary citizens deal satisfactorily with sophisticated technologies on a daily basis without fully understanding their scientific bases.

In no way does this invalidate the fact that science education can (and should) help us understand the technological devices of our world, mainly with respect to the underlying principles and theoretical models involved. However, this should not be the only or main purpose of science teaching, since such an approach risks the transmission of a *utilitarian* view of science, wherein all knowledge finds its reason for being solely in its *application*. *Technological education* remains something to be widely explored and the STSE approach is heading in this direction.

Reasons for teaching science must go *beyond* the perspective aimed at understanding the natural and technological worlds, in order to include *the ways in which this understanding is achieved*. Science provides us with a specific *view* of reality, that is, a way of seeing the world. Science teaching must be able to transmit this view, allowing students to perceive the 'game of science' and how it is being constructed. In this sense, learning *about* science is as important as mastering its concepts, models, laws, and theories (Adúriz-Bravo, 2006). This becomes more significant when we consider the general education regarding scientific knowledge provided to most of the population, who will not necessarily pursue scientific careers. Science teaching must provide ordinary citizens with possibilities for establishing an intelligent dialogue with the techno-scientific world around them.

Science is an integral part of our culture. It has undeniable *cultural value*. However, it consists of special knowledge, with its own language and epistemology, which differ from those found in other forms of knowledge. Learning *about* science also involves learning about these differences. To a certain extent, learning science entails immersion into a *new* culture – scientific culture – within our wider culture. In this sense, science learning is characterized by an *enculturation* process (Driver et al., 1994) that needs to be epistemologically oriented.

HPSS plays a crucial role in this process. Although learning about science is generally not approached exclusively from a historical-philosophical approach, the latter has proved to be an excellent path to achieve this goal. It remains for us to widen the presence of HPSS in science teaching by addressing the difficulties in and obstacles to implementing proposals with historical and philosophical content. The main challenges that we continue to face are the following: promoting greater rapprochement between historians and educators, sociologists and educators, and the latter and philosophers, with respect to the use of HPSS for didactic purposes; producing high-quality didactic material, applying it in the classroom, and evaluating such interventions, especially in primary and secondary education[11]; training in-service teachers to work with historical-philosophical perspectives, exchanging experiences in courses of both short and long duration; promoting curricular reforms implementing and recognizing the importance of HPSS in teacher education courses; among others. These are significant challenges for Brazil and the rest of the world vis-à-vis both science teaching and science teaching research.

However, for students to learn about science in effective ways, science education must also concern itself with *which* science should be taught. On the one hand, science differs in important ways from commonsense, everyday knowledge, and other cultural traditions. Yet, despite its epistemological idiosyncrasy, the science that is taught should be a science that engages in fruitful dialogues with these types of knowledge, establishing points of continuity and discontinuity. Science is a human, historical, and social construct.

On the other hand, what exactly do we want to deconstruct and what do we want to put in its place, in terms of a conception of science and its development? The relative consensus regarding this matter still appears fragile, primarily in relation to what is considered 'proper'. It may be very difficult to progress beyond that status, given the complexity of the subject. Nevertheless, it seems clear to us that science education should not be driven by the sort of extreme relativism that promotes anti-scientific positions. Criticism and problematization of science are undoubtedly healthy, as is questioning the authority of scientific discourse, which may become authoritarian. However, besides deconstructing the common view of science, our educational practices should also allow us to (once again) *admire* scientific endeavor as a human construct and a dialogue with nature. We should ask ourselves who and what would be served by a scientific education that did not offer, from the epistemological viewpoint, elements enabling us to evaluate differences between types of knowledge, the social value of science, and its underlying reasons.

Even though we may accept the consensus concerning 'which' science should be taught, a series of important questions remain open to investigation: how do individuals learn about the NOS? How does this occur in the classroom context? How useful is knowledge about the NOS in making decisions about socially relevant questions involving science? Do teachers' conceptions of the NOS really influence their pedagogical practices? Do certain epistemological positions lead to anti-scientific attitudes? (Except for the latter, these questions include some of those clearly pointed out by Lederman [2007] in his review).

We hope to have shown, throughout this chapter, the importance of HPSS in the search for answers to these and other questions involving science teaching and science teaching research. HPSS, along with other theoretical-methodological tendencies and approaches, continues to be a fertile field providing elements fundamental to epistemologically oriented scientific education.

Acknowledgement

The author wishes to thank CNPq (the National Counsel for Scientific and Technological Development/Brazil) for its financial support of the research project related to this publication.

Notes

1 Although more studies endorsing this statement are called for, arguments along this line have already been made by several authors, such as Zanetic (1989), Matthews (1992), Peduzzi (2001), and Martins (2006).

2 There is no need to assume a close parallel between historical conceptions and those of students for HPSS' contributions to teaching to remain valid.

3 In the case of Brazil, the evaluation of these school years must take into account that the mean expected high school graduation rate is slightly more than 30%, that is, just over 30% of students who enroll in first grade manage to finish high school (data from 2005, available at: http://www.edudatabrasil.inep.gov.br/, Accessed on 19 June, 2010. Although this scenario has not altered significantly in recent years, it was much worse a few decades ago).

4 In 1970, there was a conference at MIT, led by Stephen Brush and Allen King, in which Martin Klein (1972) criticized the use of HPSS for teaching purposes, stating that the only history possible in science courses would be pseudohistory, a simplified history that would fall short of the requirements and standards of historians.

Whitaker (1979) subsequently explored these arguments, featuring another significant problem: that of quasi-*history*, which would correspond to a falsification of history in order to support certain views of those who build it (for more details on this debate, see Matthews, 1992).

5 Many students express this feeling after encountering HPSS disciplines. They feel that their conception of science changed significantly and that HPSS "should be taught before" or "more often" during the course (Martins, 2007, 2012).

6 In this text, I am not concerned with establishing a 'typology of knowledge', differentiating, for example, 'common sense' from 'popular knowledge', or between knowledges specific to certain cultures, as Lopes (1999) does. For my purposes, I consider the generic reference to 'everyday knowledge' to be sufficient, following the denomination adopted by this author.

7 One can still consult the special issues of *Science Education* (Volumes 78, 83, & 85) and *Science & Education* (Volume 18, reprinted in Matthews, 2009) on this theme.

8 It is beyond the scope of this text to perform a literature review of this topic. Further information may be found in Lederman (1992, 2007).

9 Fernández et al. (2002, p. 480) have already touched on this question by stating that the critique of the 'rigid conceptions' of science could lead to other mistaken positions of methodological and conceptual relativism.

10 In addition to its large number of Catholic churches, the number of Pentecostal and neo-Pentecostal churches in Brazil has grown during recent decades. The 2000 census indicated that 15% of the Brazilian population stated that they were Protestants, which represents a growth of 100% during the 1990s (El-Hani & Sepúlveda, 2010).

11 One way in which this is being implemented is with 'historical episodes', which problematize science concepts, mainly those involving historical controversies. An example of this type of approach is the HIPST project (Höttecke, 2009).

References

AAAS – American Association for the Advancement of Science. (1990). *Science for all Americans*. New York, NY: Oxford University Press.

Adúriz-Bravo, A. (2006). *¿Qué naturaleza de la ciencia hemos de saber los profesores de ciencias? Una questión actual de la investigación didáctica* [What nature of science do science teachers have to know? A current question in didactic research]. Retrieved March 21, 2008, from http://www.unesco.cl/medios/biblioteca/documentos/

Allchin, D. (2004). Pseudohistory and pseudoscience. *Science & Education, 13*, 179–195.

Bachelard, G. (1996). *A formação do espírito científico* [The formation of the scientific spirit]. Rio de Janeiro, Brazil: Contraponto.

Brazil, Secretaria de Educação Básica. (2000). *Parâmetros curriculares nacionais para o ensino médio (PCNEM) – parte III: Ciências da natureza, matemática e suas tecnologias* [National curricular parameters for high school – Part III: Natural sciences, mathematics and their technologies]. Brasília, Brazil: MEC/SEB.

Brazil, Secretaria de Educação Básica. (2006). *Orientações curriculares nacionais para o ensino médio – Ciências da natureza, matemática e suas tecnologias* [National curricular guidelines for high school – Natural sciences, mathematics and their technologies]. Brasília, Brazil: MEC/SEB.

Campanario, J. M. (1998). Ventajas e inconvenientes de la historia de la ciencia como recurso en la enseñanza de las ciencias [Advantages and disadvantages of the history of science as a resource in science teaching]. *Revista de Enseñanza de la Física, 11*(1), 5–14.

Carvalho, A. M. P., & Gil-Pérez, D. (1998). *Formação de professores de ciências: Tendências e inovações* [Science teacher education: Tendencies and innovations]. São Paulo, Brazil: Cortez.

Cobern, W. W. (1996). Worldview theory and conceptual change in science education. *Science Education, 80*(5), 579–610.

Cobern, W. W., & Loving, C. C. (2001). Defining "science" in a multicultural world: Implications for science education. *Science Education, 85*(1), 50–67.

Colombo de Cudmani, L., & Salinas de Sandoval, J. (2004). ¿Es importante la epistemología de las ciencias en la formación de investigadores y de profesores en Física? [Is the epistemology of sciences important in the education of physics researchers and teachers?]. *Enseñanza de las Ciencias, 22*(3), 455–462.

Dawkins, R. (2006). *The God delusion.* Boston, MA: Houghton Mifflin.

Driver, R. H., Asoko, H., Leach, J., Mortimer, E. F., & Scott, P. (1994). Constructing scientific knowledge in the classroom. *Educational Researcher, 23*(7), 5–12.

Duarte, M. C. (2004). A história da ciência na prática de professores Portugueses: Implicações para a formação de professores de ciências [The history of science in Portuguese teachers' practice: Implications for science teacher education]. *Ciência & Educação, 10*(3), 317–331.

Eddington, A. (1928). *The nature of the physical world.* London: MacMillan.

El-Hani, C. N. (2006). Notas sobre o ensino de história e filosofia da ciência na educação científica de nível superior [Notes on the teaching of history and philosophy of science in higher scientific education]. In C. C. Silva (Ed.), *Estudos de história e filosofia das ciências: Subsídios para aplicação no ensino* [Studies of the history and philosophy of the sciences: contributions to teaching applications] (pp. 3–21). São Paulo, Brazil: Editora Livraria da Física.

El-Hani, C. N., & Sepúlveda, C. (2010). The relationship between science and religion in the education of protestant biology preservice teachers in a Brazilian university. *Cultural Studies of Science Education, 5*, 103–125.

Fernández, I., Gil-Pérez, D., Carrascosa, J., Cachapuz, A., & Praia, J. (2002). Visiones deformadas de la ciencia transmitidas por la enseñanza [Distorted views of science transmitted by teaching]. *Enseñanza de las Ciencias, 20*(3), 477–488.

Geelan, D. R. (1997). Epistemological anarchy and the many forms of constructivism. *Science & Education, 6*, 15–28.

Gil-Pérez, D. (1993). Contribución de la historia y de la filosofía de las ciencias al desarrollo de un modelo de enseñanza/aprendizaje como investigación [Contribution of history and philosophy of science in the elaboration of a teaching/learning model as an investigation]. *Enseñanza de las Ciencias, 11*(2), 197–212.

Hodson, D. (1993). In search of a rationale for multicultural science education. *Science Education, 77*(6), 685–711.

Höttecke, D. (2009). Developing and implementing case studies for teaching science with the help of history and philosophy. In *Proceedings of the tenth international history, philosophy, and science teaching conference*. South Bend, IN: University of Notre Dame.

Klein, M. J. (1972). The use and abuse of historical teaching in physics. In S. G. Brush & A. L. King (Eds.), *History in the teaching of physics* (pp.??–p.??). Hanover, NH: University Press of New England.

Korzybski, A. (1931). *A non-Aristotelian system and its necessity for rigour in mathematics and physics*. Paper presented before the American Mathematical Society at the New Orleans, LA meeting of the American Association for the Advancement of Science, December 28, 1931. (Reprinted in *Science and Sanity*, 1933, 747–761)

Laburu, C. E., & Carvalho, M. (2005). *Educação científica: controvérsias construtivistas e pluralismo metodológico* [Scientific education: constructivist controversies and methodological pluralism]. Londrina, Brazil: Eduel.

Lederman, N. G. (1992). Students' and teachers' conceptions of the nature of science: A review of the research. *Journal of Research in Science Teaching, 29*(4), 331–359.

Lederman, N. G. (2007). Nature of science: Past, present and future. In S. K. Abell & N. G. Lederman (Eds.), *Handbook of research on science education* (pp. 831–879). Mahwah, NJ: Lawrence Erlbaum Associates.

Lopes, A. R. C. (1999). *Conhecimento escolar: Ciência e cotidiano* [School knowledge: Scientific and everyday]. Rio de Janeiro, Brazil: EdUERJ.

Marandino, M. (2003). A prática de ensino nas licenciaturas e a pesquisa em ensino de ciências: Questões atuais [The practice of teaching in teacher training courses and science teaching research: current issues]. *Caderno Brasileiro de Ensino de Física, 20*(2), 168–193.

Marín Martínez, N., Solano Martínez, I., & Jiménez Gómez, E. (1999). Tirando el hilo de la madeja constructivista [Unraveling the thread from the constructivist skein]. *Enseñanza de las Ciencias, 17*(3), 479–492.

Martins, A. F. P. (2007). História e filosofia da ciência no ensino: Há muitas pedras nesse caminho ... [History and philosophy of science in teaching: there are many stones on the way ...]. *Caderno Brasileiro de Ensino de Física, 24*(1), 112–131.

Martins, A. F. P. (2012). História, filosofia, ensino e formação de professores: Desafios, obstáculos e possibilidades [History, philosophy, teaching and teacher education: challenges, obstacles and possibilities]. *Educação: teoria e prática, 22*(2012).

Martins, R. A. (2006). Introdução: A história das ciências e seus usos na educação [Introduction: history of the sciences and its use in education]. In C. C. Silva (Ed.), *Estudos de história e filosofia das ciências: Subsídios para aplicação no ensino* [Studies on history and philosophy of sciences: Contributions to teaching applications] (pp. xvii–xxx). São Paulo, Brazil: Editora Livraria da Física.

Matthews, M. R. (1992). History, philosophy, and science teaching: The present rapprochement. *Science & Education, 1*(1), 11–48.

Matthews, M. R. (1994a). *Science teaching: The role of history and philosophy of science.* New York, NY: Routledge.

Matthews, M. R. (1994b). Vino viejo en botellas nuevas: Un problema con la epistemología constructivista [Old wine in new bottles: A problem with constructivist epistemology]. *Enseñanza de las Ciencias, 12*(1), 79–88.

Matthews, M. R. (Ed.). (2009). *Science, worldviews and education.* Dordrecht, The Netherlands: Springer.

McComas, W. F., Almazroa, H., & Clough, M. (1998). The nature of science in science education: An introduction. *Science & Education, 7*, 511–532.

Menezes, L. C. (2005). *A matéria uma aventura do espírito: Fundamentos e fronteiras do conhecimento físico* [Matter as an adventure of the spirit: Fundamentals and frontiers of physical knowledge]. São Paulo, Brazil: Editora Livraria da Física.

NC – The National Curriculum. (2007). *Secondary curriculum key stages 3 & 4.* Retrieved June 15, 2010, from http://curriculum.qcda.gov.uk/

Ogborn, J. (1997). Constructivist metaphors of learning science. *Science & Education, 6*, 121–133.

Osborne, J. (1996). Beyond constructivism. *Science Education, 80*(1), 53–82.

Pedduzzi, L. O. Q. (2001). Sobre a utilização didática da história da ciência [On the didactic use of the history of science]. In M. Pietrocola (Ed.), *Ensino de física: Conteúdo, metodologia e epistemologia numa concepção integradora* [Physics teaching: An integrating conception of content, methodology and epistemology] (pp. 151–170). Florianópolis, Brazil: Editora da UFSC.

Posner, G. J., Strike, K. A., Hewson, P. W., & Gertzog, W. A. (1982). Accommodation of a scientific conception: Toward a theory of conceptual change. *Science Education, 66*(2), 211–227.

Shulman, L. S. (1986). Those who understand: Knowledge growth in teaching. *Educational Researcher, 15*(2), 4–14.

Villani, A. (1992). Conceptual change in science and science education. *Science Education, 76*(2), 223–237.

Whitaker, M. A. B. (1979). History and quasi-history in physics education – Parts 1 & 2. *Physics Education, 14*, 108–112, 239–242.

Zanetic, J. (1989). *Física também é cultura* [Physics is also culture] (Doctoral dissertation). Faculdade de Educação, Universidade de São Paulo, São Paulo, Brazil.

CHAPTER 16

History, Didactics, and the Transformation of Scientific Content: Epistemological Surveillance and Science Education Commitments

Maurício Pietrocola, Elio Ricardo and Thaís Forato

Abstract

This work discusses the commitments related with the use of historical contents in science teaching. We start by considering historiographic recommendations from some classical and current research. After, we present Didactic Transposition as a way of showing the transformations scientific knowledge suffers from its original context to the school context. Our conclusion is that it is impossible to produce teaching knowledge with both historiographic and didactic certitude. There are always risks during the didactic process, and we suggest preparing the teachers to evaluate them. Following Chevallard's suggestion, we name *epistemological vigilance* the skill that allows the teacher to be aware of the risks in the science education activities.

Keywords

epistemological vigilance – historiographical perspectives on science education – history of sciences – didactic transposition risks – historiographical and didactic controversies

1 Introduction

Instead of teaching science contents as 'pure science', a number of official documents and academic papers have advocated for the inclusion of a humanistic perspective on science in the curriculum as a way of contributing to the development of the skills and abilities necessary for educating the population of the 21st century (AAAS, 1990, 1993; Brazil, 2002a, b; NRC, 1996; NSTA, 1962, 1982). Several authors argue that the history and philosophy of science are suitable means of dealing with the complexity of science in the classroom

 | DOI: 10.1163/9789004409088_016

(Abd El-Khalick & Lederman, 2000; Allchin, 2004; Adúriz-Bravo & Izquierdo-Aymerich, 2009; El-Hani, 2006; Forato et al., 2012, 2009; Gil-Perez et al., 2001; Matthews, 1992; McComas et al., 1998; Lederman, 2007). This theoretical approach emphasizes the importance of holding science as a human enterprise, while considering the history of science (HS) as a teaching-pedagogical tool particularly suited to deal with its 'way of working'. For some authors, episodes from the history of science are a means by which to understand certain scientific processes that are hard to grasp, accessible only through the sort of sophisticated epistemological analysis that is appropriate work for professionals in a given area, but far from the ability of most students and teachers to replicate. Such episodes may well provide good pedagogical examples, enabling an explicit epistemological discussion of the processes of science in different cultures and times (Adúriz-Bravo & Izquierdo-Aymerich, 2009).

Reflecting on the processes that led to the formulation of concepts and laws contributes greatly to the "formation of structures of understanding that allow us to transform the real world into an intelligible world" (Pietrocola, 2003, p. 135). Any extensive examination of scientific knowledge and learning enables us to criticize distorted and stereotypical images of science which present scientific concepts as intrinsic parts of reality and ignore the complex socio-historical and cultural processes involved in the work of scientists seeking to interpret and describe the natural world.

Although it is an appropriate strategy for dealing with some of the epistemological aspects of science, the use of history also presents a wide array of challenges. The first, and perhaps most important one, is the risk of distortions in the production of historical accounts intended to be used in the school environment. Outdated historical approaches – such as those that propagate a purely empirical-inductive conception of the construction of scientific knowledge and reinforce the view of science as a producer of absolute truths – still prevail in most textbooks. Students from Elementary school through high school tend to see the work of scientists, e.g., the elaboration of models, as *discoveries* rather than *inventions*, confirming the assessment that students at this level conceive of science as a means of accessing pre-existing laws in the world (Aikenhead & Ryan, 1992; Gil-Perez et al., 2001). This view contrasts with more current understandings of the complex processes of scientific knowledge development, in which the many factors of complex cultural environments coexist with those factors normally considered scientific. Historical-epistemological analysis facilitates such understandings, thereby contributing to scientific literacy (Lederman, 2007).

In his classic paper, entitled "History and *quasi-history* in physics education", Whitaker (1979) dealt with the risks found in the use of HS in discourse intended for the lay public and students. He examined the manners in which some textbooks produced distorted historical accounts, especially linear reconstructions that seek to provide a logical and orderly framework for the emergence of concepts.

Whitaker introduces the *quasi-history* concept to define a kind of HS committed to the didactic objectives of clarity and conceptual coherence. In his words:

> In this article I shall discuss another type of material which looks Historical, but in which there is no attempt to convey history truthfully: the aim is solely to put over scientific facts, and the "history" is there to provide a framework inside which the scientific facts fit easily, appear to 'make sense' and may be easily remembered for examinations purposes. (Whitaker, 1979, p. 108)

This chapter provides a mapping of the risks and difficulties that may beset the most well-intentioned textbook author who lacks the technical training needed for such an undertaking. It ultimately offers significant warnings about the pitfalls that those wishing to make good use of historical accounts of science may encounter. The decontextualization of facts and historical episodes generated by erroneous or biased interpretations of sources constitute key errors of which to remain wary. Whitaker's insistence on assessment of the epistemological quality of the accounts found in textbooks sounds an alarm against historical accounts that are poorly connected with real scientific context.

In another work from the same period, Brush (1974) discusses debates about the historical distortions found in science education and the widely varying postures of historians of science regarding such distortions. Martin Klein cite, for example, defended fictional, linear reconstructions, deliberately arranged with historical episodes, because he feared damage to the enticing image of science if students were to discover the mistakes made by 'scientists' of the past.

Showing the wrong directions, controversies, advances, and setbacks of science, the human side of the protagonists in the HS could potentially tarnish the strong and indisputable image of scientific knowledge. In a more diagnostic than axiological analysis, Kuhn (1970) had in the early 1960s already put forth the notion that the HS in textbooks fulfills the educational function of quickly familiarizing the student with the current paradigm, and not of teaching them the NOS. The fundamental message in such works, which still prevails, points to the risks in and requirements for the effective use of history and philosophy of science (HPS) knowledge in the school environment.

Modern historiography has been the leading source for analyses of the use of the HS for educational purposes. Perhaps the greatest contribution of modern historiography lies in its demonstration that every historical account is grounded in a particular viewpoint and hence produces interpretations, even in highly specialized works (Canguilhem, 1977; Forato, 2008; Gravoglu et al., 2008; Jardine, 2003; Kragh, 1987; Martins, 2004). This points to the complex issues inherent to producing texts or activities of a historical nature to be used in classrooms. In recent decades, this has been to a certain extent the main concern of other works in this line of research (Allchin, 2006; Bell et al., 2001; El-Hani, 2006; Forato, 2009; Gil-Perez et al., 2001; Lederman, 2007; Pagliarini, 2007; Peduzzi, 2001). Avoiding distortions and ghastly blunders, such as those found in the teaching of quasi-history (as mentioned above), ends up being the yardstick for all efforts in this vein.

At first glance, it may seem that conformity to current historiography requirements should be enough to produce HS material suitable for use in the school environment. One may also naïvely believe that the role of the authors of textbooks and other didactic materials is to produce simplified historical reports while avoiding decontextualizations. But is this really enough? Can an episode from the HS be produced for educational purposes without any decontextualizations? To answer this question, one has to more closely analyze the various commitments established at the time a given aspect of scientific knowledge was produced, as well as those involved in the course of bringing that episode into the teaching and learning context, not to mention the teaching/learning context itself. What we intend to show in this chapter is that HS content prepared for instructional purposes must actively incorporate at least two aspects: i) current historiography and ii) the specific characteristics of the school environment.[1] In the next sections, we will characterize these two types of commitments.

2 First Commitment: Historiography

To deal with the aforementioned problems in HS texts, we need to consider, first, that the elaboration of narratives is inevitably impacted – whether implicitly or explicitly – by the values, beliefs, and methodological guidelines, of those producing such writings.

A historical narrative about the creation of a scientific concept or a debate between competing theories, or even about carrying out experiments, for example, carries certain conceptions about the very processes of science construction. This means that every historical account is pervaded by epistemological assumptions. Quite often, historical narratives found in the educational context incorporate a conception of science that is different from the one the author intends to present. It is not rare to find historical narratives in which the author explicitly defends a socio-historical vision of the construction of science, yet at the same time in fact presents a version of the history of science which is linear, factual, and anachronistic. Certain naïve or distorted concepts – ideas and presuppositions which an author may assume have been overcome – remain, in truth, deeply rooted, and eventually emerge in furtive comments, adjectives, or judgments, thereby affecting the outcome of the work with regard to the concepts of science and its development. It is thus essential to remain vigilant in order to minimize the influence of personal values when developing accounts of the history of science, by, for instance, confronting the latter with reports produced by experts (Canguilhem, 1977; Forato, 2008; Golinky, 1998; Gravoglu et al., 2008; Jardine, 2003; Kragh, 1987; Kuhn, 1977; Martins, 2004; Rossi, 2000).

The biggest problem in considering historical narratives for the school environment lies in the distortions that arise when the process of scientific knowledge construction is depicted. In general, producers of such educational materials – who are not necessarily historians or philosophers of science – should be aware of the need for respecting certain important basic assumptions (Allchin, 2004; Holton, 2003; Martins, 2001, 2004). Many terms that are used in narratives contribute to the distortion of history. For example, discover, built and elaboration are some time used as synonimes to designte the way laws appear in science. Although there are subtle differences between these various designations, certain these terms are linked to specifique conceptions of scientific laws.

In general, a decontextualized approach stems from the mistaken or biased interpretation of sources and/or a historical period. As a starting point, let us recall some of the most common historiographical issues, as pointed out by Allchin (2004). Anachronism is the most common error. The key principle for historians is respect for historical context, and, hence, avoiding anachronistic interpretation of the past, that is, viewing it under the lens of current rules or standards. Anachronistic views of facts or episodes of the HS interpret and judge the historical events of that period by the values, ideas, and beliefs of another era. In general, they assess the past in a biased manner, selecting and highlighting concepts and theories 'similar' to those accepted in the present. Looking into the past through the eyes of the present results either from a lack of historiographical knowledge or from intentional manipulations intended to achieve certain purposes.

One special kind of anachronistic approach is *whiggism*, which recounts historical events in order to assert the authority and heroism of certain institutions or persons. In the history of science, *whiggism* is generally used to classify a kind of narrative made to legitimate a single contribution of one scientist or group, which theoretical assumption reinforces the triumph of a specific school of thought.[2]

Allchin (2004) classifies *hagiography*[3] as a specific kind of whiggism that romanticizes some scientists and philosophers of the past as heroes by overstating specific aspects of their contributions. In a broad sense, hagiography is defined as the biography of saints and martyrs, or of any person who is idealized in order to promote their biography. Allchin uses this term to emphasize the versions of HS that not only 'sanctify' a given 'genius' thinker, but also omit their mistakes, ignore past contributions from others or current contributions from their peers, and present their conjectures as an idealized model of scientific thought. Of course, the problem is not in admiring bright individuals, but in encouraging the view that the construction of science occurs only via the work of such supposed geniuses.

Another anachronistic approach is found in the linear reconstruction of historical episodes. Although advocated for by some thinkers as useful for teaching scientific concepts and 'necessary' for avoiding conflicts with the current paradigm, it comprises yet another distortion of history.[4] In this approach, the author selects only those factors from earlier periods that contribute to an orderly reconstruction of the steps taken in the creation of the scientific concepts and theories accepted by contemporary science.[5] Ideas and events from the past are organized as if the elaboration of concepts and theories followed logically linked steps, resulting inevitably in a certain end result. Usually, this kind of anachronism makes it seem that there is a foolproof recipe for producing science: all that is needed is for an exceptional genius to consistently follow the steps of an exact, accurate, and unique universal scientific method. This perspective exemplifies precisely those erroneous ideas that we are criticizing in this chapter.

These problems are very common in historical narratives that attempt to describe a very long period, or to generalize the points of view of radical thinkers, attributing individual conceptions to a period or a school of thought (Kragh, 1987; Martins, 2001). Such misinterpretations of history ignore the complexity of scientific work, which encompasses contributions from discussions, errors, obsolete theories, and even extra-scientific factors to the development of science. Anachronistically constructed historical narratives present past theories as predecessors of current ideas. Such reports are generally filled with judgments of merit. They emphasize linearly chained facts and ignore disagreements, disputes, and alternative ideas, along with any element of the cultural context that does not reinforce the ideas intended to be highlighted

(Allchin, 2004). The essential trouble with such linear narratives is that they lead students to believe in a science that evolves infallibly, guided by a universal method and detached from any social or human influence.

These different anachronistic approaches, along with the ensuing distorted interpretations of historical documents, are detrimental to how science is perceived by students. Another problem with the presence of historical distortions in educational settings is that they discourage critical thinking in students. For example, ideas are accepted as worth entertaining simply because they were proposed by an authority or infallible genius. The purposes of understanding the human dimension of science as a cultural construction, influenced by the innumerable axiological elements of a given time, and as an activity that cannot be divorced from metaphysical or aesthetic considerations, are thereby surely impaired.

While acknowledging that it would be impossible for the authors of textbooks and other didactic materials to master the methodology of historical research and the historiographical ideas that permeate the work of experts in addition to their own areas of expertise, Allchin (2004) argues for the importance of researchers and educators being able to recognize some signs of pseudo-history, to wit, reports that present such misconceptions as perfect characters, monumental discoveries, flukes, and individualism, or which only chronicle crucial experiments, bestow a sense of the inevitable, etc. He also stresses that problematic, distorted versions of history fail to mention relevant cultural, social, and environmental aspects; human contingencies; preceding ideas; alternative ideas. Such omissions of context and scientific controversies can lead students to imagine that the course of scientific history has consisted of an uncritical acceptance of new concepts. These signs of pseudo-history are not absolutely indicative of fallacy, but are, rather, warning signs that we need to confront historical interpretations with credible sources.

3 Second Commitment: Didactic and Pedagogical

Addressing classroom challenges is a daunting task for teachers. But the interaction process in the classroom is not limited to the emotional dimensions of teachers and students. All content to be taught should be thought of as part of this interaction. Teachers, students, and knowledge comprise what the modern didactics of science defines as the *system of didactic interactions* or, briefly, the *didactic system* (Astolfi et al., 1997). What we must realize is that teachers, students, and knowledge must all conform to the constraints dictated by the didactic system. That is, teachers' actions are not only motivated by their personal convictions, but rather adapt to a social role that precedes them. The same applies to students, who seek to place themselves in the position expected of a student at a given educational level (Brousseau, 1986). But how do we consider the commitments and expectations set by knowledge adapted to the didactic system?

To answer this question, we must emancipate ourselves from fantasies and naïve ideas that lead us to disregard the classroom as a place of complex interactions. Thus, the first step is to understand that school knowledge is not purely the result of simplifications aimed at diminishing difficulties for learning. In this chapter, we adopt the didactic and epistemological perspective that knowledge brought into education, particularly into classrooms, is the result of profoundly transformational didactization processes (Chevalard, 1991).

Academic knowledge is subject to very specific rules built and adopted by communities of experts, such as logical coherence, nominalized language (Halliday & Martin, 2003), mathematical formalization, conformity to experimental results, etc. If we intend to use such knowledge just as it is, these commitments render it inadequate for primary and secondary educational systems.

Yves Chevallard introduced his expression "didactic transposition" to indicate the changes to which knowledge is subjected when transposed from the original academic context to become an educational object. In this perspective, one can distinguish between *scholarly knowledge* (produced inside a community of experts) and *school knowledge* (produced for school environments) (Chevallard, 1991). In this sense, the process of didactic transposition transforms knowledge, giving it another epistemological status (Astolfi & Devaley, 1995).

It is important to state that not all academic domain knowledge will be rendered part of everyday school life. The selection and suitability of knowledge are essential features in the process of didactic transposition. The multiple factors that influence the choices and adjustments to be made should be taken into account. Many factors, including political and commercial interests, the anxieties of a society that believes in education, and the pedagogical interests related to teaching and instruction influence didactic transposition. The characteristics that make knowledge "teachable" can be inferred from the didactic transposition framework, as we find in Chevallard's (1991), work addressing the "teachability" of knowledge. For Chevallard, the starting point lies in the very selection of knowledge, which needs to be evaluated as regards the relationship it establishes with social organization and cultural values (that is, its sociocultural agenda.) First, we must evaluate its *epistemological relevance*. Secondly, its *cultural relevance* – the measure of the cultural desire to include any knowledge content as school knowledge – must be considered. In other words, we must evaluate the degree of acceptance of any educational knowledge as a part of a broader, formative social project. Thirdly, the degree of *social exposure* of the corresponding practices in society must be addressed. This last point affirms that there are some practices in society that have more value than others in terms of their influence over curriculum development. Martinand argues along these lines in forging his notion of the social practices of references (*Pratiques Sociales de Références*, 1986, 2003).

Chevallard also considers the basic educational system requisites for promoting teaching interactions. His first such characteristic requirement is that the knowledge taught in schools should involve mostly content viewed as consensual. When a given physics subject is taught and learned in schools, for example, neither scientists, teachers, parents, nor students should have doubts about the 'value' of that subject. Thus, science content should hold the status of contemporary (or at least historical) 'truth'. This hints at why older and traditional themes are often preferred to so-called novel ones.

In addition, to be implemented, scholarly knowledge must comply with two 'types' of updating:

1. *Moral updating* regards the degree of importance assigned by society to school knowledge. It indicates the necessity for developing curricula in accordance with social expectations for formative education. It should be clear that the issue here is in terms of relevance, and not an expression of value in itself;
2. *Biological updating* concerns the degree of correlation between school knowledge and scholarly knowledge. It indicates the need to keep curricula updated in relation to knowledge produced in the academic context.

Further, school knowledge satisfies the requirement of *operability* when it can be presented in the form of activities and tasks that enable an objective assessment. Scientific contents unable to generate classroom activities and refractory to the evaluation process ultimately hinder didactic interaction and management of the school routine and tend not to remain as school knowledge.

Lastly, the process of didactic transposition involves exercising a certain didactic creativity, producing knowledge with a didactic identity.

Knowledge which conforms to the sociocultural agenda and to the requirements of teaching interactions has a greater chance of surviving in the school environment. Moreover, several additional, more-specific processes are inevitable in the didactic process, namely, the *desyncretization* of knowledge, the *depersonalization* of knowledge, the *programmability* of acquiring knowledge, the *publicity* of knowledge, and the *social control of learning* (Chevallard, 1991, p. 69).[6] These processes are consistent with the social, political, and economic interests of the educational design into which an educational system is inserted, and are manifested in its educational project (Ricardo, 2005).

In *desyncretization*, scholarly knowledge is, in the process of its reorganization and reconstruction, separated from the context in which it was built and, thus, is dislodged from the original problem to which it responds. In this process, knowledge is detached from its original epistemological environment to be reconstituted in a new one.

The *depersonalization* process is a necessary consequence of preparing knowledge to be taught in schools, since knowledge produced in the academic environment, in the internal niche of research, must be removed from this context to be understood in primary and secondary school environments (Ricardo, 2005). In depersonalizing,

> … knowledge takes on the aspect of a self-imposed, ahistorical, timeless reality that, without a producer, appears free from the whole production process, offering no opportunity to deny its origin, usefulness and relevance. Anonymous knowledge, knowledge decontextualized from its historical production, takes on the irrefutable shape of a set of natural truths. It is precisely there that one finds one of the constitutive mechanisms of the exaggerated attachment to those teaching rules that define a certain mode regarding knowledge, whereby knowledge is imposed as evident, not because it emanates from a higher authority, a master, whose word is truth, but, on the contrary, because it is totally disassociated from any person. (Chevallard & Joshua, 1982, pp. 169–170)
>
> For this knowledge to be understood and appropriated by teachers and students, it needs to have a certain degree of *publicity* (Chevallard, 1991) in the sense of being designed to be comprehensive for a broad audience. An explicit and objective discourse, as essential in rendering knowledge teachable, is sought in the construction of textbooks and other educational resources that will reach the classroom. The apparent objectivity found in textbooks and other resources provides evidence of the publicity process used in transforming academic knowledge into a more broadly spread form of knowledge. (Chevallard, 1991, p. 73)

The objectivity attained permits a certain social control of learning and a degree of programmability in knowledge acquisition, with what is to be learned generally presented in sequential series, predominantly supported by prerequisites, and compatible with allotted teaching time. This type of format is one of the most important school requirements, as well as being seen as a key learning paradigm.

However, students' subjective learning times do not necessarily conform to a single ordered, timed exposure to the knowledge to be taught. A common way to standardize the different times needed for learning is through excessive algorithmization of some areas (such as physics), in which calculations are often reduced to the application of formulas to solve exercises or problems. Such calculations aim to avoid jeopardizing the ongoing process of education, although their outcomes may not be as expected in the original teaching design. However, the belief that the reorganization of knowledge in a text is alone sufficient for coping with learning difficulties is illusory. Johsua and Dupin (1993, p. 202) emphasize that it is "the set of the characteristics of a teaching situation that can produce the effects, and not only the knowledge text". The authors add that judging a transposition extends to its practice. To put it in another way, we must take into account the internal didactic transposition which takes place between knowledge which is designed to be taught and knowledge as it is actually taught.

Thus, within this perspective, a textbook can be understood as the materialization of a complex process – one involving, at a minimum, authors, editors, and the layered contexts described above – that culminates in programmability. When analyzing a textbook, an experienced editor leaves style and emphasis as unique characteristics of the authors. However, he/she should be able to identify elements that are lacking or present in excess in teaching materials. Such editors are like 'oracles': able to both anticipate questionable facets of a work as it would be used in the classroom setting (this relates to Chevallard's "therapeutics" – see below) and to evaluate the success (or lack thereof) of the material in view of the didactic system in vogue. With this, the foundations necessary for transpositions to survive in the traditional school environment make textbooks reflect this process.

However, it is incorrect to think that any assessment of the "teachability" of knowledge is theoretically *a priori*. Knowledge taught must be subjected to empirical tests in real-life contexts, i.e., it must survive the demands of school, a process which is designated by Chevallard as "therapeutics". This is viewed as the cornerstone for the survival of knowledge. Teaching results, in the form of subsequent and collective assessment of the knowledge field involved, are essential for the maintenance (or not) of knowledge that is introduced into education. From this standpoint, the combination of knowledge in the school curricula at any given historical moment is the sum of the successes achieved in the area via the transposition process.

In a nutshell, what works under the aforementioned circumstances remains in school, while what goes wrong ends up being discarded.

Thus, the theory of didactic transposition offers some paths toward understanding the production and survival of school knowledge in the educational system. Through this lens, one can understand why pedagogical programs undergo so few changes and why new knowledge and practices are rarely incorporated into the classroom context. That being said, why would teachers abandon knowledge and practices that, rightly or wrongly, have survived in the school environment, only to implement new practices and incorporate new knowledge for which the possibilities for success remain unknown?

4 Epistemological Vigilance Regarding School Knowledge

We have shown that historiography is a benchmark for the use of the HS, while also presenting the teachability and survival of knowledge as equally important and definitive elements of education contents. At first, the requirement of using history and philosophy of science (HPS) and the conditions present in the production of school knowledge seem irreconcilable. In the previous sections, we sought to examine key problems with using the historical approach while addressing both the requirements posed by historians and the constraints of the classroom. In this broad framework, it seems clear that teachers and authors of didactic materials, when opting to use HPS, act in environments marked by commitments of very different sorts. These disparate commitments often generate tensions, such as that between avoiding linearization in historical accounts and ensuring the programmability of education. One approach to this sort of conflict would be to assign responsibility for the tensions and distortions of the content to be taught to the didactic transposition process. However, the knowledge to be taught is also necessarily subjected to didactic transposition. This is a didactic requirement and, according to Astolfi et al. (1997), is what enables the creation of teachable knowledge. According to Chevallard (1991, p. 16), "to make it merely possible to teach a particular item of knowledge, that element will undergo some distortion, which will make it fit to be taught". That is, for each level of knowledge, changes and reorganizations are inevitable throughout the process of rendering that knowledge teachable. In Chevallard's (1991, p. 17) words, "knowledge as it is taught, knowledge taught, is necessarily distinct from the knowledge that was originally designated to be taught, the knowledge to be taught. This is the terrible secret that the concept of didactic transposition places in jeopardy".

Thus, the conflicts generated by historiographical and teaching commitments surpass the teachers' and authors' spheres of decision. The action zone in question in this type of dispute contains hidden contributors, such as historians, educational advisers, curriculum designers and policies, scientists from the knowledge areas involved, etc. This action zone, defined by Chevallard (1991) as the noosphere, is understood as a site where conflicts and interests are managed and where didactic operation is thought out according to different ideologies. This is an external dimension of didactic transposition.

Solutions to such educational and historiographical conflicts must, before anything else, result from assessments of the risks involved in the transposition process. Being aware of the existence of didactic transposition allows us to "stand back, question the evidence, question simple ideas, let go of the deceptive familiarity of subject matter. Simply put, it is what enables us to use our own epistemological vigilance" (Chevallard, 1991, p. 16). The notion of epistemological vigilance draws attention to the need to examine the legitimacy of the knowledge taught and questions its *status* with regard to scholarly knowledge. Using epistemological vigilance to analyze the transformation of school knowledge reveals the differences between the knowledge taught inside the classroom and knowledge practices in the academic domain. This is one reason why epistemological vigilance is received with some hostility in the school environment.

Chevallard himself warns: "the concept of didactic transposition is stated first as a violence against the integrity of the act of teaching" (Chevallard, 1991, p. 17). To the extent that epistemological vigilance brings into question the relevance of school knowledge (in addition to its epistemological legitimacy), its cultural legitimacy will also be at stake. Why does only a certain fraction of scholarly knowledge take on the status of knowledge taught? Or, why are some historical episodes privileged in education instead of others? This apparent "violence" becomes even more evident when students themselves echo such questions, as students typically do question the need to learn what they are taught.

Being aware that – despite certain similarities – there are notable differences between, say, the physics of the physicist and physics education, or between historians' history of science and the history of science as found in schools, exposes the fact that the credibility and prestige guaranteed in both former areas is not necessarily maintained in the form of knowledge to be taught or in its teaching. We can then understand why doubting the importance of what is taught in school is problematic: It seems that one is questioning the relevance of science itself, rather than science teaching in schools. This misconception occurs when there is no clear recognition that we are in fact dealing with different kinds of knowledge. It is easy for teachers to justify the importance of teaching science because it seems sufficient to refer to the benefits and advances science offers to society. When it is revealed that those benefitting/advancing forms of science are not equivalent to the science that is found in school curricula, teachers may not feel quite so sanguine. The same can be said about the history of science in the school environment, since it cannot be legitimized by its historiographical quality. As Chevallard (1994, p. 146) warns, "no knowledge taught is empowered by itself". In the same vein, Durey and Martinand (1994, p. 77) propose the following question: "at what price is such school knowledge, which is said to be decontextualized and separated from the historical and social practices upon which it was founded, still functional today in any practice outside the school?" This is where epistemological vigilance takes on a fundamental role, because the knowledge taught must satisfy a broader educational project, with didactic choices which are made *a priori* and comply, in turn, with that broader social project. As noted, risks are inevitable in this process. In the case of the history of science, the safer path seems to be one which avoids to the greatest possible extent the historical decontextualizations discussed in Section I. It should be clear that one cannot possibly avoid all the potential traps found in the didactic transposition process of the history of science, given that, as already stated, didactization is inevitable. Didactic transposition both reveals the complexity of this process and provides analytical-theoretical tools. The need for constant epistemological vigilance concerning what is taught in schools, what is considered in designing educational tools, and what is accepted as the origins of scholarly knowledge, signals the inherent risks in the process of didactic transposition. On the one hand, epistemological vigilance seeks to answer the question of whether what one is teaching is in line with what was chosen as the object of teaching. On the other hand, it highlights the differences between these two bodies of knowledge and enables one to search for changes in and adaptations of the knowledge to be taught in order to meet the educational requirements of society. This becomes even more relevant when we consider that knowledge to be taught comes alive in the hands of teachers, i.e., when it becomes knowledge taught. It is particularly important to bear in mind that situations involving curricular innovations present teachers with many new tasks (Pintó, Couso, & Guttierrez, 2005). This is often the case when one seeks to introduce the HS in school environments, and, in this respect, further questions arise: How can concepts shaped in the past best be presented? How can activities that reveal the internal dynamics of science be adapted to classroom situations? How should the heuristic role of experimentation in the context of a supplanted theory be adjusted to the level of understanding of a given set of students? What words should be used? What analogies should be made? These are critical details, according to Pintó (2005), which may well cause deviations from the initial purposes stated in education goals when it comes to actual teaching practices. At the same time, teachers – as well as all those dedicated to the production of teaching and learning situations – inevitably make adjustments to suit their own contexts. The attendant questions are whether such adaptations are conscious and what their consequences will be.

Chevallard (1991, p. 18) emphasizes that "the knowledge produced by didactic transposition will thereafter be knowledge exiled from its origins and separated from its historical production in the sphere of scholarly knowledge, legitimizing itself as knowledge taught, as something that is not from any time or any place". Thus, if didactization of school knowledge is inevitable, the theory of didactic transposition invites us to understand these processes and organize them under constant scrutiny, overcoming appearances and questioning evidence. For Chevallard, epistemological vigilance enables the scientific analysis of school knowledge. It is the idea of didactic transposition that "allows the articulation of epistemological analysis with teaching analysis, and thus becomes a guide for the good use of epistemology for teaching" (Chevallard, 1991, p. 23). To some extent, epistemological vigilance reveals what is hidden and avoids the illusion of transparency between academic science and that which becomes the object of teaching. The inevitable condition of didactic transposition provides a means of reconciliation between teachers and historians of science, as long as it implements a permanent epistemological vigilance. And yet – how might a teacher undertake such vigilance without having access to scholarly knowledge?

We propose avoiding the placement of the historiographical and educational fields, as well as the commitments established within each, in opposing positions. That is, if the conflicts resulting from didactization and, consequently, didactic transposition, are indeed inevitable, they must be understood as part of the processes involved in teaching and learning. The only possible solution is to use epistemological vigilance by focusing on risks, avoiding pitfalls, and seeking solutions, with primary commitments to maximizing educational gains throughout. In the case of using the history of science as a science-teaching tool, certain risks, traps, and gains may be likely to emerge. For the remainder of this chapter, we will take a particular HS topic and discuss how science teachers in training can be prepared to efficiently use epistemological vigilance.

5 Epistemological Vigilance and Newton's Theory of Colors in Science Teacher Education

Teachers' beliefs and educational practices are initially formed in pre-service education courses and develop further throughout their experiences (both personal and shared with colleagues) over the course of their training. Success in dealing with the complexity of the school context is the result of participating in the challenging process of transforming theoretical knowledge into practical knowledge. However, such practical knowledge can generate genuine obstacles to curricular changes. In addition, teachers' considerable reliance on textbooks and, in some cases, their lack of sufficient analytical perspective on the relevance of content and social-educational design, provide further obstacles. These factors entail the need to offer teachers complementary theoretical and methodological tools, often after their initial training, in order to enable them to use epistemological vigilance regarding both what they are teaching and the learning materials they are using.

However, offering just the theoretical aspects underlying epistemological vigilance does not seem adequate. We need to help future teachers pursue critical analyses of the materials they will use in their profession, and of textbooks in particular. Thus, we developed and implemented a practical activity for teachers in training[7] which was aimed at examining the didactization route of specific physics content in textbooks, with historical articles or extracts functioning as empirical references for the analysis of the transformations and simplifications that occurred in the subjects to be taught. After an initial discussion about the didactic transposition of the chosen subjects, our pre-service teachers received the textbooks to be analyzed. The activity was carried out in groups. The following analytical categories, extracted from Chevallard's approach, were offered to guide their work:

- Time reduction:
- Associated with what Chevallard calls the programmability of the acquisition of knowledge, which defines the progress of didactic relations by the construction of rational sequences, which enable the acquisition and control of learning.
- Dogmatization – maintenance of products and exclusion of processes:
- Emphasizes the products of science and, at the same time, the exclusion of the processes of knowledge production. Scientific controversies and debates disappear, and, in general, a linear and empiricist conception of science is promoted.
- Depersonalization – loss of the contexts of discovery and individual actions:
- Separates knowledge and the subjects of its production. Leads also to the disappearance of the collective dimensions of knowledge production, in some cases keeping historical figures isolated.
- Desyncretization – loss of the epistemological context to which knowledge is bound:
- Produces a reorganization of the knowledge to be taught, including chronological inversions, allowing for the definition and division of fields of structured knowledge, for example, in sections and chapters of a textbook.

- Decontextualization – loss of historicity:
- Entails the loss of the historical context of the production of knowledge. According to Chevallard, it is the "disarticulation of network issues and problems disarticulation of knowledge from the network issues and problems that would be presented in a full account (1991, p. 71). That is, the knowledge to be taught will be extracted from its context of historical production.
- Inevitables Risks:
- The risks are the result of the processes described above, which should be didactically treated. The student teachers were also informed that these processes result in certain risks, which should be didactically treated.The main risks include:
- Temporal impossibility/inadequacy:
- Involves problems with temporal reduction and the programmability of knowledge, which can conflict with the time and/or classes allotted for working with a particular knowledge to be taught. This bears the risk of a gap between the objects of teaching and the expected educational project. Recognizing the changes incurred in knowledge to be taught should not just consist of analyzing its distance from scholarly knowledge, but also the educational implications of the choices made when compared to the initial educational design.
- Mischaracterization of school knowledge:
- Relates to the adequacy of the initially proposed educational project, especially as regards the relationship between the material to be taught and the material which is taught. For example, a teaching material that could take the form of a scientific concept might instead be turned into the mere application of equations to solve exercises: say, the material to be taught, electrical resistance of materials, could become, when actually taught, simply about the association of resistors.

In this case, it was decided to use contents related to the nature of light, a theme found in all high school physics textbooks. Furthermore, this subject often reinforces empiricist and inductive views of science. Newton's color theory appears free of controversy in most textbooks and appears to have emerged as a direct result of a few experiments, thereby giving students a false idea of the simplicity of scientific work. In some cases, as highlighted by Silva and Martins (2003), Newton's theory of color is treated as an example of the application of scientific method.

As a historical reference, the article "Isaac Newton's 'New Theory about Light and Colors': A Translation with Commentary" (Silva & Martins, 1996) was used. The authors point out that this work, having been published in 1672 in the *Philosophical Transactions of the Royal Society,* was the first in which Newton presented his conceptions of the nature of white light and colors. Acceptance of Newton's hypotheses was mixed, and they generated intense discussions among his contemporaries, including familiar names such as Hooke and Huygens. Newton's explanation of the formation of the color spectrum after white light passes through a prism was based on the assumption that light is a heterogeneous mixture formed by rays of all colors. However, the authors contend that this hypothesis could not have been proven by his prism experiment alone, since other interpretations could also satisfactorily explain the phenomena. In fact, some of their criticisms were directed at Newton's experiment itself. Thus, this chapter verifies the complex nature of the development and acceptance of scientific hypotheses and theories. It also presents other issues that deserve attention, such as the *experimentum crucis* and the globular nature of light, among others. Newton's 'personal style' of writing, and his expectations are also highlighted. At the start of his description of his prism experiment, Newton says, "it was initially a very enjoyable diversion" (Silva & Martins, 1996, p. 315). He then stresses his surprise at the problem that drew his attention toward a more careful investigation of the phenomenon: "but after a while dedicated to considering them more seriously, I was surprised to see them in an oblong shape that, according to the accepted laws of Refraction, was expected to have been circular" (Silva & Martins, 1996, p. 315). Before presenting his hypothesis, Newton reports some of his suspicions and goes on to describe new experiments attempted in order to refine his explanations. In another study, Silva and Martins (2003) emphasize Newton's intellectual efforts toward furthering acceptance of his theory, including his use of epistemological arguments and theoretical definitions. For the authors, even the *experimentum crucis* was not decisive. Other experiments and further discussions were necessary to bolster Newton's hypothesis, since a major point of controversy was the composition of white light; in this matter, Newton's explanation opposed, for example, the view taken by Robert Hooke, for whom white light was of more simple composition. Having read the chosen text, the student teachers (referred to as A1, A2, and so on for the purposes of our study) were then asked to consider the guiding steps listed above as they examined several standard textbooks

Regarding the reduction of time, the students observed in their analyses that:

> The textbook begins the study of light in Greece with Plato, who tried to answer the following question: why do we see an object? Soon after, the book "lands" in our time. (A7)
>
> The textbook presents condensed concepts, with the explicit purpose of solving the exercises that are proposed in the material. (A11)

> The processes of questioning and comparisons between the various theories that emerged before those which are accepted today loses ground to the exercises. (A5)

Textbooks considerably reduce the dimension of time between various previous scientific theories. They prioritize an operational approach to science subjects by conveying key information in the form of exercises, giving the impression that the very same research questions occupied the minds of philosophers from Plato to Newton, as Student A7 seems to suggest.

With regard to the dogmatization of science, students stated that:

> No book describes the experiences that led Newton to formulate his theory of light and color and much less the reasoning sequences used to arrive at it. As a rule, only a sketch of the prism separating the components of white light is presented. (A1)
>
> The book [in analyzis]treats the subject in a totally dogmatic form, without presenting the process that created the theory of light dispersion. Desyncretization is also clear, from the moment that the authors present the prepared concepts without showing what problems led to his study. (A6)

Coupled with the operational form in which knowledge is presented in teaching, that is, one which favors solving problems without exploring more conceptual aspects, science takes on a dogmatic form in textbooks, as the analyses of A1 and A6 indicate. This leads to misunderstandings about science as a human and historical endeavor.

The process of depersonalization identified by Chevallard also contributes to these misunderstandings:

> Newton's text is very personal, written in first-person. The textbook says … informs us, that the separation of colors is like that, as if the truth was beyond question or experiment. (A3)
>
> The experiment, history, date, character, all of that is left out and instead there is a brief theoretical description about refraction and dispersion, and then, right away, it begins working on the problems and exercises. (A9)
>
> In the approach to content, a clear loss of individual action, of discovery, is evident. However, there is a simplified explanation of how Newton arrived at his results in the appendix. (A12)

To meet the requirements of the school context, knowledge to be taught is generally presented in an order that is grounded in pre-requisites and in forms that omit its related knowledge construction processes as well as the dynamics of the actors in that process, resulting in knowledge to be taught which is depersonalized. Personal investments, conflicts, and previous contributions disappear to make way for a theoretical model that seems to be ready made. The students identified these characteristics in the textbooks examined, as may be observed in their comments. According to A12, one of the textbooks tries to rescue some historical aspects in a supplementary text. Dogmatic qualities are revealed as well in A3's analysis. And, again, the operational dimension of the content, in the form of practical exercises, is verified by A9.

Desyncretization was also identified in the textbooks analyzed:

> This high school text, with its brief historical contextualization, gives students the impression that the contents of the chapter developed linearly and rapidly over time, resulting in a final product: the explanation of colors. (A10)

Desyncretization indicates the necessary delimitation of knowledge areas. It consists of separating areas of theoretical knowledge into specific learning areas, dividing theories into school subjects, chapters, and sections, in order to conform to the teaching project. It entails a reorganization of knowledge, separating it from its chronological sequence, so that it can be presented in a new structure. Concepts that are part of the body of a theory are often treated as separate and independent. A10 draws attention to the linear form assumed by/forced upon knowledge to be taught. This excess of systematization can convince teachers that the sequences of teaching contents presented in textbooks represent the one and only course of events, when, according to Chevallard, such texts only reflect a local variant of contextualized knowledge.

Similarly, students found that decontextualization was also present in the textbooks examined:

> Another point[in the historical reference article regarding Newton's work] refers to some problems that occurred during the [original Newton's] experiment, for instance, seeing oblong images, when they were expected to be circular. In the textbook, the experiment appears perfect, problem-free. (A2)
>
> The textbook provides false information about Newton's studies, saying that Newton discovered that a beam of sunlight, when passing through a glass prism, breaks up into a colored beam. But Newton himself

> stated: "I procured a triangular glass prism to try therewith the celebrated phenomena of colors". (A6)

The epistemological exile described by Chevallard is evident. The original problems disappear and the final products are presented as if they answered an initial question, in an apparently smooth and linear progression of ideas and hypotheses. When knowledge to be taught is separated from the research problems and conceptual network in which it originated, it loses much of its meaning. Or, it has meaning only in idealized situations with purely didactic purposes, as if there are two different sciences: science for school and real science

These transposition processes occur simultaneously and their separation into categories serves only to more clearly illustrate each one and to facilitate analysis by future teachers.

Characteristics such as these make this historical episode of Newton's work with color a good subject for the use of epistemological vigilance, as the changes taking place in school knowledge in a given set of textbooks are easy to identify. Newton's initial hypothesis, original research problems, and personal discourse, as well as the related controversies, disappear, with color theory taking on, in the textbooks, an impersonal, unquestionable, and conclusive form which is amenable to set experiments. The processes of depersonalization, desyncretization, decontextualization and others described by Chevallard, along with a dogmatic approach to science, are clearly identified in the pre-service teachers' analyses. The epistemological exile of school knowledge, as discussed by Chevallard, is unmistakable.

6 Conclusions

Our discussion of the commitments to be honored when using the history of science in science teaching makes it clear that it is not possible to produce learning contents that meet all historiographical and didactic-pedagogical determinants in their entirety. The didactization process inevitably involves knowledge being transformed and taken out of the context of its original production (whether by the community of scientists or by historians of science). There is an aggravating factor in the use of HS when facing this dilemma between original and school environments: the lack of preparation for dealing with the transformation of knowledge that characterizes most pre-service teaching courses at the undergraduate level. It is unrealistic to expect school science teachers to be, in addition, historians or philosophers of science in order to make good use of the HS within school environments. We believe that one solution would be to select, by means of research, those HS-related educational contents which face the highest risk in the didactization process, 'highest risk' being understood here as showing the greatest potential for 'scientific miseducation'; in other words, by identifying contents that compromise the scientific education desired by the current educational project. Once such contents are found, epistemological vigilance, as grounded in historical analysis, would be able to reveal hidden and/or corrupted aspects of the process of didactic transposition, as demonstrated via our research using Newton's studies on the nature of light as a historical episode. Incorporating such analysis into both initial and continuing science teacher education would prepare teachers to be less naïve and more vigilant with regard to the learning contents presented to them in school curricula. The strength of our proposal rests, therefore, on educating teachers who are able, given sufficient autonomy, to implement epistemological vigilance to the benefit of their educational practices and, above all, their students.

Acknowledgements

Thaís Forato thanks FAPESP for its support. Maurício Pietrocola thanks FAPESP and CNPq. The authors would like to thank Charbel Niño El-Hani for his careful revisions and suggestions for the first draft of this chapter.

Notes

1 Parts of these theoretical considerations were previously presented at conferences (e.g., Forato et al., 2010).
2 The term "Whig" was used by historians in the early twentieth century to characterize the heroification of English institutions. It became popular among historians of science when Butterfield used it in his book *The Whig Interpretation of History* (1931), to refer to a kind of anachronistic and selective factual history, glorifying the genius of some characters and fields of study. Jardine (2003) discusses Herbert Butterfield's controversial proposal for the use of whiggism in the context of the history of science.
3 For an example of hagiography, see Adúriz-Bravo and Izquierdo-Aymerich (2009). The authors discuss the use of a well-known episode from the history of science involving the 'discovery' of radium by the Curies in a pre-service science teacher

course. Besides investigating ideas about the nature of science (around methodology, theory change, scientific inference and explanation, values, gender issues), they debate the extended hagiographic treatment of Madame Curie in the 1997 French commercial film, 'Les Palmes de Monsieur Schutz'.

4 Stephen Brush (1974) presents this debate among physicists, historians of science, and physics educators in the early 1970s as one with two sides: those who defended anachronistic and hagiographic versions of the history of science, aiming at reinforcing rational and impartial experimental control and validation of hypotheses – and seeking acceptance of the prevailing paradigm; and others who advocated for the idea of presenting science within its socio-historical context of production. In the latter case, the scientist would be framed as immersed in and influenced by the values of his or her time.

5 Kuhn (1970, pp. 173–181) recognizes the pedagogical nature of the distortions present in the history of science in scientific manuals (which aimed at perpetuating normal science), but warned that such historical narratives do not take into account the relationship between scientific research and its historical context. He emphasizes that historical errors make experts and laypeople alike delude themselves about the nature of the scientific enterprise (Kuhn, 1997, p. 180). See details in Phorate (2009, pp. 13–23).

6 It is important to note that these terms, "desyncretization", "depersonalization", and "decontextualization", were originally in French, and could not be translated directly into English. But we concluded that it was better to keep the English words as close to the French as possible to maintain their original meanings.

7 A full description of this activity is available in Ricardo and Pietrocola (2010).

References

AAAS-American Association for the Advancement of Science. (1990). *Science for all Americans*. New York, NY: Oxford University Press.

AAAS-American Association for the Advancement of Science. (1993). *Benchmarks for science literacy: A Project 2061 report*. New York, NY: Oxford University Press.

Abd-El-Khalick, F., & Lederman, N. (2000). The influence of history of science courses on students' views of nature of science. *Journal of Research in Science Teaching, 37*(10), 1057–1095.

Aikenhead, G., & Ryan, A. (1992). The development of a new instrument: Views on Science-Technology-Society (VOSTS). *Science Education, 76*(5), 477–491.

Allchin, D. (2004). Pseudohistory and pseudoscience. *Science & Education, 13*, 179–195.

Allchin, D. (2006). Why respect for history – and historical error – matters. *Science & Education, 15*(1), 91–111.

Aduríz-Bravo, A., & Izquierdo-Aymerich, M. (2009). A research-informed instructional unit to teach the nature of science to pre-service science teachers. *Science & Education, 18*, 1177–1192.

Astolfi, J.-P., Darot, E., Ginsburger-Vogel, Y., & Toussaint, J. (1997). *Mots-clés de la didactique des sciences: Repères, definitions, bibliographies*. Paris: De Boeck & Larcier.

Astolfi, J.-P., & Develay, M. (1995). *A didática das ciências*. Campinas, Brazil: Papirus. (La didactique des sciences. Paris: Presses Universitaires de France, 1989)

Bell, R., Abd-El-Khalick, F., Lederman, N. G., McComas, W. F., & Matthews, M. R. (2001). The nature of science and science education: A bibliography. *Science & Education, 10*(1–2), 187–204.

Bosch, M., & Gascón, J. (2006). Twenty-five years of didactic transposition. *ICMI Bulletin, 58*, 51–64.

Brazil. (2002a). *Parâmetros curriculares nacionais: ensino médio*. Brasília, Brazil: MEC/SEMTEC.

Brazil. (2002b). *Orientações educacionais complementares aos parâmetros curriculares nacionais: ensino médio. Ciências da natureza, matemática e suas tecnologias*. Brasília, Brazil: MEC/SEMTEC.

Brockington, G., & Pietrocola, M. (2005). Serão as regras da transposição didática aplicáveis aos conceitos de física moderna? *Investigações em Ensino de Ciências* (Online), *10*(3), 1–17.

Brush, S. G. (1974). Should the history of science be rated X? *Science, 183*, 1164–1172.

Brush, S. G. (1979). Comments on "on the distortion of the history of science in science education". *Science Education, 63*, 277–278.

Canguilhem, G. (1977). *Ideologia e Racionalidade nas Ciências da Vida* (E. Piedade, Trans.). Lisboa, Portugal: Edições 70.

Chevallard, Y. (1991[1985]). *La transposition didactique du savoir savant au savoir enseigné* (2nd ed.). Grenoble, France: La Pensée Sauvage.

Chevallard, Y. (1994). Les processus de transposition didactique et leur theorization. In G. Arsac, Y. Chevallard, J. L. Martinand, & A. Tiberghien (Eds.), *La transposition didactique à l'épreuve*. Grenoble, France: La Pensée Sauvage.

Chevallard, Y., & Joshua, M.-A. (1985). Un exemple d'analyse de la transposition didactique: La notion de distance (1982). In Y. Chevallard (Ed.), *La transposition didactique du savoir savant au savoir enseigné*. Grenoble, France: La Pensée Sauvage.

Durey, A., & Martinand, J.-L. (1994). Un analyseur pour la transposition didactique entre pratiques de référence et activités scolaires. In G. Arsac, Y. Chevallard, J.-L. Martinand, & A. Tiberghien (Eds.), *La Transposition Didactique à l'Épreuve*. Grenoble, France: La Pensée Sauvage.

El-Hani, C. N. (2006). Notas sobre o ensino de história e filosofia da ciência na educação científica de nível superior. In C. C. Silva (Ed.), *Estudos de história e filosofia das ciências: subsídios para aplicação no ensino* (pp. 3–21). São Paulo, Brazil: Ed. Livraria da Física.

Ferreira, M. S., Gomes, M. M., & Lopes, A. C. (2001). Trajetória histórica da disciplina escolar ciências no colégio de aplicação da UFRJ (1949–1968). *Pro-posições, 12*(1), 9–26.

Forato, T. C. M. (2008). A filosofia mística e a doutrina Newtoniana: uma discussão historiográfica. [Mythical philosophy and Newtonian doctrine: an historiographycal discussion]. *ALEXANDRIA Revista de Educação em Ciência e Tecnologia, 1*(3), 29–53.

Forato, T. C. M. (2009). *A natureza da ciência como saber escolar: Um estudo de caso a partir da história da luz* [The nature of science as school knowledge: a case study from the history of light], 2 vols. (Doctoral dissertation). FEUSP, São Paulo, Brazil.

Forato, T. C. M., Martins, R. de A., & Pietrocola, M. (2010). Alguns debates históricos sobre a natureza da luz: Discutindo a natureza da ciência no ensino [Some historical debates about the nature of light: discussing the nature of science in science teaching]. In R. de A. Martins, L. Lewowics, J. H. Ferreira, C. C. Silva, & L. A. P. Martins (Eds.), *Filosofia e história da ciência no Cone Sul Seleção de trabalhos do 6º Encontro* (1st ed.; pp. 616–626). Campinas, Brazil: Associação de Filosofia e História da Ciência do Cone Sul.

Forato, T. C. M., Martins, R. de A., & Pietrocola, M. (2012). History and nature of science in high school: Building up parameters to guide educational materials and strategies. *Science & Education, 21*(5), 657–682.

Gil-Perez, D., Montoro, I. F., Alis, J. C., Cachapuz, A., & Praia, J. (2001). Para uma imagem não deformada do trabalho científico. *Ciência & Educação, 7*(2), 125–153.

Golinsky, J. (1998). *Making natural knowledge: Constructivism and the history of science.* Cambridge: Cambridge University Press.

Gravoglu, K., Patiniotis, M., Papaneloupoulou, F., Simões, A., Carneiro, A., Diogo, M., Sánchez, J., Belmar, A., & Nieto-Galan, A. (2008). Science and technology in the European periphery: Some historiographical reflections. *History of Science, 46*, 153–175.

Halliday, M. A. K., & Martin, J. R. (1993). *Writing science: Literacy and discursive power.* Pittsburgh, PA: University of Pittsburgh Press.

Holton, G. (2003). What historians of science and science educators can do for one another. *Science Education, 12*(7), 603–616.

Jardine, N. (2003). Whigs and stories: Herbert Butterfield and the historiography of science. *History of Science* [part 2], *41*(132), 125–140.

Johsua, S., & Dupin, J.-J. (1993). *Introduction à la didactique des sciences et des mathématiques.* Paris: Presses Universitaires de France.

Klein, M. (1972). The use and abuse of historical teaching in physics. In S. G. Brush & A. L. King (Eds.), *History in the teaching of physics* (pp. 12-27). Hanover NH: The University Press of New England.

Koyré, A. (1982). *Estudos da história do pensamento científico* (M. Ramalho, Trans.). Rio de Janeiro, Brazil: Ed. Forense Universitária.

Kragh, H. (1987). *An introduction to the historiography of science.* Cambridge: Cambridge University Press.

Kuhn, T. S. (1970). *The structure of scientific revolutions* (2nd ed.). Chicago, IL: University of Chicago Press.

Kuhn, T. S. (1977). *The essential tension: Selected studies in scientific tradition and change*. Chicago, IL & London: University of Chicago Press.

Lederman, N. G. (2007). Nature of science: Past, present, and future. In S. K. Abell & N. G. Lederman (Eds.), *Handbook of research on science education* (pp. 831–880). Mahwah, NJ: Lawrence Erlbaum Associates.

Martinand, J.-L. (1986). *Connaître et transformer la matière: des objectifs pour l'initiation aux sciences et techniques*. Bern, Switzerland: Peter Lang.

Martinand, J.-L. (2003). La question de la référence en didactique du curriculum. *Investigações em Ensino de Ciências, 8*(2), 125–130.

Martins, R. de A. (2001). Como não escrever sobre história da física: Um manifesto historiográfico [How one should not write about history of physics: A historiographical manifesto]. *Revista Brasileira de Ensino de Física, 23*(1), 113–129.

Martins, R. de A. (2004). Ciência versus historiografia: os diferentes níveis discursivos nas obras sobre história da ciência [Science versus historiography: the several levels of discourse in history of science writings]. In A. M. Alfonso-Goldfarb, M. H. R. Beltran, & A. G. Debus (Eds.), *Escrevendo a história da ciência: Tendências, propostas e discussões historiográficas* (pp. 115–145). São Paulo, Brazil: Educ; Fapesp; Ed. Livraria da Física.

Martins, R. de A. (2006). Introdução: a história da ciência e seus usos na educação [Introduction: history of science and its uses in education]. In C. C. Silva (Ed.), *Estudos de história e filosofia das ciências: Subsídios para aplicação no ensino* (pp. 3–21). São Paulo, Brazil: Ed. Livraria da Física.

Martins, R. de A., & Silva, C. C. (2001). Newton and colour: The complex interplay of theory and experiment. *Science & Education, 10*(3), 287–305.

Matthews, M. R. (1989). A role for history and philosophy in science teaching. *Interchange, 20*(2), 3–15.

Matthews, M. R. (1992). History, philosophy and science education: The present rapproachment. *Science & Education, 1*(1), 11–47.

McComas, W., Almazroa, H., & Clough, M. P. (1998). The nature of science in science education: An introduction. *Science & Education, 7*, 511–532.

NRC-National Research Council. (1996). *National science education standards*. Washington, DC: National Academic Press.

NSTA-National Science Teachers Association. (1962). The NSTA position on curriculum development in science. *The Science Teacher, 29*(9), 32–37.

NSTA-National Science Teachers Association. (1982). *Science-technology-society: Science education for the 1980s* (An NSTA position statement). Washington, DC: National Science Teachers Association.

Pagliarini, C. R. (2007). *Uma análise da história e filosofia da ciência presente em livros didáticos de física para o ensino médio* (Master's thesis). Instituto de Física da Universidade de São Paulo/São Carlos.

Pietrocola, M. (2003). A história e a epistemologia no ensino de ciências: Dos processos aos modelos de realidade na educação científica (pp. 133–149). In A. M. R. Andrade (Ed.), *Ciência em perspectiva: estudos, ensaios e debates*. Rio de Janeiro, Brazil: MAST/SBHC.

Pietrocola, M. (2008). A transposição da física moderna e contemporânea para o ensino médio: Superando obstáculos epistemológicos e didático-pedagógicos. In R. Borges (Ed.), *Propostas interativas na educação científica e tecnológica*. Porto Alegre, Brazil: EDUC.

Peduzzi, L. (2001). Sobre a utilização didática da história da ciência. In M. Pietrocola (Ed.), *Ensino de física: Conteúdo, metodologia e epistemologia numa concepção integradora*. Florianópolis, Brazil: Editora UFSC.

Pintó, R. (2005). Introducing curriculum innovations in science: Identifying teachers' transformations and the design of related teacher education. *International Journal of Science Education, 89*, 1–12.

Pintó, R., Couso, D., & Gutierrez, R. (2005). Using research on teachers' transformations of innovations to inform teacher education: The case of energy degradation. *International Journal of Science Education, 89*, 38–55.

Pumfrey, S. (1991). History of science in the National Science Curriculum: a critical review of resources and their aims. *British Journal of History of Science, 24*, 61–78.

Ricardo, E. C. (2005). *Competências, interdisciplinaridade e contextualização: Dos parâmetros curriculares nacionais a uma compreensão para o ensino de ciências* (Doctoral dissertation). UFSC, Florianópolis.

Ricardo, E. C., & Pietrocola, M. (2010). *Epistemological vigilance and didactic textbooks: Demonstrating the didactic transposition process of physics knowledge*. III International Conference on the Anthropological Theory of the Didactic. Girona – Barcelona, Spain.

Rossi, P. (2000). *Naufrágios sem espectador: A idéia de progresso* (A. Lorencini, Trans.). São Paulo, Brazil: Unesp.

Silva, C. C., & Martins, R. de A. (2003). A teoria das cores de Newton: um exemplo do uso da história da ciência em sala de aula [Newton's theory of colors: an instance of history of science use in the classroom]. *Ciência & Educação, 9*(1), 53–65.

Silva, C. C., & Martins, R. de A. (1996). A "Nova teoria sobre luz e cores" de Isaac Newton: Uma tradução comentada [Isaac Newton's "New theory of light and colors": A translation with commentary]. *Revista Brasileira de Ensino de Física, 18*(4), 313–327.

Whitaker, M. A. B. (1979). History and quasi-history in physics education – Part 1. *Physics Education, 14*, 108–112.

CHAPTER 17

Contributions to the Nature of Science: Scientific Investigation as Inquiry, Modeling, and Argumentation

Agustín Adúriz-Bravo

Abstract

In this chapter, we will look into some of the diverse contributions that the history and philosophy of science can make to science education. Focus is on how these meta-sciences have been used to construct an educationally valuable image of what is usually called the 'nature of science', that is, an epistemological characterization of science as a product (scientific knowledge) and as a process (scientific activity). We deal with a didactical characterization of the nature of the scientific enterprise that conceptualizes science-in-the-making from a perspective that resorts to the idea of evidence. From this point of view, we claim that science can be seen as the collection, transformation and use of evidence to model the natural world. We direct the inquiry-based characterization of science towards the education of pre- and in-service science teachers. One of the main aims of the chapter is to provide a comprehensive picture of recent Latin American scholarship devoted to this issue within the field of didactics of science (i.e. science education as an academic discipline).

Keywords

nature of science – inquiry – modelling – argumentation – Latin-American scholarship

1 Purposes and Emphases of This Chapter

In this chapter, I look into contributions from the history and philosophy of science to science education. I focus on how these disciplines have been used to construct an educationally valuable image of what is called the 'nature of science', that is, an epistemological characterization of scientific knowledge

 | DOI: 10.1163/9789004409088_017

and activity. The aim of this chapter is to provide an overview of recent Latin American scholarship devoted to this topic within the field of didactics of science (i.e., science education as an academic discipline).

The history and philosophy of science are sometimes labeled *metasciences*, since they are 'second-order' disciplines that take science as their object of analysis, and study it with different intentions and perspectives. This 'meta-science' label is also taken to include – among others – the sociology, psychology, and linguistics of science. As Kuznetsov and Balzer put it (2011),

> [i]t is a truism that many sciences, often called metasciences, investigate science. Philosophy of science [...] is only one of these. Others are history of science, economics of science, ethics of science, axiology of science, psychology of science, sociology of science, organization and management of science, ecology of science, etc. Each of the metasciences studies a specific aspect or dimension of science, i.e. has a specific domain. This allows us to distinguish between a all these metasciences.

The metasciences have been inspected by the didactics of science to various degrees, in order to identify their different contributions to science education. The use of the history and philosophy of science is now a long-standing tradition in our scholarly community (Matthews, 1994); these disciplines have also been extensively employed to shape our understanding of the nature of science (McComas, 1998). In contrast, the other metasciences have only recently begun to be incorporated into our field.

The second-order, 'meta-', perspective is also shared by other academic endeavours that are less 'disciplined', such as science studies, science-technology-society (STS), feminist epistemology, public understanding of science, complexity science, etc. These theorizations have also, to a greater or lesser extent, incited the interest of didacticians of science (i.e., science education researchers).

One objective of my research activity is to examine how meta-theoretical fields contribute to the didactic portrayal of the nature of science; most of my work is concentrated on the contributions of the *philosophy of science* (see references). In this chapter, I present an educational conceptualization of science focusing on scientific activity (i.e. 'science as a process'), and using a theoretical perspective that is grounded in the idea of *evidence*. From this point of view, I propose that the activity of scientific investigation – at least at schools – may be seen as the collection, structuring, and use of evidence to understand the natural world. I argue that these three processes can be identified, respectively, with scientific *inquiry*, *modeling*, and *argumentation* (Figure 17.1).

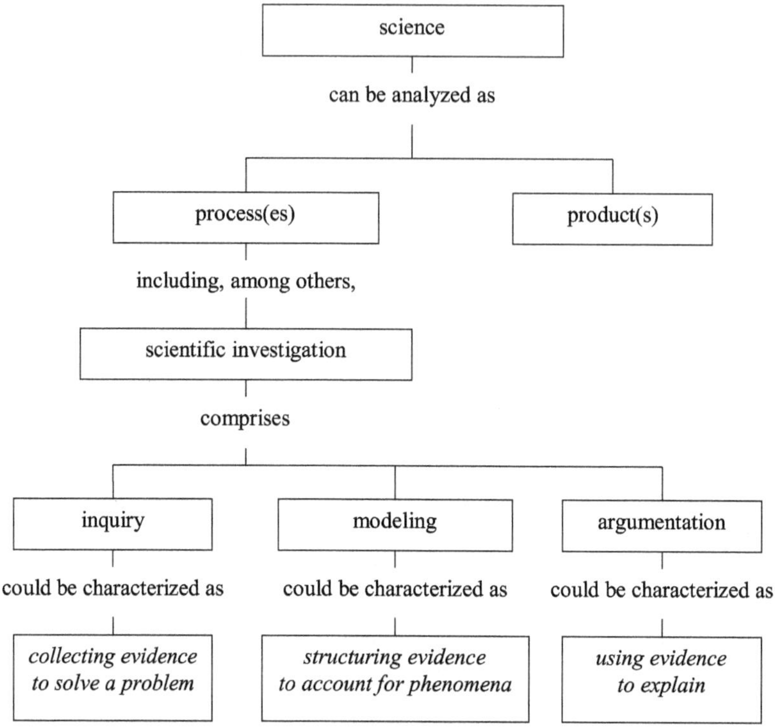

FIGURE 17.1 A view on the nature of science as a process, with inquiry, modeling, and argumentation as key components of school scientific investigation

For the purposes of this chapter, I must first clarify certain terms. It may be argued that, in the world of science education, the two widely used expressions, *history and philosophy of science* (sometimes abbreviated under the acronym HPS; see Matthews, 1994) and *nature of science* (usually referred to by the acronym NOS; see McComas, 1998) each bear two different, intimately connected, but still conceptually separable meanings. In the most general sense, the referents of these constructs could both be considered, borrowing the expression coined by Flick and Lederman (2004a, p. ix), as "major and interrelated foci of the reforms in science education"; these foci admit interpretations from the points of view of *teaching* (when we look at 'reformed' science education practices) and of *research* (when we look at the academic input supporting the proposal and implementation of reforms).

From the viewpoint of teaching, the HPS and NOS constitute new content to be taught as well as a new pedagogy (i.e., a new method of teaching it). They may be primarily identified as "a major element of the science curriculum" (Lederman & Abd-el-Khalick, 1998, p. 83), intended to provide "an accurate description of the function, processes and limits of science" (McComas et al.,

1998, p. 6). From the viewpoint of research, HPS and NOS can be conceptualized as intellectual tools used to organize the results of investigations and innovations in the didactics of science (Adúriz-Bravo, 2008a). In this context, HPS constitutes a broad research *area* dealing with "the interplay of disciplines that inform [didactics of science] about the character of science itself" (McComas et al., 1998, p. 4). Thus, HPS would comprise studies on the epistemological status of our discipline, the philosophical foundations of science curricula, the components of science teachers' professional knowledge, the 'family resemblance' between students' naïve conceptions and constructs from the history of science, the production of historical narratives for teaching, the role of scientific and social values in decision-making, and a host of others.

Meanwhile, the NOS represents a specific research *line* within HPS, which deals with producing 'didactic transpositions' (i.e., educationally valid versions) of metascientific ideas that can be taught and learned in classrooms. The NOS is identified with

> the intersection of issues addressed by the philosophy, history, sociology, and psychology of science as they apply to and potentially impact science teaching and learning. As such, the nature of science is a fundamental domain for guiding science educators in accurately portraying science (…). (McComas et al., 1998, p. 5)

In this chapter, I will refer to these two constructs/meanings of HPS and NOS; I will indicate which is more appropriate in each occurrence.

If we look at the HPS and NOS from the standpoint of the first, teaching-related meaning, the last 15 years of science curricula in Latin America have increasingly been incorporating innovative components carrying a metascientific perspective: along with the usual requirement of knowing science, curricula are now adding a new goal, that of knowing *about* science. Incorporating the question of 'how we know what we know' (Duschl, 2008; Osborne, 2010) into curricula has in turn affected teaching, textbooks and materials, assessment, and teacher education. It could be argued that these changes have only been comprehensively instituted in a certain number of the countries in our region (notably Argentina, Brazil, Colombia, and Mexico), as well as, with more or less delay, in the US, Canada, and many countries in Europe.

If we approach the HPS and NOS via the second, research-based meaning, the Latin American didactics of science community has been increasingly directing its attention to these issues. The interest in HPS and NOS in our region was apparent in the successful *1st Latin American Conference of the International History, Philosophy, and Science Teaching Group* (*IHPST-LA*), which was held in

Maresias, Brazil, in August of 2010. At that conference, around 120 scholars from several countries in the region presented work related – in various ways – to the contributions of the metasciences to the theory and practice of science education. The second such conference, held in Mendoza, Argentina in October of 2012, confirmed the existence of an active and thriving Latin American community around these issues.

My own research group – the *GEHyD, Grupo de Epistemología, Historia y Didáctica de las Ciencias Naturales* (Philosophy and History of Science and Science Education Group) – has been exploring these issues for the past ten years, with a special interest in the pre- and in-service education of science teachers for all educational levels, kindergarten to university. One of my main concerns as a didactician of science is with finding ways to acquaint teachers with theoretical frameworks that propose particular characterizations of the nature of science as adjusted to the objectives of current science curricula.

One such characterization – which has emerged as fruitful in our group investigations and analyses (Adúriz-Bravo, 2001, 2002, 2004, 2005a, 2008b, 2011b, in press) – is briefly discussed in this chapter. This characterization adds three more restricted foci, all of which are related in various degrees, to 'mainstream' HPS/NOS research. The first focus is *scientific models and modeling*. Although models have generally been addressed from psychological and cognitive perspectives in the didactics of science, there is some emergent literature analyzing the *epistemic* nature of scientific models, and drawing implications for science education (Erduran & Duschl, 2004; Sensevy et al., 2008; Adúriz-Bravo, in press). The second focus is what I call *school scientific argumentation* (i.e. the use of arguments in school science). Here, the connections with HPS/NOS research have been scarce (apart from the use of Stephen Toulmin's argumentation pattern); nevertheless, some authors have recently begun to inspect epistemic criteria for scientific debates in the classroom, or the role of scientific models in constructing and using arguments (cf. Duschl, 2008; Böttcher & Meisert, 2011). The third and last focus is *inquiry-based science teaching*. In this case, the relationships with HPS/NOS research have been ambiguous, ranging from the lack of meta-theoretical input in the literature to 'parallel', and rather independent, characterizations of the nature of science and the so-called 'nature of scientific inquiry' (or NOSI, cf. Schwartz et al., 2008).

With all of the previous considerations in mind, this chapter has three main purposes. On the one hand, it aims to present some of the NOS research findings and innovations that have been produced in the last ten years by the GEHyD. Secondly, and in order to illuminate that presentation, it retrieves the core notions of models, argumentation, and inquiry, which constitute the pillars of the theoretical framework underlying those research and innovation results. And, thirdly, it uses these foci as a lens through which to review the ever-increasing didactical literature on HPS in Latin America, which also serves as a context for my work. In order to give readers a sense of what is being accomplished, I have endeavored to include key writings in Spanish and Portuguese which are only very poorly known outside our region.

For rhetorical reasons, I invert the order of these three points. I start by briefly reviewing, in Section 2, the literature produced on HPS and NOS in Latin America within the field of didactics of science. To this review, I add models, argumentation, and inquiry. Then, in the third section, I address my second purpose, above, by striving to underline substantive *connections* between these three main themes, under the umbrella of NOS. At the end of Section 3, I expand upon GEHyD findings and those of other Latin American scholars, sketching my theoretical framework on the nature of science – which conceptualizes science as the collection, structuring, and use of evidence – and offering proposals for science teacher education.

2 A Review of the Latin American Literature that Uses Metascientific Perspectives for Science Education

2.1 *Scope and Topics of the Review*

This section is devoted to a rather rapid 'literature review' of Latin American production around issues that involve the more or less rigorous uses of metascientific studies for science education. As I noted in the first section, the review is organized around five 'structuring' topics that are not completely separable: HPS, NOS, models, argumentation, and inquiry. I identified appropriate publications belonging to each topic using titles, keywords, and abstracts; I then classified them into different taxons. In the case of HPS and NOS, I resorted to already existing theoretical categories (taken mainly from Matthews, 1994; McComas, 1998; Adúriz-Bravo, 2008a) for classification. For the rest of topics, since the number of relevant texts was very low, I utilized standard categories: theoretical and empirical studies, investigations centered on teachers or students, questions concerning curriculum and materials, etc.

My review does not constitute a *meta-analysis*: I have not attempted to create new categories, and I have not looked for patterns or generalizations in the results reported. My aim here is to provide a comprehensive panorama of what is being done in our region, and is especially directed towards English-speaking scholars, who are often unacquainted with Latin American scholarship. I intend to depict the state of the art of Latin American efforts in didactics of science centered on 'meta-' contributions. Nevertheless, I do not purport to have produced an exhaustive review, since the literature is now vast and the number of sources constantly increases. Thus, one derived value of my effort is that my classifications can be modified and my corpus expanded in order to include ideas, texts, and authors left out of this chapter.

As I stated above, the literature around the contributions of the meta-sciences – and other second-order views on science – to the didactics of science has come to constitute a significant research and innovation area in our discipline: HPS in the second sense. Several lines can be recognized within this area (Matthews, 1994), NOS, also in the second sense, being one of those lines. Most of those HPS lines are represented, albeit timidly, in Latin American literature. As far as I know, there are only a very few reviews of that literature (see Adúriz-Bravo, 2008a; Erthal & Linhares, 2009; Pansera-de-Araújo et al., 2009; Villani et al., 2010; Amador Rodríguez & Adúriz-Bravo, 2012; Arriassecq, 2012, for partial reviews with specific aims and scopes).

As for the other three topics (models, argumentation, and inquiry), these are only partly included in HPS/NOS. My review takes into account all research perspectives around these topics, although I have concentrated especially on works that use a 'meta-' perspective.

For this chapter, the central corpus is comprised of *journal articles* authored by scholars working in Latin America, i.e., scholars born either in our region or somewhere else, but who have conducted at least part of their research in Latin American institutions. My review encompasses works in the three main languages of publication of research in didactics of science in the region: Spanish, Portuguese, and English.

For the literature in English, I review *all* papers published from 2001–2011 in the four major journals in our field: *International Journal of Science Education, Journal of Research in Science Teaching, Science & Education*, and *Science Education*. This corpus (Table 17.1) is rather small: only 48 papers out of the over 1,200 published in those journals over that period focused on one or more of the five topics under consideration. This shows that the participation of our scholarly community in the English-speaking arena is still very limited, due both to the relatively recent growth of Latin American scholarship in the area, and to the difficulties that we non-native speakers face when writing in English.

Of the 48 papers in English, 31 appear in *Science & Education*. This asymmetry may of course be explained by the fact that this is the only one of the four reviewed journals entirely devoted to HPS and connected issues.

Due to the much larger number of published texts in Spanish and Portuguese, I only review *some* of the papers that appeared in very prestigious

TABLE 17.1 Papers published in English by Latin American scholars on five topics informed by a 'meta-' perspective between 2001–2011

Journals topics	International journal of science education	Journal of research in science teaching	Science & education	Science education	Total per topic
HPS (excluding NOS, models, argumentation, and inquiry)	3	2	24	1	30
NOS	0	0	2	2	4
Models	6	0	4	2	12
Argumentation	0	0	1	0	1
Inquiry	1	0	0	0	1
Total per journal	10	2	31	5	48

journals from our region and Spain (a country where many Latin American scholars regularly publish) between 2001 and 2011. These journals include *Ciência & Educação*, *Educación Química*, *Enseñanza de las Ciencias*, *Investigações em Ensino de Ciências*, *Revista Brasileira de Pesquisa em Educação em Ciências*, *Revista de Educación en Ciencias/Journal of Science Education*, *Revista de Enseñanza de la Física*, *Revista Electrónica de Enseñanza de las Ciencias*, *Revista Eureka de Enseñanza y Divulgación de las Ciencias*, and *Tecné, Episteme y Didaxis*.

A proviso should be added here: I do not claim that these ten journals are the most important, distributed, or cited of our region. I have chosen them for a series of reasons that are convenient for my review. First, they have all attained the status of being 'regional' journals, publishing texts by scholars from many different countries. Second, most of them are multilingual, accepting papers in Spanish or Portuguese, and sometimes in English or other languages. Third, they are all journals of didactics of science, and not of general educational research (another area to which Latin American didacticians of science sometimes submit their work). And, finally, these ten journals are published in Argentina, Brazil, Colombia, Mexico, and Spain, the five countries with the highest dissemination of regional literature in our field.

To this corpus, I also add a few more papers (from other journals and conference proceedings), book chapters, and books in order to complement my review with other ideas that I think are valuable, and which would not be included if I restrained myself to the previous criteria.

I now present the results of the review, along with some brief comments, organized in three sub-sections.

2.2 *The State of the Art of* HPS *in Latin America (Excluding* NOS*, Models, Argumentation, and Inquiry)*

According to my conceptualization, HPS entirely encompasses NOS, but only partially includes models (e.g., from the perspective of the nature of models), argumentation (e.g., regarding the use of scientific models in arguments), and inquiry (e.g., from the perspective of NOSI). This sub-section reviews HPS literature on topics that exclude the four mentioned above.

Latin American literature on these 'other' topics of HPS comprises a wide variety of issues. My own classification of these issues follows:

1. *The philosophy and history of didactics of science:* epistemological characterizations of didactics of science as a discipline (Peme-Aranega, 1997; Eder & Adúriz-Bravo, 2001; Adúriz-Bravo & Izquierdo-Aymerich, 2002, 2005); critical appraisal of constructivism as an epistemology for science education (Niaz et al., 2003); examination of the presence of STS perspectives in Brazilian didactics of science (Pansera-de-Araújo et al., 2009); analysis of the historical evolution of didactics of science in Brazil (Villani et al., 2010); philosophical analysis of the evolution of one research line in didactics of science (Soto et al., 2011).
2. *The role of the philosophy of science in science education:* philosophy of science in didactics of science (Adúriz-Bravo et al., 2002; Paruelo, 2003; Gallego Torres & Gallego Badillo, 2007; Amador Rodríguez & Adúriz-Bravo, 2011; Ariza & Adúriz-Bravo, 2012); philosophy of science in science curricula (Colombo de Cudmani, 2003); philosophy of science in science teacher education (Colombo de Cudmani & Salinas de Sandoval, 2004; Adúriz-Bravo, 2005b); Toulmin's epistemology for science education (Chamizo, 2007a; Rengifo Gallego & Claret Zambrano, 2009); Popper's epistemology for science education (Rufatto & Carneiro, 2009).
3. *The role of the history of science in science education:* historico-philosophical studies of science with implications for teaching and learning (Martins & Silva, 2001; Freire Jr., 2003; Waldegg, 2005; García Martínez, 2007; Camelo Bustos & Rodríguez Sotelo, 2008; Duarte et al., 2008; Padilla & Furió-Mas, 2008; Gallego Badillo et al., 2009; Henao et al., 2009; Uribe et al., 2010; Viana & Porto, 2010); historical reconstructions and their implications for science textbooks (Niaz, 2001a; Brito et al., 2005;

Niaz et al., 2010); the role of the history of science in promoting scientific competencies (Camacho & Quintanilla, 2008); the role of the history of science in teachers' professional knowledge (Cuéllar et al., 2010); the role of the history of science in teaching and learning (Silva et al., 2008; Garay Garay, 2011).

4. *Theoretical frameworks for using the history and philosophy of science:* discussions of foundational concepts in chemistry and their implications for teaching (Niaz, 2001b; García Martínez et al., 2002); the importance of 'specific' philosophies of science for science education (Adúriz-Bravo & Erduran, 2003); frameworks to support the teaching of HPS to science teachers (Adúriz-Bravo, 2004, 2011a; Quintanilla et al., 2007); a framework to support the teaching of HPS to science students (Guridi & Arriassecq, 2004); a proposal to teach chemistry via an historiographic model (Chamizo, 2007b, 2009).
5. *Teacher education in HPS: Studies of teacher education in HPS:* research findings about a graduate studies program on HPS (Freire Jr. & Tenório, 2001) and an undergraduate course on HPS (Rosa & Martins, 2009); experiences of implementation of HPS materials for teachers (Reis et al., 2001; Oki & Moradillo, 2008).
6. *Misconceptions, epistemological obstacles, and learning difficulties through the lens of HPS:* the relationships between students' conceptions and the history of concepts in different scientific fields (Waldegg, 2001; Cotignola et al., 2002; Pocoví & Finley, 2002; Speltini & Dibar Ure, 2002; Greca & Freire Jr., 2003); analyses of concepts and concept evolution using models from the philosophy of science (Flores-Camacho et al., 2007; Tamayo & Sanmartí Puig, 2007).
7. *HPS-supported instruction, assessment, and materials for students:* the contextual teaching of modern physics (Arriassecq & Greca, 2002); history- and philosophy-based materials and their results for students (Pocoví, 2007).
8. *The presentation of science in textbooks:* analyses of historical accounts in textbooks (Rodríguez & Niaz, 2002; Pocoví & Finley, 2003); an analysis of the presentation of an historical experiment in textbooks (Niaz & Rodríguez, 2005); analyses of the presentation of scientific theories in textbooks (Arriassecq & Greca, 2007; Pitombo et al., 2008; Krapas, 2008; Niaz & Fernández, 2008).
9. *STS and socio-scientific issues:* STS approaches to science teaching (Guerra et al., 2008; Siqueira-Batista et al., 2010; Firme & Amaral, 2011); moral reasoning and decision-making in socio-scientific issues (Guimarães et al., 2010); teachers' conceptions of STS (Kist & Ferraz, 2010).

10. *Scientific knowledge versus other worldviews:* science and religion in university education (Sepúlveda & El-Hani, 2004; El-Hani & Sepúlveda, 2010; an intervention promoting 'dialogue' between scientific and traditional knowledge in the classroom (Baptista & El-Hani, 2009).

It can be seen that the problems that have been given the most attention in our region are those linking the evolution of scientific ideas in the history of science and in students; the treatment of such problems adheres, to a greater or lesser extent, to the classic 'parallelism' between *phylogeny* and *ontogeny* (à la Piaget, and others). Research into this issue has been conducted from the point of view of the history of science (developing historico-philosophical reconstructions of scientific ideas and then drawing implications for students' learning), and from the point of view of students' learning (making sense of misconceptions by means of exploring scientific conceptions that were considered valid over a certain period of the history of science).

2.3 *Latin American Literature on 'Mainstream'* NOS

This sub-section focuses on NOS, which is currently the line of research within HPS with the highest number of publications. I will, however, leave the treatment of models, argumentation, and inquiry from the perspective of their participation in the nature of science for the next sub-section.

I divide NOS research in Latin America into the following categories:

1. *Theoretical and meta-analytical studies:* theoretical characterizations of the nature of science (Niaz, 2001c; Adúriz-Bravo, 2005b, 2008b; Ayala et al., 2011; Inzillo & Adúriz-Bravo, 2011).
2. *NOS for teachers and teaching:* primary, secondary, and university teachers' NOS views (Scandroli & Rocha, 2002; Vilela-Ribeiro & Benite, 2009; Chinelli et al., 2010; Guerra-Ramos et al., 2010); the goals and values that teachers attribute to lab work (Antúnez et al., 2008); changes in teachers' understandings of the nature of science (Niaz, 2009); bringing the experimental nature of science into teacher education (Inzillo et al., 2010).
3. *Students' perspectives on NOS:* students' NOS views analyzed using Laudan's epistemology (Guridi et al., 2006); students' NOS views around science and pseudo-science (Viau et al., 2006); students' NOS views after an intervention utilizing a contextual approach (El-Hani et al., 2004; Teixeira et al., 2009); students' NOS views on scientific knowledge construction (Rezende et al., 2010); students' NOS views as derived from textbook usage (Briccia & Carvalho, 2011).
4. *Images of science and scientists:* images of science and scientists among undergraduate students (Mengascini et al., 2004); images of science in cartoons (Mesquita & Soares, 2008); the influence of gender in attitudes

towards science (Porro & Acevedo, 2011); images of science among science teachers (Pujalte et al., 2011).

5. *Didactical* (i.e. instructional) *units and materials for teaching NOS:* didactical units for teaching NOS ideas to science teachers (Adúriz-Bravo, 2002; Adúriz-Bravo & Izquierdo-Aymerich, 2009); a didactical unit for teaching NOS ideas to science students (Couló & Adúriz-Bravo, 2010).

The literature here is so scarce that there are no recognizable patterns. Nevertheless, it should be noted that *all* studies in 1, and some in 2, 3, and 5, propose or refer to NOS characterizations that strongly differ from the hegemonic, *tenet-based,* approach (McComas, 1998). For instance, they suggest that NOS should be framed as a form of didactical selection between [pre-existing?] *models* from the philosophy of science. This could be taken as an indication that there are theoretical concerns in our region about the applicability of NOS views that have been generated in other research traditions.

2.4 *Overview of Studies Focusing on Models, Argumentation, and Inquiry*

This sub-section is devoted to three topics that, although considered peripheral to canonic HPS and NOS research, can, I contend, establish valuable connections with these fields. I later use these three topics in my portrayal of the nature of science as a process that can be profitably discussed with science teachers.

In Latin America, models and modeling are the most studied of these topics. The literature concerning models surpasses that on argumentation and inquiry combined. In addition, it can be safely said that this is the only topic that has by now accumulated a recognizable regional corpus of work that is regularly published in English, in the major journals of our field.

Latin American literature on models may be classified as follows:

1. *The nature of models:* the nature of models and implications for teaching and learning (Lombardi, 1998; Greca & Moreira, 2000, 2002; Gallego Badillo, 2004; Chamizo, 2006, 2010; Sanabria Rojas et al., 2009; Adúriz-Bravo, in press); teachers' conceptions of the nature of models (Justi & Gilbert, 2002a, 2002b, 2003); the epistemological nature of teachers' mental models (Amador Rodríguez et al., 2008).
2. *Model-based teaching:* teaching models proposed in a museum exhibition (Falcão et al., 2004); modeling in school science laboratories (Medina et al., 2004; Merino Rubilar & Izquierdo-Aymerich, 2011); model-based secondary and post-secondary science education (Justi, 2006; Maia & Justi, 2009; Souza & Justi, 2010); model-based primary and secondary science education (Acher et al., 2007; López y Mota & Sanmartí, 2011); models in

science teacher education (Gallego Torres et al., 2007); scientific models in textbooks (Matus et al., 2011).

3. *Model-based learning:* the construction of students' mental models through model-based teaching (Núñez-Oviedo et al., 2008); students' understandings of a physics model (Wainmaier & Salinas, 2008); students' modeling of a physics problem (Massa et al., 2010).
4. *Models, analogies, and metaphors in science education:* the role of models in analogies (Miguel, 1999; Galagovsky & Adúriz-Bravo, 2001; Linares & Izquierdo-Aymerich, 2006; Viau et al., 2008; Raviolo, 2009; Marcelos & Nagem, 2010); models in scientific explanation (Concari, 2001); semantic and syntactic aspects of modeling (Ragout de Lozano & Cárdenas, 2002); models in problem solving (Gangoso et al., 2008); models in textbooks (Santos & El-Hani, 2009; Santos et al., 2012).

It should be noted that Latin American work on models is (slightly) more abundant than that on NOS. This does not wholly reflect international trends, where NOS is a 'hot topic' that has been extensively investigated in the last two decades.

Scientific argumentation in school settings is steadily becoming a point of interest in the Latin American community, with several significant projects in development (most to be published in Spanish and Portuguese). Issues around argumentation could be classified as follows:

1. *Toulmin's argumentation pattern:* the use of TAP in science education (Henao & Stipcich, 2008); the use of TAP in the analysis of classroom situations (Nascimento & Vieira, 2008; Sasseron & Carvalho, 2011).
2. *Analyzing and fostering argumentation in students:* sharing assessment criteria for school scientific argumentation with students (Adúriz-Bravo & Revel Chion, 2005); argumentative elements in scientific explanations in schools (Gómez Galindo, 2006); fostering students' argumentation in environmental education (Campaner & De Longhi, 2007); argumentation as proposed in mathematics textbooks (Llanos et al., 2007; Llanos & Otero, 2009); students' argumentation around a physical phenomenon (Stipcich, 2008); argumentative elements in meaning-making and negotiation in the classroom (Mortimer & Scott, 2003; Domínguez & Stipcich, 2009); the place of argumentation in undergraduate chemistry (Linhares Queiroz & Passos Sá, 2009); reflecting on argumentation with science students (García Romano & Valeiras, 2010); the relationship between students' conceptual knowledge and argumentation practices (Tavares et al., 2010).
3. *Analyzing and fostering argumentation in teachers:* argumentation in science teacher education (Islas et al., 2009; Adúriz-Bravo, 2011b); argu-

mentation patterns in a teacher education journal (Martins, 2009); argumentative elements in the analysis of teachers' practices (Vieira & Nascimento, 2009; Fagúndez et al., 2011).

Scientific inquiry is the least investigated of the three topics. Most studies around inquiry that can be found in our region are directly derived from large-scale, publicly or privately funded projects on 'inquiry-based science education', or IBSE (the American English acronym). It should be noted that, although such projects are now generating growing interest, formal didactical research around them is still scarce. The very few studies that are readily available can be grouped as follows:

1. *Theoretical studies:* the theoretical foundations of IBSE (Hernández et al., 2004; Devés & Reyes, 2007); theoretical frameworks to support IBSE (Gellon et al., 2005; Bahamonde, 2008).
2. *Empirical studies on inquiry and 'investigative' approaches:* student performance in 'investigative activities' (Trani Gomes et al., 2008); an inquiry-based teaching experience in college agricultural sciences (Gajardo et al., 2008); challenges that teachers face when implementing scientific inquiry (González Weil et al., 2009); the impact of IBSE programs on students (Meisel Donoso et al., 2010; Muñoz Rojo, 2010); an assessment of a Colombian IBSE program (Tamayo & Restrepo, 2012).
3. *Practical proposals:* a workshop for teachers on the use of inquiry in science education (Nicaretta & Bueno, 2011).
4. *Technical reports on the implementation of IBSE programs and projects:* a report on an IBSE project in Venezuela (Bifano et al., 2006); a report on an IBSE project using information and communication technologies (ICTs) in Chile (Moënne Rivas et al., 2008).

Regarding this last topic, two distinctive traits can be highlighted. In the first place, almost all of the texts reviewed come from Chile and Colombia, the two countries in the region where IBSE projects have been most widely implemented. In the second place, many of the texts reviewed *are not authored by specialists in the didactics of science*, but by active (natural) scientists who have turned towards education in order to encourage the use of IBSE methodologies in classrooms.

3 The Nature of Scientific Investigation: A Didactical Model for Science Teachers

In this last section, I seek to identify fruitful connections between the nature of science, on the one hand, and models, argumentation and inquiry, on the other, using theoretical contributions from scholars outside Latin America. I then use these ideas as lenses through which to identify similar formulations in our region, and to present and support my own proposal for science teacher education.

3.1 *Contributions of Models, Argumentation, and Inquiry to the Construction of the Nature of Science*

In recent *model-based* or *model-theoretic* portrayals of the nature of science, models and NOS are substantively linked. Such portrayals, usually transposed from the so-called *semanticist family* of contemporary philosophy of science, *explicitly* depict school scientific theories as families of 'theoretical models', and school scientific investigation as a process of modeling (i.e. constructing and using models to represent and intervene) (Izquierdo-Aymerich & Adúriz-Bravo, 2003; Develaki, 2007; Koponen, 2007).

In Latin America, Chamizo (2006) and Justi (2006) have elaborated conceptualizations along this line: they suggest that model-based teaching supposes not only the use of models in classrooms, but also an effort to discuss the epistemological nature of scientific models with students. In science education, models can then be understood as representations constructed by scientists, which are adequate for the set of aims and values shared in a specific moment of the history of science.

Argumentation and NOS are connected in the meta-theoretical characterization of school science as an argumentative activity (cf. Duschl, 2008; Osborne, 2010). According to this characterization, the key epistemic ideal of science would be *explaining* the natural world. Argumentative processes, in turn, would provide support to scientific explanations, and help convince others of their pertinence and robustness. As Richard Duschl (2008, p. 159) highlights, the language of science "is a discourse that critically examines and evaluates the numerous and at times iterative transformations of evidence into explanations".

In Latin America, several scholars who completed their Ph.D. studies at the Universitat Autònoma de Barcelona in Spain (Adrianna Gómez Galindo, Rita Linares, Cristian Merino Rubilar, Mario Quintanilla, and Oscar Tamayo) adhere to a model-based theoretical conceptualization of school scientific argumentation (see also Izquierdo-Aymerich & Adúriz-Bravo, 2003). They see argumentation as an 'act of speech' in which scientific models give meaning to the natural world and make sense to the 'cognitive agents' who use them. These authors infuse this view of argumentation into their didactical studies of conceptualization, explanation, analogies, experiments, etc. in science classrooms (see references).

Inquiry and NOS can establish a cogent dialogue when it is understood that the nature of scientific knowledge (one of the aspects of 'science as a product') is conditioned by the fundamental composition of the culturally conquered processes employed by scientists to develop such knowledge (see Flick & Lederman, 2004b; Schwartz et al., 2008). At least part of what is done in school science can be located under the umbrella of scientific inquiry, "a variety of processes and ways of thinking that support the development of new knowledge in science" (Flick & Lederman, 2004a, p. ix).

In Latin America, some 'adaptations' of international IBSE projects (such as the French *La main à la pâte*) to local conditions and expectations have implied the expansion of the original theoretical frameworks. That expansion has been accomplished using elements taken from didactical research that have an epistemological, historical, linguistic, or cognitive character (see Gellon et al., 2005; Bahamonde, 2008; Tamayo & Restrepo, 2012). The use of this research-based input has strengthened the epistemological foundations of the original projects, adding new facets to the nature of inquiry so that it goes beyond traditional investigative projects and hands-on science.

3.2 *School Scientific Investigation as the Collection, Structuring, and Use of Evidence*

Didactics of science has contemplated the nature of science from a broad range of theoretical perspectives. Both philosophers and didacticians have repeatedly pointed out that there is no such thing as 'one' nature of science. Even the small corpus of literature reviewed here shows that NOS can be construed in quite dissimilar ways: as a list of general tenets or a set of philosophical models, either of which may focus on specific issues such as demarcation, experiments, values, explanation, gender, etc. For this chapter, I have chosen a semantic, model-based approach, taken from the philosophy of science of the last three decades, which has helped me in identifying fruitful links between NOS and models, argumentation, and inquiry.

Figure 17.1, at the very beginning of this chapter, was intended to express the extremely rich and complex nature of scientific endeavour. On the one hand, science is constituted by a set of products: not only knowledge, but also methods, values, data, artifacts, publications, etc. (not shown in the figure). On the other hand, it comprises a manifold of processes carefully designed and adjusted in order to generate those products. I include innovation, justification, systematization, application, evaluation, and communication among the key scientific processes.

I locate scientific investigation under the umbrella of processes, while not conflating it with science-in-the-making as a whole. The term 'investigation' comes from the Latin 'investigare', meaning *to follow trails*; it is connected to the English word 'vestiges': traces, marks, signs, and – more concretely – footprints. This etymological connection can be profitably used from a didactical point of view: when teaching science, it can be interesting to highlight the problem-solving, evidence-based nature of science. Scientific activity says things about the world, and demands reasons to uphold what is being said. *Investigation can be intimately connected to the idea of evidence*: when doing science, scientists produce or obtain – via different 'methods' – 'compelling' elements in favor of certain understandings of the real world. In my proposal, school scientific investigation is defined as the collection, structuring, and use of evidence in order to *give support* to a scientific view on a natural phenomenon.

Flick and Lederman (2004a) recognize that, "[a]lthough intimately related, scientific inquiry and nature of science are different constructs" (p. ix). I consider the nature of science as the most encompassing of those two constructs, referring to what 'this thing called science' is. Accordingly, scientific inquiry is one particular *aspect* of the nature of science: I subsume it as a distinct element of scientific investigation. Inquiry, like investigation, can be conceptualized as an evidence-related construct, since the noun comes from the Latin verb 'inquaerere', with the meaning of *to seek* or *ask*. I define inquiry as the model-driven collection of evidence: an activity in which theoretical models guide, and even 'load', our observations, explanations, predictions, and interventions regarding phenomena (Figure 17.2). This is why I say that modeling gives structure, coherence, and direction to a set of evidence.

The word *inquiry* has Spanish and Portuguese equivalents that differ slightly from the English. In Portuguese, inquiry is sometimes translated as 'pesquisa', from the Latin verb 'perquirere' (which shares the same root, 'quaerere', but is intensified with the prefix 'per-'). Perquirere means *to seek earnestly for something*. In Spanish, on the other hand, inquiry is usually translated as 'indagación', from the Latin verb 'indagare', with the meaning of searching out or tracking something down; the choice of this word stresses the idea of *pursuit*, which is often used in analogies about the scientific enterprise.

My view on the nature of science gives centrality to scientific explanation, locating it at the top of the hierarchy of the epistemic purposes of science. It can be argued that the convergent use of models and argumentation is in fact necessary for explanation. Models give *intelligibility* to evidence; such highly systematized evidence is then used in elaborate scientific arguments to show how certain carefully selected phenomena make 'good cases' for a particular theoretical view.

My work on these three processes (inquiry, modeling, and argumentation) assumes them to be founded on evidence-based *inferences* (i.e. modes

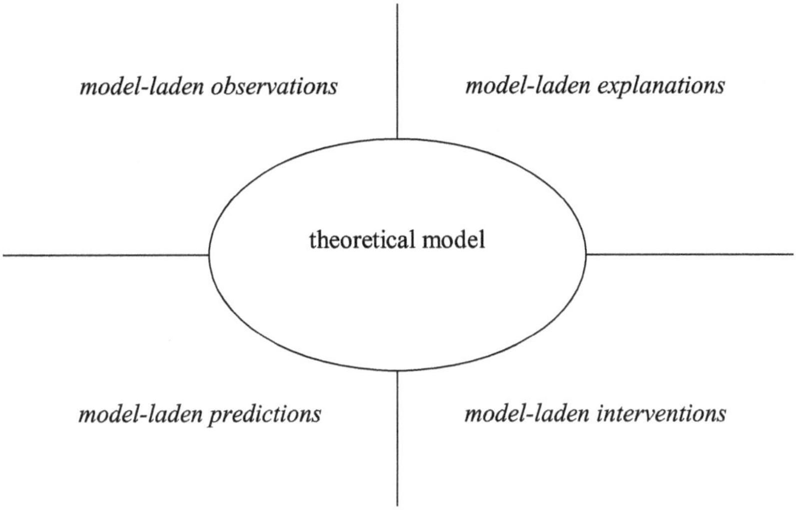

FIGURE 17.2 In a model-based account of the nature of science, school theoretical models serve as a guide for observation, prediction, explanation, and intervention

of reasoning or thinking); such inferences can be grouped under the label of *abduction* in its most general sense (Adúriz-Bravo, 2001, 2002). That is why, in my theoretical and practical proposals directed toward teachers, I am particularly interested in drawing analogies between scientific reasoning and detective, medical, forensic, and 'gossipographic' (garden-variety) thinking. In all these fields, a parsimonious collection of 'facts', gathered under the guidance of a strong model, can be used as the premises for a logically 'ampliative' reasoning pattern, which, with its explanatory power, may 'ascend' to general, abstract, and extremely audacious conclusions.

References

Acher, A., Arcà, M., & Sanmartí, N. (2007). Modeling as a teaching learning process for understanding materials: A case study in primary education. *Science Education, 91*(3), 398–418.

Adúriz-Bravo, A. (2001). A proposal to teach the abductive argumentation pattern through detective novels. In D. Psillos, P. Kariotoglou, V. Tselfes, G. Bisdikian, G. Fassoulopoulos, E. Hatzikraniotis, & M. Kallery (Eds.), *Proceedings of the Third International Conference on Science Education Research in the Knowledge Based Society* (Vol. 2, pp. 715–717). Thessaloniki, Greece: Aristotle University of Thessaloniki.

Adúriz-Bravo, A. (2002). Aprender sobre el pensamiento científico en el aula de ciencias: Una propuesta para usar novelas policiacas. *Alambique, 31*, 105–111.

Adúriz-Bravo, A. (2004). Methodology and politics: A proposal to teach the structuring ideas of the philosophy of science through the pendulum. *Science & Education, 13*(7–8), 717–731.

Adúriz-Bravo, A. (2005a). 'Los descubrimientos del radio': Una unidad didáctica para enseñar sobre la naturaleza de la ciencia a futuros profesores de ciencias naturales. In D. Couso, E. Badillo, G. A. Perafán, & A. Adúriz-Bravo (Eds.), *Unidades didácticas en ciencias y matemáticas* (pp. 317–336). Bogotá, Colombia: Editorial Magisterio.

Adúriz-Bravo, A. (2005b). ¿Qué naturaleza de la ciencia hemos de saber los profesores de ciencias?: Una cuestión actual de la investigación didáctica. *Tecné, Episteme y Didaxis*, extra issue, 23–33.

Adúriz-Bravo, A. (2008a). Áreas de investigación en la didáctica de las ciencias experimentales: La naturaleza de la ciencia. In C. Merino, A. Gómez Galindo, & A. Adúriz-Bravo (Eds.), *Áreas y estrategias de investigación en la didáctica de las ciencias experimentales* (pp. 111–125). Bellaterra, Spain: Servei de Publicacions de la UAB.

Adúriz-Bravo, A. (2008b). ¿Existirá el 'método científico'? In L. Galagovsky (Ed.), *¿Qué tienen de 'naturales' las ciencias naturales?* (pp. 47–59). Buenos Aires, Argentina: Biblos.

Adúriz-Bravo, A. (2011a). Epistemología para el profesorado de física: Operaciones transpositivas y creación de una 'actividad metacientífica escolar'. *Revista de Enseñanza de la Física, 24*(1), 7–20.

Adúriz-Bravo, A. (2011b). Fostering model-based school scientific argumentation among prospective science teachers. *US-China Education Review, 8*(5), 718–723.

Adúriz-Bravo, A. (in press). A 'semantic' view of scientific models for science education. *Science & Education.*

Adúriz-Bravo, A., & Erduran, S. (2003). La epistemología específica de la biología como disciplina emergente y su posible contribución a la didáctica de la biología. *Revista de Educación en Biología, 6*(1), 9–14.

Adúriz-Bravo, A., & Izquierdo-Aymerich, M. (2002). Acerca de la didáctica de las ciencias como disciplina autónoma. *Revista Electrónica de Enseñanza de las Ciencias, 1*(3), 130–140.

Adúriz-Bravo, A., & Izquierdo-Aymerich, M. (2005). Utilising the '3P-model' to characterize the discipline of didactics of science. *Science & Education, 14*(1), 29–41.

Adúriz-Bravo, A., & Izquierdo-Aymerich, M. (2009). A research-informed instructional unit to teach the nature of science to pre-service science teachers. *Science & Education, 18*(9), 1177–1192.

Adúriz-Bravo, A., Izquierdo-Aymerich, M., & Galagovsky, L. (2002). Relationships between the philosophy of science and didactics of science. *Revista de Educación en Ciencias/Journal of Science Education, 3*(1), 6–7.

Adúriz-Bravo, A., & Revel Chion, A. (2005). Sharing assessment criteria on school scientific argumentation with secondary science students. In R. Pintó & D. Couso (Eds.), *Proceedings of the Fifth International ESERA Conference on Contributions of Research to Enhancing Students' Interest in Learning Science, CD-ROM* (pp. 589–592). Barcelona, Spain: ESERA.

Amador Rodríguez, R. Y., & Adúriz-Bravo, A. (2011). A qué epistemología recurrir para investigar sobre la enseñanza de las ciencias. *Revista Electrónica EDUCyT, 2*(2), 3–18.

Amador Rodríguez, R. Y., & Adúriz-Bravo, A. (2012). Consensos y disensos en torno al concepto de *naturaleza de la ciencia* (NOS) en la comunidad iberoamericana de didáctica de las ciencias. *Revista Científica, 15*, 30–46.

Amador Rodríguez, R. Y., Gallego Badillo, R., & Pérez Miranda, R. (2008). Desde qué versiones epistemológicas construyen modelos mentales los profesores en formación inicial: Una investigación didáctica. *Tecné, Episteme y Didaxis, 24*, 8–22.

Antúnez, G. C., Pérez, S. M., & Petrucci, D. (2008). Concepciones de los docentes universitarios sobre los trabajos prácticos de laboratorio. *Revista Brasileira de Pesquisa em Educação em Ciências, 8*(1), n/pp.

Ariza, Y., & Adúriz-Bravo, A. (2012). La 'nueva filosofía de la ciencia' y la 'concepción semántica de las teorías científicas' en la didáctica de las ciencias naturales. *Revista de Educación en Ciencias Experimentales y Matemática, 2*(2), n/pp.

Arriassecq, I. (2012). *Contributions from history and philosophy of science in physics education: The role both in the curriculum of high school as in teacher training courses analyzed in different educational reforms in Argentina* (Unpublished manuscript). UNICEN, Tandil, Argentina.

Arriassecq, I., & Greca, I. M. (2002). Algunas consideraciones históricas, epistemológicas y didácticas para el abordaje de la teoría de la relatividad especial en el nivel medio y polimodal. *Ciência & Educação, 8*(1), 55–69.

Arriassecq, I., & Greca, I. M. (2007). Approaches to the teaching of special relativity theory in high school and university textbooks of Argentina. *Science & Education, 16*(1), 65–86.

Ayala, M. M., Malagón, J. F., & Sandoval, S. (2011). Magnitudes, medición y fenomenologías. *Revista de Enseñanza de la Física, 24*(1), 43–54.

Bahamonde, N. (2008). Un desafío de la alfabetización científica: Hacer ciencia a través del lenguaje. *El Monitor de la Educación, 16*, 28–31.

Baptista, G. C. S., & El-Hani, C. N. (2009). The contribution of ethnobiology to the construction of a dialogue between ways of knowing: A case study in a Brazilian public high school. *Science & Education, 18*(3–4), 503–520.

Bifano, C., Hernández-Szczurek, C., & Valdivieso, R. (2006). Educación en Ciencias Basada en la Indagación: La experiencia de Venezuela. In C. Bosch & C. Bifano (Eds.), *Taller Educación en Ciencias Basada en la Indagación: Logros y tropiezos* (Online). Caracas, Venezuela: Academica de Ciencias Físicas, Matemáticas y Naturales.

Böttcher, F., & Meisert, A. (2011). Argumentation in science education: A model-based framework. *Science & Education, 20*(2), 103–140.

Briccia, V., & Carvalho, A. M. P. (2011). Visões sobre a natureza da ciência construídas a partir do uso de um texto histórico na escola média. *Revista Electrónica de Enseñanza de las Ciencias, 10*(1), 1–22.

Brito, A., Rodríguez, M. A., & Niaz, M. (2005). A reconstruction of development of the periodic table based on history and philosophy of science and its implications for general chemistry textbooks. *Journal of Research in Science Teaching, 42*(1), 84–111.

Camacho, J. P., & Quintanilla, M. (2008). Resolución de problemas científicos desde la historia de la ciencia: Retos y desafíos para promover competencias cognitivo lingüísticas en la química escolar. *Ciência & Educação, 14*(2), 197–212.

Camelo Bustos, F. J., & Rodríguez Sotelo, S. J. (2008). Una revisión histórica del concepto de calor: Algunas implicaciones para su aprendizaje. *Tecné, Episteme y Didaxis, 23*, 67–77.

Campaner, G., & De Longhi, A. L. (2007). La argumentación en educación ambiental: Una estrategia didáctica para la escuela media. *Revista Electrónica de Enseñanza de las Ciencias, 6*(2), 442–456.

Chamizo, J. A. (2006). *Los modelos de la química. Educación Química, 17*(4), 476–482.

Chamizo, J. A. (2007a). Las aportaciones de Toulmin a la enseñanza de las ciencias. *Enseñanza de las Ciencias, 25*(1), 133–146.

Chamizo, J. A. (2007b). Teaching modern chemistry through 'recurrent historical teaching models'. *Science & Education, 16*(2), 197–216.

Chamizo, J. A. (2009). El aprendizaje de la historia experimental de la química. *Tecné, Episteme y Didaxis, 26*, 82–96.

Chamizo, J. A. (2010). Una tipología de los modelos para la enseñanza de las ciencias. *Revista Eureka de Enseñanza y Divulgación de las Ciencias, 7*(1), 26–41.

Chinelli, M. V., Ferreira, M. V. S., & Aguiar, L. E. V. (2010). Epistemologia em sala de aula: A natureza da ciência e da atividade científica na prática profissional de professores de ciências. *Ciência & Educação, 16*(1), 17–35.

Colombo de Cudmani, L. (2003). ¿Qué puede aportar la epistemología a los diseños curriculares en física? *Ciência & Educação, 9*(1), 83–91.

Colombo de Cudmani, L., & Salinas de Sandoval, J. (2004). ¿Es importante la epistemología de las ciencias en la formación de investigadores y de profesores en física? *Enseñanza de las Ciencias, 22*(3), 455–462.

Concari, S. (2001). Las teorías y modelos en la explicación científica: Implicancias para la enseñanza de las ciencias. *Ciência & Educação, 7*(1), 85–94.

Cotignola, M. I., Bordogna, C., Punte, G., & Cappannini, O. M. (2002). Learning difficulties of thermodynamics concepts: Are they linked to the historical development of the field? Science & Education, 11(2), 279–291.

Couló, A., & Adúriz-Bravo, A. (2010). 'La idea más minúscula': Unidad didáctica para aprender sobre modelos en torno a la estructura atómica de la materia. In F. Angulo Delgado & M. Quintanilla (Eds.), *Unidades didácticas en ciencias naturales y educación ambiental: Su contribución a la promoción de competencias de pensamiento científico* (Vol. 2, pp. 159–186). Medellín, Colombia: Universidad de Antioquia.

Cuéllar, L., Quintanilla, M., & Marzàbal, A. (2010). La importancia de la historia de la química en la enseñanza escolar: Análisis del pensamiento y elaboración de material didáctico de profesores en formación. *Ciência & Educação, 16*(2), 277–291.

Develaki, M. (2007). The model-based view of scientific theories and the structuring of school science programs. *Science & Education, 16*(7), 725–749.

Devés, R., & Reyes, P. (2007). Principios y estrategias del programa de educación en ciencias basada en la indagación (ECBI). *Pensamiento Educativo, 41*(2), 115–131.

Domínguez, M. A., & Stipcich, M. S. (2009). Buscando indicadores de la negociación de significados en clases de ciencias naturales. *Revista Electrónica de Enseñanza de las Ciencias, 8*(2), 539–551.

Duarte, S. U., Gallego Badillo, R., Pérez Miranda, R., & Gallego Torres, A. P. (2008). Una construcción histórico-epistemológica del modelo del octeto para el enlace químico. *Tecné, Episteme y Didaxis, 23*, 52–66.

Duschl, R. A. (2008). Quality argumentation and epistemic criteria. In S. Erduran & M. P. Jiménez-Aleixandre (Eds.), *Argumentation in science education: Perspectives from classroom-based research* (pp. 159–175). Dordrecht, The Netherlands: Springer.

Eder, M. L., & Adúriz-Bravo, A. (2001). Aproximación epistemológica a las relaciones entre la didáctica de las ciencias naturales y la didáctica general. *Tecné, Episteme y Didaxis, 9*, 2–16.

El-Hani, C. N., & Sepúlveda, C. (2010). The relationship between science and religion in the education of protestant biology preservice teachers in a Brazilian university. *Cultural Studies of Science Education, 5*(1), 103–125.

El-Hani, C. N., Tavares, E. J. M., & Rocha, P. L. B. (2004). Concepções epistemológicas de estudantes de biologia e sua transformação por uma proposta explícita de ensino sobre história e filosofia das ciências. *Investigações em Ensino de Ciências, 9*(3), 265–313.

Erduran, S., & Duschl, R. (2004). Interdisciplinary characterizations of models and the nature of chemical knowledge in the classroom. *Studies in Science Education, 40*, 111–144.

Erthal, J. P. C., & Linhares, M. P. (2009). História da ciência em sala de aula: O que tem aparecido em nossas revistas? In *VII ENPEC-Encontro Nacional de Pesquisadores em Educação em Ciências* (Online). Florianópolis, Brazil: ABRAPEC.

Fagúndez, T., Rangel, N., & Castells, M. (2011). El 'qué hacer' docente en clases universitarias de física: Una perspectiva semiótico-comunicativa-multimodal de la construcción del conocimiento. *Revista de Enseñanza de la Física, 24*(1), 55–79.

Falcão, D., Colinvaux, D., Krapas, S., Queiroz, G., Alves, F., Cazaelli, S., Valente, M. E., & Gouvea, G. (2004). A model-based approach to science exhibition evaluation: A case study in a Brazilian astronomy museum. *International Journal of Science Education, 26*(8), 951–978.

Firme, R. N., & Amaral, E. M. R. (2011). Analisando a implementação de uma abordagem CTS na sala de aula de química. *Ciência & Educação, 17*(2), 383–399.

Flick, L. B., & Lederman, N. G. (2004a). Introduction. In L. B. Flick & N. G. Lederman (Eds.), *Scientific inquiry and nature of science: Implications for teaching, learning, and teacher education* (pp. ix–xviii). Dordrecht, The Netherlands: Kluwer Academic Publishers.

Flick, L. B., & Lederman, N. G. (Eds.). (2004b). *Scientific inquiry and nature of science: Implications for teaching, learning, and teacher education.* Dordrecht, The Netherlands: Kluwer Academic Publishers.

Flores-Camacho, F., Gallegos-Cázares, L., Garritz, A., & García-Franco, A. (2007). Incommensurability and multiple models: Representations about structure of matter in undergraduate chemistry students. *Science & Education, 16*(7–8), 775–800.

Freire Jr., O. (2003). A story without an ending: The quantum physics controversy 1950–1970. *Science & Education, 12*(5), 573–586.

Freire Jr., O., & Tenório, R. M. (2001). A graduate program in history, philosophy and science teaching in Brazil. *Science & Education, 10*(6), 601–608.

Gajardo, O. A., Bezic, C., Dall'Armellina, A., Avilés, L., & Cañón, S. (2008). Una experiencia educativa en la enseñanza de las ciencias a partir de un sistema productivo: La indagación como metodología de estudio en carreras universitarias de agronomía. *Revista Pilquén, Sección Agronomía, 10*(9), 1–11.

Galagovsky, L., & Adúriz-Bravo, A. (2001). Modelos y analogías en la enseñanza de las ciencias naturales: El concepto de *modelo didáctico analógico*. *Enseñanza de las Ciencias, 19*(2), 231–242.

Gallego Badillo, R. (2004). Un concepto epistemológico de modelo para la didáctica de las ciencias experimentales. *Revista Electrónica de Enseñanza de las Ciencias, 3*(3), 301–319.

Gallego Badillo, R., Pérez Miranda, R., & Gallego Torres, A. P. (2009). Una aproximación histórico epistemológica a las leyes fundamentales de la química. *Revista Electrónica de Enseñanza de las Ciencias, 8*(1), 359–375.

Gallego Torres, A. P., & Gallego Badillo, R. (2007). Historia, epistemología y didáctica de las ciencias: Unas relaciones necesarias. *Ciência & Educação, 13*(1), 85–98.

Gallego Torres, A. P., Gallego Badillo, R., & Pérez Miranda, R. (2007). Algunas consideraciones metodológicas en torno a la didáctica de la modelación. *Revista de Enseñanza de la Física, 20*(1–2), 7–15.

Gangoso, Z., Truyol, M. E., Brincones, I., & Gattoni, A. (2008). Resolución de problemas, comprensión, modelización y desempeño: Un caso con estudiantes de ingeniería. *Latin American Journal of Physics Education, 2*(3), 233–240.

Garay Garay, F. R. (2011). Perspectivas de historia y contexto cultural en la enseñanza de las ciencias: Discusiones para los procesos de enseñanza y aprendizaje. *Ciência & Educação, 17*(1), 51–62.

García Martínez, Á. (2007). Prácticas experimentales e instrumentos científicos en la construcción del conocimiento científico escolar. In M. Quintanilla (Ed.), *Historia de la ciencia: Propuestas para su divulgación y enseñanza* (Vol. 2, pp. 13–45). Santiago, Chile: Editorial Arrayán.

García Martínez, Á., Mosquera Suárez, C. J., & Mora Penagos, W. M. (2002). Bases para la construcción de un cuerpo conceptual didáctico del desarrollo histórico-epistemológico de los conceptos estructurantes de la química. *Revista Científica, 5*, 259–286.

García Romano, L., & Valeiras, N. (2010). *Lectura y escritura en el aula de ciencias: Una propuesta para reflexionar sobre la argumentación. Alambique, 63*, 57–64.

Gellon, G., Rosenvasser Feher, E., Furman, M., & Golombek, D. (2005). *La ciencia en el aula: Lo que nos dice la ciencia sobre cómo enseñarla.* Buenos Aires, Argentina: Paidós.

Gómez Galindo, A. A. (2006). Construcción de explicaciones científicas escolares. *Revista Educación y Pedagogía, 18*(45), 73–83.

González Weil, C., Martínez Larraín, M. T., Martínez Galaz, C., Cuevas Solís, K., & Muñoz Concha, L. (2009). La educación científica como apoyo a la movilidad social: Desafíos en torno al rol del profesor secundario en la implementación de la indagación científica como enfoque pedagógico. *Estudios Pedagógicos, 35*(1), 63–78.

Greca, I. M., & Freire Jr., O. (2003). Does an emphasis on the concept of quantum state enhance student's understanding of quantum mechanics? *Science & Education, 12*(5–6), 541–557.

Greca, I. M., & Moreira, M. A. (2000). Mental models, conceptual models, and modeling. *International Journal of Science Education, 22*(1), 1–11.

Greca, I. M., & Moreira, M. A. (2002). Mental, physical, and mathematical models in the teaching and learning of physics. *Science Education, 86*(1), 106–121.

Guerra, G., Alvarado, C., Zenteno, B. E., & Garritz, A. (2008). La dimensión ciencia-tecnología-sociedad del tema de ácidos y bases en un aula del bachillerato. *Educación Química, 19*(4), 277–288.

Guerra-Ramos, M. T., Ryder, J., & Leach, J. (2010). Ideas about the nature of science in pedagogically relevant contexts: Insights from a situated perspective of primary teachers' knowledge. *Science Education, 94*(2), 282–307.

Guimarães, M. A., Carvalho, W. L. P., & Oliveira, M. S. (2010). Raciocínio moral na tomada de decisões em relação a questões sociocientíficas: O exemplo do melhoramento genético humano. *Ciência & Educação, 16*(2), 465–477.

Guridi, V., & Arriassecq, I. (2004). Historia y filosofía de las ciencias en la educación polimodal: Propuesta para su incorporación al aula. *Ciência & Educação, 10*(3), 307–316.

Guridi, V., Salinas, J., & Villani, A. (2006). Contribuciones de la epistemología de Laudan para la comprensión de concepciones epistemológicas sustentadas por estudiantes secundarios de física. *Investigações em Ensino de Ciências, 11*(1), 97–117.

Henao, B. L., & Stipcich, M. S. (2008). Educación en ciencias y argumentación: La perspectiva de Toulmin como posible respuesta a las demandas y desafíos contemporáneos para la enseñanza de las ciencias experimentales. *Revista Electrónica de Enseñanza de las Ciencias, 7*(1), 47–62.

Henao, B. L., Stipcich, M. S., & Moreira, M. A. (2009). 'Sustancia' en el devenir la química: Dime cómo te buscan y te diré qué eres. *Ciência & Educação, 15*(3), 497–514.

Hernández, J. T., Figueroa, M., Carulla, C., Patiño, M. I., Tafur, M., & Duque, M. (2004). Pequeños científicos, una aproximación sistémica al aprendizaje de las ciencias en la escuela. *Revista de Estudios Sociales, 19*, 51–56.

Inzillo, L. N., & Adúriz-Bravo, A. (2011). La naturaleza experimental de la biología: Una reflexión indispensable para los docentes. *Quehacer Educativo, 106*, 30–39.

Inzillo, L. N., Rodríguez, E. M., & Adúriz-Bravo, A. (2010). Introducir la naturaleza de la biología en la formación inicial del profesorado. *Revista Electrónica Iberoamericana de Educación en Ciencias y Tecnología, 2*(1), 141–152.

Islas, S. M., Sgro, M. R., & Pesa, M. A. (2009). La argumentación en la comunidad científica y en la formación de profesores de física. *Ciência & Educação, 15*(2), 291–304.

Izquierdo-Aymerich, M., & Adúriz-Bravo, A. (2003). Epistemological foundations of school science. *Science & Education, 12*(1), 27–43.

Justi, R. (2006). La enseñanza de ciencias basada en la elaboración de modelos. *Enseñanza de las Ciencias, 24*(2), 173–184.

Justi, R., & Gilbert, J. K. (2002a). Modelling, teachers' views on the nature of modeling, and implications for the education of modellers. *International Journal of Science Education, 24*(4), 369–387.

Justi, R., & Gilbert, J. K. (2002b). Science teachers' knowledge about and attitudes towards the use of models and modelling in learning science. *International Journal of Science Education, 24*(12), 1273–1292.

Justi, R., & Gilbert, J. K. (2003). Teachers' views on the nature of models. *International Journal of Science Education, 25*(11), 1369–1386.

Kist, C. P., & Ferraz, D. F. (2010). Compreensão de professores de biologia sobre as interações entre ciência, tecnologia e sociedade. *Revista Brasileira de Pesquisa em Educação em Ciências, 10*(1), n/pp.

Koponen, I. T. (2007). Models and modelling in physics education: A critical re-analysis of philosophical underpinnings and suggestions for revisions. *Science & Education, 16*(7–8), 751–773.

Krapas, S. (2008). El *tratado sobre la luz* de Huygens y su transposición didáctica en la enseñanza introductoria de óptica. *Revista de Enseñanza de la Física, 21*(2), 49–60.

Kuznetsov, V., & Balzer, W. (2011). From philosophy of science to theories of knowledge systems. In *14th Congress of Logic, Methodology and Philosophy of Science* (Online). Nancy, France: Université de Nancy 2.

Lederman, N. G., & Abd-el-Khalick, F. (1998). Avoiding de-natured science: Activities that promote understandings of the nature of science. In W. F. McComas (Ed.), *The nature of science in science education: Rationales and strategies* (pp. 83–126). Dordrecht, The Netherlands: Kluwer Academic Publishers.

Linares, R. M., & Izquierdo-Aymerich, M. (2006). El rescate de la princesa encerrada en lo más alto de la más alta torre: Un episodio para aprender sobre analogías, símiles y metáforas. *El Hombre y la Máquina, 27*, 24–37.

Linhares Queiroz, S., & Passos Sá, L. (2009). O espaço para a argumentação no ensino superior de química. *Educación Química, 20*(2), 104–110.

Llanos, V. C., & Otero, M. R. (2009). Argumentación matemática en los libros de la enseñanza secundaria: Un análisis descriptivo de las caracterísitcas de los libros de texto y de la argumentación. *Revista Electrónica de Investigación en Educación en Ciencias, 4*(1), 37–50.

Llanos, V. C., Otero, M. R., & Banks Leite, L. (2007). Argumentación matemática en los libros de texto de la enseñanza media. *Revista Electrónica de Investigación en Educación en Ciencias, 2*(2), 39–53.

Lombardi, O. (1998). La noción de modelo en ciencias. *Educación en Ciencias, 2*(4), 5–13.

López y Mota, Á. D., & Sanmartí, N. (2011). ¿Desde dónde y con qué perspectiva enseñar ciencias? In *Las ciencias naturales en educación básica: Formación de ciudadanía para el siglo XXI* (pp. 43–94). México: Secretaría de Educación Pública/Universidad Pedagógica Nacional.

Maia, P. F., & Justi, R. (2009). Learning of chemical equilibrium through modelling-based teaching. *International Journal of Science Education, 31*(5), 603–630.

Marcelos, M. F., & Nagem, R. L. (2010). Comparative structural models of similarities and differences between vehicle and target in order to teach Darwinian evolution. *Science & Education, 19*(6–8), 599–623.

Martins, I. (2009). Argumentation in texts from a teacher education journal: An exercise of analysis based upon the Bakhtinian concepts of genre and social language. *Educación Química, 20*(2), 26–36.

Martins, R. A., & Silva, C. C. (2001). Newton and colour: The complex interplay of theory and experiment. *Science & Education, 10*(3), 287–305.

Massa, M., Llonch, E., & D'Amico, H. (2010). Cómo los estudiantes modelizan un problema de encuentro: Un estudio de caso. In *III Jornadas de Experiencias Innovadoras en Educación en la FCEIA* (Online). Rosario, Argentina: Universidad Nacional de Rosario.

Matthews, M. R. (1994). *Science teaching: The role of history and philosophy of science.* New York, NY: Routledge.

Matus, L., Benarroch, A., & Nappa, N. (2011). La modelización del enlace químico en libros de texto de distintos niveles educativos. *Revista Electrónica de Enseñanza de las Ciencias, 10*(1), 178–201.

McComas, W. F. (Ed.). (1998). *The nature of science in science education: Rationales and strategies.* Dordrecht, The Netherlands: Kluwer Academic Publishers.

McComas, W. F., Clough, M. P., & Almazroa, H. (1998). The role and character of the nature of science in science education. In W. F. McComas (Ed.), *The nature of science in science education: Rationales and strategies* (pp. 3–39). Dordrecht, The Netherlands: Kluwer Academic Publishers.

Medina, C., Velazco, S., & Salinas, J. (2004). Experimental control of simple pendulum model. *Science & Education, 13*(7–8), 631–640.

Meisel Donoso, J. D., Bermeo Andrade, H. P., Saavedra Moreno, C., & Patiño Garzón, L. (2010). El éxito de la enseñanza de las ciencias basada en indagación (ECBI): Una cuestión más allá del aula de clase. *Pedagogía y Saberes, 32*, 111–124.

Mengascini, A., Menegaz, A., Murriello, S., & Petrucci, D. (2004). "... Yo así, locos como los vi a ustedes, no me lo imaginaba": Las imágenes de ciencia y de científico de estudiantes de carreras científicas. *Enseñanza de las Ciencias, 22*(1), 65–78.

Merino Rubilar, C., & Izquierdo-Aymerich, M. (2011). Aportes a la modelización según el cambio químico. *Educación Química, 22*(3), 212–223.

Mesquita, N. A. S., & Soares, M. H. F. B. (2008). Visões de ciência em desenhos animados: Uma alternativa para o debate sobre a construção do conhecimento científico em sala de aula. *Ciência & Educação, 14*(3), 417–429.

Miguel, H. (1999). La analogía como herramienta en la generación de ideas previas. *El Caldero de la Escuela, 73*, 85–97.

Moënne Rivas, G., Filsecher Wagner, M., Flores Clerfeuille, L., Runge, E., & Verdi Rademacher, M. (2008). *Enseñanza de las ciencias basada en la indagación (ECBI) con TIC: Informe final.* Temuco, Chile: Universidad de la Frontera.

Mortimer, E., & Scott, P. H. (2003). *Meaning making in secondary science classrooms.* Maidenhead: Open University Press.

Muñoz Rojo, M. I. (2010). Análisis del impacto en la implementación del programa enseñanza de las ciencias basada en la indagación (ECBI), en las escuelas municipalizadas de la V región, Valparaíso, Chile. In *Congreso Iberoamericano de Educación*

'*Metas 2021*' (Online). Buenos Aires, Argentina: Organización de Estados Iberoamericanos.

Nascimento, S. S., & Vieira, R. D. (2008). Contribuições e limites do padrão de argumento de Toulmin aplicado em situações argumentativas de sala de aula de ciências. *Revista Brasileira de Pesquisa em Educação em Ciências, 8*(2), n/pp.

Niaz, M. (2001a). A rational reconstruction of the origin of the covalent bond and its implications for general chemistry textbooks. *International Journal of Science Education, 23*(6), 623–641.

Niaz, M. (2001b). How important are the laws of definite and multiple proportions in chemistry and teaching chemistry?: A history and philosophy of science perspective. *Science & Education, 10*(3), 243–266.

Niaz, M. (2001c). Understanding nature of science as progressive transitions in heuristic principles. *Science Education, 85*(6), 684–690.

Niaz, M. (2009). Progressive transitions in chemistry teachers' understanding of nature of science based on historical controversies. *Science & Education, 18*(1), 43–65.

Niaz, M., Abd-el-Khalick, F., Benarroch, A., Cardellini, L., Laburú, C. E., Marín, N., Montes, L. A., Nola, R., Orlik, Y., Scharmann, L. C., Tsai, C. C., & Tsaparlis, G. (2003). Constructivism: Defense or a continual critical appraisal?: A response to Gil-Pérez et al. *Science & Education, 12*(8), 787–797.

Niaz, M., & Fernández, R. (2008). Understanding quantum numbers in general chemistry textbooks. *International Journal of Science Education, 30*(7), 869–901.

Niaz, M., Klassen, S., McMillan, B., & Metz, D. (2010). Reconstruction of the history of the photoelectric effect and its implications for general physics textbooks. *Science Education, 94*(5), 903–931.

Niaz, M., & Rodríguez, M. A. (2005). The oil drop experiment: Do physical chemistry textbooks refer to its controversial nature? *Science & Education, 14*(1), 43–57.

Nicaretta, G. M., & Bueno, M. A. P. (2011). O uso de mídias no ensino de ciências por investigação. In *X Congresso Nacional de Educacão-Educere e I Seminário Internacional de Representações Sociais, Subjetividade e Educação-SIRSSE* (Online) (pp. 1716–1727). Curitiba, Brazil: Pontifícia Universidade Católica do Paraná.

Núñez-Oviedo, M. C., Clement, J., & Rea-Ramirez, M. A. (2008). Developing complex mental models in biology through model evolution. In J. Clement & M. A. Rea-Ramirez (Eds.), *Model based learning and instruction in science* (pp. 173–119). Dordrecht, The Netherlands: Springer.

Oki, M. C. M., & Moradillo, E. F. (2008). O ensino de história da química: Contribuindo para a compreensão da natureza da ciencia. *Ciência & Educação, 14*(1), 67–88.

Osborne, J. (2010). Arguing to learn in science: The role of collaborative, critical discourse. *Science, 328*(5977), 463–466.

Padilla, K., & Furió-Mas, C. (2008). The importance of history and philosophy of science in correcting distorted views of 'amount of substance' and 'mole' concepts in chemistry teaching. *Science & Education, 17*(4), 403–424.

Pansera-de-Araújo, M. C., Gehlen, S. T., Mezalira, S. M., & Scheid, N. M. J. (2009). Enfoque CTS na pesquisa em educação em ciências: Extensão e disseminação. *Revista Brasileira de Pesquisa em Educação em Ciências, 9*(3), n/pp.

Paruelo, J. (2003). Enseñanza de las ciencias y filosofía. *Enseñanza de las Ciencias, 21*(2), 329–335.

Peme-Aranega, C. (1997). El carácter epistemológico interdisciplinar de la didáctica de las ciencias. *Educación en Ciencias, 1*(2), 5–13.

Pitombo, M. A., Almeida, A. M. R., & El-Hani, C. N. (2008). Conceitos de gene e idéias sobre função gênica em livros didáticos de biologia celular e molecular do ensino superior. *Contexto e Educação, 22*(77), 81–110.

Pocoví, M. C. (2007). The effects of a history-based instructional material on the students' understanding of field lines. *Journal of Research in Science Teaching, 44*(1), 107–132.

Pocoví, M. C., & Finley, F. (2002). Lines of force: Faraday's and students' views. *Science & Education, 11*(5), 459–474.

Pocoví, M. C., & Finley, F. (2003). Historical evolution of the field view and textbook accounts. *Science & Education, 11*(5), 459–474.

Porro, S., & Acevedo, C. (2011). Una perspectiva de género en los temas de ciencia, tecnología y sociedad. *Revista de Educación en Ciencias/Journal of Science Education, 12*(1), 21–24.

Pujalte, A. P., Porro, S., & Adúriz-Bravo, A. (2011). Las imágenes de ciencia del profesorado: Su relación con una educación científica de calidad para todos y todas. *Tecné, Episteme y Didaxis*, extra issue, 410–415.

Quintanilla, M., Izquierdo-Aymerich, M., & Adúriz-Bravo, A. (2007). Discusión en torno a una propuesta para introducir la historia de la ciencia en la formación inicial del profesorado de ciencias. In M. Izquierdo-Aymerich, A. Caamaño, & M. Quintanilla (Eds.), *Investigar en la enseñanza de la química: Nuevos horizontes: Contextualizar y modelizar* (pp. 173–196). Bellaterra, Spain: Servei de Publicacions UAB.

Ragout de Lozano, S., & Cárdenas, M. (2002). Some learning problems concerning the use of symbolic language in physics. *Science & Education, 11*(6), 589–599.

Raviolo, A. (2009). Modelos, analogías y metáforas en la enseñanza de la química. *Educación Química, 20*(1), 55–60.

Reis, J. C., Guerra, A., Braga, M., & Freitas, J. (2001). History, science and culture: Curricular experiences in Brazil. *Science & Education, 10*(4), 369–378.

Rengifo Gallego, L. A., & Claret Zambrano, A. (2009). La epistemología de Toulmin como referente para la selección de contenidos en la enseñanza de la biología. *Tecné, Episteme y Didaxis, 26*, 123–141.

Rezende, F. S., Ferreira, L. N. A., & Linhares Queiroz, S. (2010). Concepções a respeito da construção do conhecimento científico: Uma análise a partir de textos produzidos por estudantes de um curso superior de química. *Revista Electrónica de Educación en Ciencias, 9*(3), 596–617.

Rodríguez, M. A., & Niaz, M. (2002). How in spite of the rhetoric, history of chemistry has been ignored in presenting atomic structure in textbooks. *Science & Education, 11*(5), 423–441.

Rosa, K., & Martins, M. C. (2009). Approaches and methodologies for a course on history and epistemology of physics: Analyzing the experience of a Brazilian university. *Science & Education, 18*(1), 149–155.

Rufatto, C. A., & Carneiro, M. C. (2009). A concepção de ciência de Popper e o ensino de ciências. *Ciência & Educação, 15*(2), 269–289.

Sanabria Rojas, Q. A., Pérez Miranda, R., & Gallego Badillo, R. (2009). Modelos científicos y algunas implicaciones en la formación inicial de profesores de química. *Tecné, Episteme y Didaxis, 25*, 85–98.

Santos, V. C., & El-Hani, C. N. (2009). Idéias sobre genes em livros didáticos de biologia do ensino médio publicados no Brasil. *Revista Brasileira de Pesquisa em Educação em Ciências, 9*(1), a6.

Santos, V. C., Joaquim, L. M., & El-Hani, C. N. (2012). Hybrid deterministic views about genes in biology textbooks: A key problem in genetics teaching. *Science & Education, 21*(4), 543–578.

Sasseron, L. H., & Carvalho, A. M. P. (2011). Construindo argumentação na sala de aula: A presença do ciclo argumentativo, os indicadores de alfabetização científica e o padrão de Toulmin. *Ciência & Educação, 17*(1), 97–114.

Scandroli, N., & Rocha, A. (2002). Las concepciones de ciencia de los docentes de enseñanza general básica (EGB): Un diagnóstico. *Revista de Educación en Ciencias/ Journal of Science Education, 3*(1), 38–41.

Schwartz, R. S., Lederman, N. G., & Lederman, J. S. (2008). An instrument to assess views of scientific inquiry: The VOSI questionnaire. In *2008 Annual International Conference of the National Association for Research in Science Teaching* (Online). Baltimore, MD: NARST.

Sensevy, G., Tiberghien, A., Santini, J., Laubé, S., & Griggs, P. (2008). An epistemological approach to modeling: Cases studies and implications for science teaching. *Science Education, 92*(3), 424–446.

Sepúlveda, C., & El-Hani, C. N. (2004). Quando visões de mundo se encontram: Religião e ciência na trajetória de formação de alunos protestantes de uma licenciatura em ciências biológicas. *Investigações em Ensino de Ciências, 9*(2), 1–49.

Silva, C. P., Figueirôa, S. F. M., Newerla, V. B., & Mendes, M. I. P. (2008). Subsídios para o uso da história das ciências no ensino: Exemplos extraídos das geociências. *Ciência & Educação, 14*(3), 497–517.

Siqueira-Batista, R., Silva, C. A., Maria-da-Silva, L., Rôças, G., Souza, R. R. M., Oliveira, A. L., Pires-do-Prado, H. J., & Helayël-Neto, J. A. (2010). Nanociência e nanotecnologia como temáticas para discussão de ciência, tecnologia, sociedade e ambiente. *Ciência & Educação, 16*(2), 479–490.

Soto, C., Angulo, F., & Sanjosé, V. (2011). Applying the epistemological theories of Kuhn and Lakatos to the analysis of scientific progress in conceptual change research. *Revista de Educación en Ciencias/Journal of Science Education, 12*(2), 62–67.

Souza, V. C. A., & Justi, R. (2010). Estudo da utilização de modelagem como estratégia para fundamentar uma proposta de ensino relacionada à energia envolvida nas transformações químicas. *Revista Brasileira de Pesquisa em Educação em Ciências, 10*(2), n/pp.

Speltini, C., & Dibar Ure, M. C. (2002). Conservation in physics teaching, history of science and in child development. *Science & Education, 11*(5), 475–486.

Stipcich, S. (2008). Las argumentaciones de estudiantes de polimodal sobre la interacción eléctrica. *Caderno Brasileiro de Ensino de Física, 25*(3), 397–423.

Tamayo, O. E., & Restrepo, F. (Eds.). (2012). *Niños y maestros: El caso de pequeños científicos.* Manizales, Colombia: Editorial Universidad Autónoma de Manizales.

Tamayo, O. E., & Sanmartí Puig, N. (2007). High-school students' conceptual evolution of the respiration concept from the perspective of Giere's cognitive science model. *International Journal of Science Education, 29*(2), 215–248.

Tavares, M. L., Jiménez-Aleixandre, M. P., & Mortimer, E. F. (2010). Articulation of conceptual kowledge and argumentation practices by high school students in evolution problems. *Science & Education, 19*(6), 573–598.

Teixeira, E. S., Freire Jr., O., & El-Hani, C. N. (2009). A influência de uma abordagem contextual sobre as concepções acerca da natureza da ciência de estudantes de física. *Ciência & Educação, 15*(3), 529–556.

Trani Gomes, A. D., Borges, A. T., & Justi, R. (2008). Students' performance in investigative activity and their understanding of activity aims. *International Journal of Science Education, 30*(1), 109–135.

Uribe, M., Quintanilla, M., Izquierdo-Aymerich, M., & Solsona i Pairó, N. (2010). Aplicación del modelo de Stephen Toulmin a la evolución conceptual del sistema circulatorio: Perspectivas didácticas. *Ciência & Educação, 16*(1), 61–86.

Viana, H. E. B., & Porto, P. A. (2010). The development of Dalton's atomic theory as a case study in the history of science: Reflections for educators in chemistry. *Science & Education, 19*(1), 75–90.

Viau, J., Moro. L., Zamorano, R., & Gibbs, H. (2008). La transferencia epistemológica de un modelo didáctico analógico. *Revista Eureka sobre Enseñanza y Divulgación de las Ciencias, 5*(2), 170–184.

Viau, J., Zamorano, R. O., Gibbs, H., & Moro, L. (2006). Ciencia y pseudociencia en el aula: El caso del "bosque energético". *Revista Electrónica de Enseñanza de las Ciencias, 5*(3), 451–465.

Vieira, R. D., & Nascimento, S. S. (2009). Uma visão integrada dos procedimentos discursivos didáticos de um formador em situações argumentativas de sala de aula. *Ciência & Educação, 15*(3), 443–457.

Vilela-Ribeiro, E. B., & Benite, A. M. C. (2009). Concepções sobre natureza da ciência e ensino de ciências: Um estudo das interações discursivas em um núcleo de pesquisa em ensino de ciências. *Revista Brasileira de Pesquisa em Educação em Ciências, 9*(1), n/pp.

Villani, A., Silva Dias, V., & Melgaco Valadares, J. (2010). The development of science education research in Brazil and contributions from the history and philosophy of science. *International Journal of Science Education, 32*(7), 907–937.

Wainmaier, C., & Salinas, J. (2008). Incomprensiones sobre el modelo de punto material en estudiantes universitarios que regularizaron Física I. *Revista de Enseñanza de la Física, 21*(1), 29–42.

Waldegg, G. (2001). Ontological convictions and epistemological obstacles in Bolzano's elementary geometry. *Science & Education, 10*(4), 409–418.

Waldegg, G. (2005). Bolzano's approach to the paradoxes of infinity: Implications for teaching. *Science & Education, 14*(6), 559–577.

CHAPTER 18

The History of Science and Science Education: Tools for Practice and Research

Nelio Bizzo

Abstract

This chapter discusses different perspectives in which the field of the History of Science (HS) can be related to science education, and brings a new perspective for further collaborative work in both areas. Considering different ways in which the importance of the social context is recognized in the conceptualization process, new heuristic tools are discussed and applied to real situations. The case of the relations between Evolution and Genetics is focused, showing how learning difficulties can be taken as literal clues for historiographical investigation. Another case is presented, namely, the way Charles Darwin had considered the inclusion of humans since the first edition of "Origin of Species". There was a long debate on this issue and historians had been divided, but classroom work gave important contributions to this issue. The chapter closes with the discussion of the possible relations between cognitive activity, as a socially defined process, and the reappraisal of key concepts for HS.

Keywords

history of biology – evolution education – history of Mendelism – social reconceptualization of science

History of science may not mean the same thing to educators and social scientists. In fact, Brush (1989) discusses criticisms of the use of history of science in science classrooms which are based on the idea that students might be bored by experiments and ideas which are no longer valid. Such critiques, Brush argued, entail a caricature of science lessons, one which obviously cannot be genuinely informative to social scientists, since science teachers' jobs cannot in reality be reduced to performing a list of experiments or reading conceptual definitions to the students in their classrooms.

 | DOI: 10.1163/9789004409088_018

These complaints involve a certain image of schools or, more precisely, of science classes, which can be found outside the community of science educators, and may explain to some extent the tensions between these two communities, namely, educators and social scientists. I intend to summarize some ways in which history of science is considered useful for educators and then suggest some potential interchanges between educational and historiographic research programs.

1 History of Science as a New Field of Knowledge

What we call history of science (HS) at the present time is a very specialized field of knowledge, a tributary to the longstanding field of history of philosophy, which became an academic discipline in the post-1945 era, along with a range of developments in higher education and the rapid expansion of science and technology. In fact, it has been framed as a consequence of the astonishing growth of scientific knowledge and the need to retrieve useful information in a quick and precise way. For many scientists of the post-WWII period, HS provided a huge repository of information which, while not cutting-edge science, could be taken as "epistemological cul-de-sacs" which might eventually give way to a new flow of creative thinking. In this restricted view, HS could provide a repository not only of factual scientific knowledge itself, but also of ways of constructing and communicating scientific representations. From this standpoint, HS could allow scientists to avoid the burden of every new investigation being an *ad initio* endeavor by extending the collective memory of the scientific community.

It is worthwhile to recall the seminal words – written during a time of war – of the U.S. Director of the Office of Scientific Research and Development, Dr. Vannevar Bush, from his well-known article from 1945, "As We May Think":

> Professionally our methods of transmitting and reviewing the results of research are generations old and by now are totally inadequate for their purpose. If the aggregate time spent in writing scholarly works and in reading them [...], the ratio between these amounts of time might well be startling. Those who conscientiously attempt to keep abreast of current thought, even in restricted fields, by close and continuous reading might well shy away from an examination calculated to show how much of the previous month's efforts could be produced on call. [...] The difficulty seems to be, not so much that we publish unduly in view of the extent and variety of present day interests, but rather that publication has been

> extended far beyond our present ability to make real use of the record. The summation of human experience is being expanded at a prodigious rate, and the means we use for threading through the consequent maze to the momentarily important item is the same as was used in the days of square-rigged ships. (Bush, 1945)

Bush proposed the concept of "memex", meaning "memory extension", given how important it was to change the way information, including visual information, was stored and retrieved. He wrote, "it would be advantageous to be able to snap the camera and to look at the picture immediately", a situation very familiar today, but one that almost nobody could imagine half a century ago. The rapid growth of science from the early twentieth century through the post-war era and beyond brought a new perspective to HS, as it was clear that the sources of information for future historical research would increase dramatically in the short term.

This situation brought some fresh air to debates about reconstructions of science, some of which were considered simplistic, biased, and *whiggish*. On the one hand, some scientists from the hard sciences claimed that the logic of past science was not properly addressed in the social sciences, and tended to look for purely "logical" – or "epistemic" – forms of HS. Others, on the other hand, complained that the social context of science was not properly addressed in the "epistemic" approach. These different views eventually gave rise to the so called *internalist* vs. *externalist* debates in the historiography of science, *before* the aforementioned post-war expansion. *Externalism*, in the context of the historiography of science, is the view that scientific knowledge changes mainly due to its social context – the sociopolitical climate and the surrounding economy determine scientific progress. *Internalism* stresses the intellectual component of the scientific enterprise as distinct from the social acceptance of a given statement, such as, "humans and apes have a common ancestor" or "Earth is not the center of the universe".

Between 1904 and 1909 Emanuel Rádl (1873–1942) published his "Geschichte der biologischen Theorien", which was translated into English in 1930 (The History of Biological Theories, Oxford, 1930). Rádl recognized the importance of social context in the conceptualization process and devised different versions of the same system of ideas according to the theoretical framework adopted. For instance, the second volume of his book is almost entirely devoted to evolutionary biology, with Rádl describing two versions of Darwinism: one by Ernst Haeckel and the other one by Charles Darwin himself (!). He held that Darwin's work was, in fact, a sort of 'sociology of nature', in which he simply applied the ruling political ideas in England at the time to the natural world.

For instance, there was an evident parallel between Adam Smith's economic theories and Darwin's view of nature as governed by competing agents. However, this was not an attempt to denounce some sort of "impurity" detected in Darwin's theorizing process. On the contrary, Rádl welcomed the rise of a "democratic" notion of biological nature, one governed (in the tradition of "matter in motion") by the laws of nature, and without the constant interventions of God present in the old archaic systems of ideas.

Although many historians described the biological sciences as self-consciously 'historical', J. B. Bury (1909), writing at the beginning of the 20th century, observed:

> From the more general influence of Darwinism on the place of history in the system of human knowledge, we may turn to the influence of the principles and methods by which Darwin explained development. It had been recognized even by ancient writers (such as Aristotle and Polybius) that physical circumstances (geography, climate) were factors conditioning the character and history of a race or society. In the sixteenth century Bodin emphasized these factors, and many subsequent writers took them into account. The investigations of Darwin, which brought them into the foreground, naturally promoted attempts to discover in them the chief key to the growth of civilization. Comte had expressly denounced the notion that the biological methods of Lamarck could be applied to social man. Buckle had taken account of natural influences, but had relegated them to a secondary plane, compared with psychological factors. But the Darwinian theory made it tempting to explain the development of civilization in terms of "adaptation to environment", "struggle for existence", "natural selection", "survival of the fittest", etc. (Bury, 1909)

Perhaps the most well-known event of this debate can be found in the Second International Congress of Science and Technology, which took place in London in at the end of June, 1931. The USSR delegation made a major impact with a paper delivered by Boris M. Hessen (1893–1936),[1] entitled "The Social and Economic Roots of Newton's Principia", which is recognized as a landmark in so-called *externalist* historiography. Hessen and his group emphasized the economic and sociopolitical forces that would have shaped Newton's science, which had long been taken as emblematic of the *internalist* perspective. He wrote:

> Earlier historical theories examined only the ideological motives in the historical activities of human beings. Consequently, they were unable

> to reveal the true origins of those motives and regarded history as being driven by the ideological impulses of individual human beings, thereby blocking the way to recognition of the objective laws of the historical process. "Opinion governed the world". The course of history depended on the talents and the personal impulses of man. The individual created history. (Hessen, 1931)

In contrast, in this new line of reasoning regarding Newton's oeuvre, the development of economic forces would have driven science at the time; his work could therefore be explained in terms of needs for increased production and distribution of goods. Other works tried to show an opposite view, stressing the intellectual content and rational nature of science, along with its cumulative nature, none of which would have parallels with social change. Science development was seen as predominantly gradual.

This explains, to some extent, the reasons for the weakening of this debate after the publication of Thomas Kuhn's seminal book, *The Structure of Scientific Revolutions* (1962). Despite the fact that he himself had been a student – and was considered a disciple – of a well-known *internalist* (Alexandre Koyré), who had published a very influential book on the internal factors and metaphysical aspects of the development of Galileo's works (*Études galiléennes*, 1939), Kuhn recognized the role of social influences in the development of science. In the Kuhnian perspective, the very bases of scientific knowledge are replaced during brief revolutionary periods. These periods are followed by periods of stability, involving a more gradual and cumulative growth of knowledge, which then eventually face increasing instability before another revolutionary period. Kuhn directly contested the idea of the gradual development of science.

Despite the criticisms faced by the Kuhnian model, it provided room for a balanced approach to HS. Although non-historians often refer to the *internalist* vs. *externalist* debate, the social sciences tend to consider both explanations as complementary ways of understanding modifications of science. Science representations became an important field of research, since critics questioned the Kuhnian view of the integrity of what was then called the *paradigm*, the group of interconnected concepts which were the very basis of scientific knowledge. Some noted that groups of concepts constituting paradigms would not necessarily stand or fall as a whole, as in the Copernican revolution (see, for instance, Toulmin, 1972). At the same time, extra-scientific influences on the origins of scientific ideas were pointed out, as seen in the Darwinian revolution (see, for instance, Gale, 1972).

The closely related field of the history of ideas provided more evidence of the need for a balanced approach to HS. The Kuhnian perspective recognized that concepts were related to perceptions, since people who share certain concepts tend to have similar perceptions; in other words, different social groups may not agree as to what constitutes relevant evidence to support a given belief about nature, since they may not perceive evidence in the same way. This means that "worldviews" are as important as rational explanations, since the production of meaning depends on both. For instance, one of the major problems faced by the Darwinian perspective was the idea that the world of nature possesses a natural harmony, which was an important component of natural theology. Perfect adaptation to existing circumstances was long taken as a proof of the perfection of creation and was even one of St. Thomas Aquinas' proofs of the existence of God. Darwin himself acknowledged that his views on adaptation were, at first, deeply influenced by William Paley's model of natural theology, which could see perfection everywhere. The notion of an imperfect world, governed by chance, was not part of Darwin's contemporary Western cultural milieu (see, for instance, Greene, 1981).

HS became closely linked to the sociology of science as a consequence of this dual approach, which renders both internal and external factors equally important in knowledge change, taking the personal and social into account along with other relevant dimensions of scientific endeavor. As the late John Ziman (1978) puts it:

> (…) no scientist is a disembodied observing and conceptualizing instrument; he is a conscious human being, born and reared in the common life of his era. Long before he is taught about electrons, and genes, and exogamous fratries, he has acquired practical experience of pots and pans, cats and dogs, uncles and aunts. Although such mundane objects are seldom discussed as such in high science, they are not excluded from its realm. However fantastic it may appear on its wilder shores, the scientific consensus includes, by definition, the matter-of-fact, and must be coherent with everyday reality. (Ziman, 1978, p. 9)

This "mundane" dimension, as Ziman calls it, is part of the cultural milieu of a given time. It is an important factor in the scientific process since science is governed to some extent by consensus. The epistemological challenge to science refers not only to the ways scientists acquire their views of the world, but also to the extent to which they "objectively" see the same sets of evidences as being in support of their common views, and, in addition, to whether there may be conceivable alternatives that are strong enough to force scientists to question such views.

Scientific knowledge cannot be taken as "absolute truth" for several reasons (including the social production of consensus), especially as it is inherently vulnerable to two sorts of error. First, the achievement of intersubjective agreement is seldom rigorous and may admit mistaken beliefs which, although maintained collectively, eventually prove false when they cannot explain new evidence brought to the stage. Second, and more importantly, beliefs held by the scientific community may be "self-sustained delusions" (as Ziman put it), in the sense that scientists are almost always deliberately trained to hold a particular attitude toward natural phenomena, and are not as open to alternative views as some would like them to be (Ziman, 1978, p. 8). Therefore, change is inherent to the scientific enterprise: scientific paradigms, or "wordviews", inevitably change over time.

2 Science and History: Tensions

This history of science S *scenario* emerged at the end of the 20th century, as a result of tensions between different communities. As Brush stated, "professional historians of science, seeing themselves as historians, rather than scientists, criticized scientists for promulgating '*Whiggism*', and some of them overemphasized the social context at the expense of the technical content of science", but HS would have "started to rebuild the bridges to science" (Brush, 1989, p. 70).

Science education is related to the expansion of the common views shaped and held by scientists to a larger group of people. That being said, it was only toward the end of the 20th century that history and philosophy of science seem to have finally begun to come to terms with science education (Matthews, 1990). The so-called "traditional view", as found in the well-known writings of John Dewey, Jerome Bruner, and James Conant, long maintained that students should have contact with the roots of scientific knowledge, as this could provide motivation and context for understanding more complex ideas. However, the question was: "Whose history of science?" There was no apparent consensus among educators, "hard science" practitioners, and social scientists.

These "traditional views" differed significantly from more recent "recapitulation theories", which are used to draw parallels between the construction of concepts in the past and in the minds of students in the present. While the latter have produced some interesting perspectives, they have long faced serious restrictions (Bizzo, 1992a). For instance, Piaget and Garcia, who investigated such parallels, did not discern a correspondence in the *content* of what scientists thought in the past and what students think today, but rather in the *processes* involved in changes in ways of thinking (Piaget & Garcia, 1987, p. 39). In any case, recapitulating the historical process is not *possible, feasible,* or *desirable* (Nersessian, 1992, p. 54). This does not mean, however, that HS is of no importance at all for teaching. On the contrary, it has been seen as highly relevant to the three main themes of science education, generally called "ideas about science" (IAS). Together with the practice of science, in the sense of applying science to real situations, and science inquiry, as related to understanding the methods of science, HS has much to offer for teaching the nature of science (Osborne et al., 2003), as we will see below.

3 Science Teachers and Professional Historians

In one more recent perspective, historical developments in scientific knowledge should be presented to students in an explicit manner, in order to nurture their appreciation of scientific processes and to demonstrate "the ways and extent to which such developments have been affected by the demands and expectations of society at different points in history" (Osborne et al., 2003, p. 706). This indicates that the sociological approach to history and philosophy appears to be as important to educators as to historians.

It is important to acknowledge that the entire sequence of past scientific knowledge is not necessarily relevant to understanding science today. For instance, understanding the Ptolemaic system is not necessary for every classroom lesson about the heliocentric model. However, if one reads Copernicus from primary sources (for instance, *De revolutionibus orbium caelestium* or *Comentariolus*), one will have to contend with Ptolemaic terminology; therefore, knowing the Ptolemaic system may in fact become both relevant and necessary in more in-depth learning contexts. Copernicus and his fellow astronomers followed the only system known at the time to enable the prediction of astronomical events, and, as Roberto Martins remarked in the introduction to his (Portuguese) translation of *Comentariolus,* "He wrote his works to contemporary astronomers and not to us" (Martins, 2003, p. 27). Even the explanations of the past can seem baffling today. For instance, the reasons for Copernicus' devotion to uniform circular motion cannot be properly understood in the 21st century, since they date back to the ancient Greek philosophers, and were linked to their idea of perfection: "In the case of the movement of the Earth around the Sun, for instance, Copernicus described an eccentric circle, without an epicycle (as Ptolemy and, before him, Hiparcus had done to explain the movement of the Sun)" (Martins, 2003, p. 82).

Another example can be found in Charles Darwin's works. It is not necessary to understand the theories of Lamarck or Buffon, for instance, in a lesson on evolution today. However, if one reads Darwin's own works one will certainly find more than natural selection as the basis of his explanations for evolutionary change. When he himself wrote that his views were different from those of Lamarck, he was not referring to the inheritance of acquired characteristics, but rather to the tendency of gradual change toward improving characters, the "progression of animals", as it was called (Martins, 2007). In the Lamarckian perspective, the heart of a mammal would be far better than the one of a lizard or of a frog.[2]

Historians have to deal with a wide range of ways of conceiving of the natural world in order to produce historical narratives. However, in a science lesson, this range must necessarily be narrowed. This 'bottleneck effect' reduces the range of HS questions in use in classrooms, and, therefore, can lead to false history, or, more precisely, to the sort of "pseudohistory" Douglas Allchin has described in the context of science education (Allchin, 2004). He wrote:

> Not every science teacher can become a professional historian, of course. While teachers should certainly be concerned about historical accuracy, my primary focus here is not false history [...] per se, but *pseudohistory*. Pseudohistory conveys false ideas about the historical process of science and the nature of scientific knowledge, even if based on acknowledged facts. Fragmentary accounts of real historical events that omit context can mislead, even while purporting to overemphasize the contributions of one individual, minimize the role of accident or errors, simplify the investigative process, disguise less than noble motivations, hide the effect of personal or cultural values [...]. They turn real science into an imaginary idealized science. Such a misleading selective history masquerading as responsible history is justly called *pseudohistory*. (Allchin, 2004, p. 186)

Condensing a long story about Darwin's work – or Copernicus' or Galileo's – for a science lesson is not an easy task, and carries the risk of presenting a fragmentary account that omits certain key aspects of context or oversimplifies the investigative process. The production of historical narratives for school use is a process in which history has to be re-written, and, further, one in which it is necessary to (deliberately) leave out at least some relevant events. 'Real science' and 'imaginary idealized science' may not be as clear cut entities as some might believe, after all.

Recent bibliographies on the subject tend to identify 'real science' with a nature of science (NOS) perspective, wherein science education is not limited to facts and concepts, but also involves the understanding and practice of the processes involved in the production of scientific knowledge. This means that science has to be related to observation and inference in order to establish theories, without disregarding the empirical basis of scientific inquiry, which also involves imagination and creativity. However, from this approach, science must be seen as a tentative and subjective form of knowledge, which is socially and culturally embedded (Lederman, 2004; Lederman & Lederman, 2005).

History of science is always linked to wordview. Thus, educators may be in trouble not because they are propagating idealized science, but rather because their "pictures" are not in accordance with the ones upon which scientists are currently agreeing. Recently, researchers suggested that HS should be abandoned in the teaching of genetics, as the most modern approaches to heredity present a view of how genes govern biological characteristics which is essentially distinct from the traditional Mendelian one (Dougherty, 2009). The latter traditional, deterministic framework will be of no use in understanding the enormous quantity of genetic information to which people will increasingly have access. The practice of human genetics is fast becoming a real and concrete task for the average person, who will have to deal with complex and extensive information about their own genetic makeup. The picture of the world geneticists see today is very different from the one Gregor Mendel conceived in 1865. In other words, non-Mendelian genetics should be seriously considered by those who develop school curricula.[3]

This example shows that even the most solid scientific knowledge – knowledge which is as reliable as any scientific truth could possibly be – may nonetheless change dramatically during a relatively brief span of time. This underlines the importance of HS for education, as different "worldviews" are continuously being produced and constitute a repository of alternative ways in which the natural world can be conceived. Of course, the simple act of presenting students with the ways scientists resolved relevant problems in a given social and cultural context does not mean that those learners will consequently figure out novel ways to construct a scientific representation. However, as Nersessian (1992) has argued, those historical processes may provide a model for the learning activity itself, for instance, by promoting the idea that inherited traits do not appear in every generation. This new perspective throws light not only on the relevance of HS to educational research but also to classroom practice.

4 History of Science and Curriculum Design: The Case of Genetics

There is another area in which HS is relevant to education, that of curriculum design. Several aspects of science curricula are informed and clarified by history, for instance, the sequence of the presentation of concepts. In physics, for example, "Physics I" will likely cover the work of Galileo and Newton, and probably not that of Einstein. As we have argued vis-à-vis the Ptolemaic and Copernican systems, it is not always necessary to retrieve old ideas in order to learn modern ones. Similar reasoning also indicates that the past cannot be seen as a time when things were more simple and easy. As argued elsewhere, the idea of a simpler past is a very *whiggish* one (Bizzo, 1992a); as well, it is not fair or wise to take HS as a simple chronology of cumulative facts.

Another example can be seen in the connections between genetics and evolution. Despite the fact that Gregor Mendel may not be seen as a scientist who tried to solve the problem of heredity – his foci were in fact hybridization and the investigation of the distinctions between species and varieties (Lorenzano, 1998) – school biology courses worldwide adopt a historical approach to teaching genetics. In an influential work by William Bateson that was published at the beginning of the 20th century, the author lamented that Charles Darwin had never heard of Mendel: "Had Mendel's work come into the hands of Darwin, it is not too much to say that the history of the development of evolutionary philosophy would have been very different from that which we have witnessed" (1902, p. 39). In the following edition of the book, seven years later, Bateson not only repeated the phrase, but also stressed the same argument, adding another similar comment (Bizzo & El-Hani, 2009). This was certainly the first time that this historical version was presented to the public, and it has informed educators since, especially in the area of curriculum design. As we have argued elsewhere:

> In many sources, one can learn that Darwin "lacked" a theory of heredity; and, therefore, he has not been able to achieve a more sophisticated view of biological evolution. According to this stance, school could provide a prior basis for the study of heredity, so that students – unlike Darwin, and thus, avoiding the difficulties he had – could begin to study evolution with a proper background of genetics. (Bizzo & El-Hani, 2009, p. 108)

This version is not restricted to lay or school audiences, but can also be found among academics. For instance, Mayr (1991) stated that "Darwin never heard of Mendel's work and was never able to solve the problem" (p. 109), a vision echoed by Giordan (1987). More recently, another version of the same idea appeared, which revealed that an "uncut" – and therefore unread – copy of a Mendel paper was found in Darwin's archives; moreover, the author claimed that if Darwin had read it carefully "evolutionary biology would have been advanced by at least three decades" (Rose, 2000[1998], p. 43).

We investigated the above argument and concluded that HS did not support this account. We therefore contend that the notion that the traditional curriculum was based on historically accurate accounts of the development of science should be abandoned. The traditional argument runs as follows:

> Darwin lacked the theoretical framework of Mendelian genetics and, therefore, was not able to develop the so-called evolutionary synthesis, something which has eventually become possible in the 1930s. School would then offer an epistemological bypass to students, showing Mendel's work as directly connected with Darwin's. Hence, when students begin to study evolution, they would have already achieved what Darwin lacked in his time, namely Mendelian genetics. The school environment would be therefore presented as a 'fast lane' in terms of curriculum design aiming at understanding biological evolution. We believe that this argument is a fallacious one, with no sound evidence, in either historical or cognitive psychological grounds. (Bizzo & El-Hani, 2009, p. 111)

As a result of the teaching process, students' understanding of biological evolution is very poor in many different places of the world (for a revision of literature, see Hokayem & BouJaoude, 2008). It is likely that curricular organization plays a significant role in this situation. Schools present students with the works of these two scientists framed as complementary projects. However, Mendel and Darwin could not themselves see how compatible their perspectives were in their own time. Darwin himself had experimented on several species, including sweet peas, and published very similar results to Mendel's, having reached the 3:1 ratio in F2 generations. His experiments with snapdragons showed that the first generation produced common plants; and, in the second generation, "out of a hundred and twenty-seven seedlings, eighty-eight proved to be common snapdragons", as expected in Mendelian terms, "and thirty-seven were perfectly peloric,[4] having reverted to the structure of their one grandparent" (Darwin, 1868, pp. 70–71). The recessive trait reappeared in the F2 offspring, in Darwin's views, as a result of "reversion", a sort of change in the particle mediating inheritance that would "jump back" to a previous condition, as found in the grandparent. This means that the hereditary particles in Darwin's conceptual framework were very flexible; Mendel's, on the contrary, was based on the idea that particles traveled unchanged through generations. In other words, Darwin believed in "soft inheritance", while Mendel worked under a strict "hard inheritance" model.

> When we flutter the pages of the 'Proceedings' which contain Mendel's paper [*referring to the 1866 one*], we discover an additional reason

> (probably the chief reason) why his work was so little understood either in the Brünn Society for the Study of Natural Science or by the wider scientific public of those days. We note that at the very same meeting at which Mendel's paper was read, Professor Alexander Makowsky, one of the leading members of the society, refers with the utmost enthusiasm to Darwin's theory of the origin of species (…) its theme must have held the minds of biologists captive (…) since the consciousness of the epoch was entirely filled by the flood of ideas contained in Darwinian theory and its consequences, people would not trouble themselves to make a place in their minds for the profound and peculiar ideas of Mendel, even though these were concerned with a kindred field. (Iltis, 1966, p. 178)

This discussion is not aimed at exploring the fate of Mendel's work in detail, but rather meant to stress how HS can be relevant to curriculum design. The two scientists in question did not see how their work could be complementary; a larger, qualified audience could not immediately see the connections between them. However, school teachers are told that understanding this very connection should be an easy task for young students – a statement which may well be historically, factually, and epistemologically inaccurate. Thus, HS can be used by educators in planning science lessons which incorporate some of the original problems theories faced in the past, which are likely to appear – in some sense at least – when they are taught today.

In addition, it is worthwhile to note that, in this case, HS has produced divergent views on the work of the very same scientists. History of science does not provide 'final judgments' of the scientists of the past, or an 'objective' report of a scientist's lifelong work, but, rather, continuously produces new insights and new ways of interpreting the past. This dynamic feature of HS depends, to a certain extent, on the research effort carried out. For instance, 18th-century science has been far less studied than that of the 19th and 20th centuries, not to mention Arabic sciences, etc. As new studies progress, light is thrown on unknown fields and may reveal novel ways of thinking about the past. The bridge between HS and education may be broader than originally thought and may admit transit in both directions, as we discuss in the following section.

5 History of Science and Education: Analogies and Metaphors

In the post-1945 period – the early days of HS as an academic discipline – HS was seen as a sort of repository of factual knowledge for scientists, and, as a result, for educators as well. In fact, this idea of HS is often linked to a short-sighted educational focus on profitable skills, which results in a pedagogical approach that may erode students' ability to criticize authority. As Martha Nussbaum has argued, the humanities should bring a broader philosophical perspective to school curricula (Nussbaum, 2010). However, another role for HS in science education has been recently proposed, in which not only facts and theories, but also the processes for constructing, changing, and communicating scientific representations are framed as important for educators. As Nancy Nersessian puts it:

> In coming to understand the constructive practices of scientists, educators will be in a better position to devise explicit strategies for leading students through their own constructions of extant scientific representations. Examinations of the history of scientific change will enable us to discern the specific processes that generate new scientific conceptual structures and the means by which those who have created these structures communicate them to others, that is, instruct other scientists how to construct them for themselves. The recommendation is, thus, to "mine" the historical data – publications, diaries, notebooks and correspondence – for these practices and then integrate into our instructional procedures what we learn about how scientists have brought about conceptual change. (Nersessian, 1992, p. 54)

Instructional procedures could be further informed by HS in the search for parallels between the ways scientists changed their ways of conceiving phenomena and the changes which are expected in students' minds. Nancy Nersessian reported that her own investigations of the major conceptual "revolutions" in physics revealed the repeated use of heuristic procedures, including analogies. Such practices allow scientists to abstract from existing conceptual structures, creating genuinely novel ones. A long list of publications on the uses of analogies in science education recommends some caution, although it is clear that scientists' understandings of abstract phenomena can benefit from the concrete references generated by analogies.

Science teachers frequently use analogies to render unfamiliar matters comprehensible to their students, but – since many important social actors will not necessarily subscribe to the exact same viewpoint – an agreed-upon interpretation of the particular phenomena under scrutiny cannot be taken for granted. Nonetheless, while being of less use as a tool for generating concrete references than perhaps previously hoped, analogies may serve as an important tool for engagement in the learning process. From this perspective, meaning in science would in fact be derived from discourse by both teachers and learners, rather than being independent of it, as is usually assumed (Heywood, 2002). Science is, as John Ziman (2000) argued, basically a group of theoretical models which operate through analogy and metaphor.

> It is clear that scientific maps, models, metaphors, themata, and other analogies are not just tools of thought, or figures of speech. They are of the very substance of scientific theory. As sources of meaning and understanding, they stand on equal footing with explicit verbal and symbolic representations. (Ziman, 2000, p. 150)

As mentioned before, the scientific community works on the basis of "consensibility" and "consensuality", through forms of "unambiguous" communication (Ziman, 1978) which seldom encompass the educational community, which has its own specific rules and traditions.

Thus, analogies may play significantly different roles for scientists and educators. As argued elsewhere (Bizzo, 1993), analogies can bring completely novel conceptual structures into the minds of scientists, especially when linking different semantic fields. According to this view, the meaning of a word or group of words in a discourse is partly dependent on its relation to other words in the same conceptual area. In creative scientific thinking, linking macro- and micro-world references may lead to completely new insights; for instance, in comparing a planetary system to the atomic structure of a chemical element. However, in science education, an analogy which brings together concrete references from different semantic fields may, partly due to oversimplification, cause problems in students' understanding. For instance, it is very common to illustrate plate tectonics by depicting solid pieces of earth crust, floating buoyantly on liquid lava in convective cells, an image as common in science textbooks as it is inaccurate (King, 2010). Another example of mixed fields may be found in certain classroom experiments: any prediction of what will happen to the level of water in a small plastic cup when pierced by a needle will certainly fail if based on larger-scale receptacles. In this case, a concrete reference is of no help, as the key elements for understanding the phenomena belong to different conceptual areas.

6 Educational Research "Mining": Levels of Scrutiny

Nersessian's (1992) recommendations for the "mining" of historical data, including sparse materials such as publications, unpublished manuscripts, correspondence, etc., may well open too large a field both to educators and historians. This indicates the necessity of identifying more specific methods for prioritizing key areas of research. In the same way that no miner looks for gold in random places, but instead searches in those areas more likely to yield gold, simply drawing from all existing scientists' diaries, published and unpublished works, etc., would be of little use compared to focused research. Since the late 1980s, our research group has been developing a method for discerning which works of the past are more relevant to education today, based on three levels of scrutiny. We have approached this project via studies in Brazil and Italy encompassing a range of cultural and scientific subjects intended to refine the elements of these three levels.

The first level is students' academic performance. As mentioned above, if students cannot understand the instructional procedures in a given area, we must examine the underlying reasons for their lack of success. It is quite possible that teachers' professional skills, students' cognitive abilities, school settings, and inadequate support materials (such as textbooks), play important roles in students' low academic performance. However, it is important to also consider the organization of the subject matter itself, as a cultural artifact. As shown in the case of the connections between Mendel and Darwin, genetics and evolution, pedagogical processes often conform to a very biased, if not *whiggish*, version of the historical development of scientific knowledge. There has been a long tradition, based on various criteria for accountability in education, of blaming teachers for students' low academic performance. Before that, it was common to blame the students themselves, whether for psychological traits, family influences, malnutrition, etc. A sociological approach would recommend analyzing the situation from a broader perspective, taking into account a range of different influences, including the shaping of the body of knowledge as presented to students. The latter may be framed as a significant variable to be considered as part of a wider process of "social reconceptualization of science" (Bizzo, 1994).

The second level of scrutiny is students' discourse, which reflects their understanding of a subject area. Students construct logical models which articulate sociocultural influences, including formal and informal instructional procedures (such as texts, images, and discourses), as well as the available direct and indirect evidence. If the researcher listens carefully to the learner, it is very possible that the former may perceive the novel production of meanings and relations among concepts and evidence. The vast body of literature on students' conceptions (see Limón et al., 2009) is deeply relevant to teaching practices and educational research and has revealed a rich "intellectual ecology" in the classroom. In the educational community, there is a consensus that eliciting students' conceptions is a first step in the design of lesson plans, etc.

Our adoption of a sociocultural perspective entails that we recognize different sorts of contributions to students' ideas. A student with no direct contact with a certain phenomenon will construct representations based on indirect evidence, such as other people's reports, literature, movies, etc. However, a student with direct contact with evidence will produce what we call a "first-hand discourse", in the form of rich and persuasive representations which generate concrete examples and are based on personal experience. As we have seen in the case of models of inheritance, members of families with inherited diseases tend to have very sophisticated models by which to explain the ways a certain characteristic passes through generations; while those with healthier family histories may have much simpler ones.

The third level of scrutiny gives significant importance to the sociocultural perspective, looking at students' representations not as individual outcomes, but rather as social constructs. This means that it is necessary to study the various influences to which students are subjected in order to understand their conceptions. For instance, many of the "student misconceptions" described in the science education literature were found in science textbooks (Franzolin & Bizzo, 2008; King, 2010), TV ads, and popular films (Jordão, 2006). Although there is no evidence that sound science shown in films can correct common misconceptions about scientific research and public health issues (see, for instance, Secker, 2001), it is likely that movies and TV can play an important role in reinforcing and consolidating common beliefs.[5] A longer exposition of this third level of scrutiny will show its importance in the context of our research program.

7 The Sociocultural Perspective

In this level of scrutiny, we include research on the ways a community may conceive of a certain phenomenon, which may be a generally unrecognized form of influence on students' representations. In order to investigate this possibility, we conducted interviews with young students [ages?] in two areas with plenty of fossils. We looked for the ways used by each community to represent animals from the past in order to elucidate similar – or divergent – sociocultural factors in both locations.

A region in northeastern Brazil known as the Crato formation, where there are very important fossil remains from the Cretaceous, is mined commercially for cement manufacture and paving stones. The quarries from which the slab limestone is extracted contain an enormous quantity of rare and beautiful fossils, rendering world-class status to the site. "Fossil Konservat Lagerstätte" is a deposit known for its exceptional preservation of fossilized organisms, wherein the soft parts of those organisms are preserved in the form of impressions or casts. It has proven to be "one of the most diverse assemblages of a Cretaceous terrestrial biota known anywhere" (Martill et al., 2007, p. 4).

During a period of research in that area, we discovered previously unnoticed evidence of social views about paleontology in the local culture, as evinced in the work of artisans selling paving stones with engraved images of fossils. Contrary to first impressions, which might take these engravings as simple counterfeit souvenirs intended to defraud tourists, they are in fact artistic representations which reproduce the artisans' understandings of the meaning of fossils. Many of them could be considered valuable for their intrinsic beauty of form. These engravings are sold out of artisans' homes; and their creators are proud to say that they are handmade. They bear images not only of locally known fossils, such as scorpions, dragonflies and insects, but also of *Jurassic Park's* dinosaurs, and even parrots, disregarding scale in all figures (Figure 18.1).

These keepsakes express the local understanding of the meaning of fossils, which are seen not as proof of life 100 million years ago, in the high Cretaceous, but rather as beautiful forms engraved in solid stone. A fossil of a feather, which was found in the area and displayed in the local Paleontology museum as a tantalizing fact, was not necessarily understood as proof of the antiquity of the birds in the region (and the whole continent), but was expressed in the form of a present-day parrot (Figure 18.1b). We found that our observations of students' discourse in the region could be contextualized as part of the socially shared meaning given to fossil evidence by their community. Such "first-hand discourses" are continually reinforced by the local culture and constitute a basis for the social construction of local reality (Berger & Luckmann, 1966).

Thus, from a sociocultural perspective, cognitive skills are not fixed or fully predictable as clear cut results of teaching tasks, but rely heavily on *contexts*

(a) (b)

FIGURE 18.1 Engraved images produced in the region of the Crato Formation (Brazil)

(Rogoff & Lave, 1999). Although this understanding is not new, it is necessary to note that there are many different ways in which "context" can be defined, since it cannot be seen simply as a structure or the features of a task or domain of knowledge. Interpersonal relations and cultural aspects, such as values and beliefs, are important parts of the contexts in which actions are embedded.

Social interaction includes cognitive activity, which implies sharing socially provided tools and *schemas* for acting on reality, since

> Cognitive activity is socially defined, interpreted and supported. People, usually in conjunction with each other and always guided by social norms, set goals, negotiate appropriate means to reach the goals, and assist each other in implementing the means and resetting the goals as activities evolve. (Rogoff & Lave, 1999, p. 4)

From a sociocultural perspective, cognitive activity is not a lonely and purely logical mental task, but rather a two-leveled process which makes use, on the one hand, of tools for cognitive activity (such as scientific theories) and socially negotiated practices. On the other hand, immediate social interaction creates a material basis for individual cognitive activity, such as the artistic representations of fossils as seen above.

There is no reason to believe that these processes related to the production of "first-hand discourses" did not occur in the past. Certain frequently repeated actions may create a sort of pattern, which can then be reproduced with an economy of effort and be apprehended by its performer *as* that pattern. This process has been called "habituation", which implies that the action may be performed again in the future in the same manner and with the same economical effort (Berger & Luckmann, 1966). Therefore, the "mining" done by the educational researcher encompasses not only HS, as a tool for examining cognitive activity in the past, but also the products of the immediate social interactions related to the culture of the social group in question. Some of these products may be relegated to the past, and the "mining" thereof may involve revealing new meanings for traditional views. "Habituated" actions retain set meanings related to specific social groups and become routine, narrowing the possibilities for future actions, and providing a stable background for social activity. This means that critical cultural studies may reveal new aspects of cultures and environments which are not evident to the embedded social agents themselves, who are accustomed to certain local traditions. This phenomenon may be observed when the meanings of past artistic representations prove to be less straightforward than expected and we may extrapolate similar shifts in the realms of of science and education. In the

FIGURE 18.2 *Fanciullo con disegno di un pupazzo* (*Child with a puppet drawing*), a painting by Giovanni Francesco Caroto (1515)

case of art, a well-known painting by Giovanni Francesco Caroto (1480–1555), *Fanciullo con disegno di un pupazzo* (Figure 18.2), displayed at the Castelvecchio Museum in Verona, offers an opportunity for this sort of hermeneutical exercise.

In an article recently published in the British daily journal *The Independent*, Tom Lubbock wrote:

> Around the time when Raphael was at the height of his powers, a minor Veronese painter made his great one-off. Giovanni Francesco Caroto painted the Portrait of a Young Boy Holding a Child's Drawing (circa 1515). The boy's eager, slightly toothsome smile gives this picture a place in the history of portraiture. But the page he holds upstages it. It has the first depiction of child art in a European painting. Whoever the boy is, this stickman is presumably meant to be his own work, proudly presented. But study the sheet more closely. Lower right, notice the profile eye, drawn with an expert hand. We can imagine the boy hanging around the studio, picking up bits of paper used by the artist or his pupils for sketches, adding his own. But what of the stickman itself? It's an attempt by an experienced artist to imitate a child's handiwork. It's uneven. The scratchy, wobbly lines are persuasive. Some of the formations seem too complex – see its right eye, constructed from curved eyebrow and eyelid. Indeed the incomplete head in the corner suggests a grown-up approach.

> Children of this age push ahead, don't have a second try. And of course, this drawing is not a drawing. It's a painting of a drawing, made in the infinitely correctable medium of oil paint. Caroto has closely observed how children draw. He probably hasn't tried to unteach his own hand. He has faked it. And his careful copying has preserved for us evidence that while art styles change, children 500 years ago failed much as they do today. (Lubbock, 2010)

For many years, this painting was taken as a representation of a child's drawing, a rare element in European paintings. However, the drawing itself does not seem to have been produced by the boy depicted, who is not a child but an adolescent. In fact, around 1964, an English pediatrician named Harry Angelman (1915–1996) noticed the boy's peculiar laughing face while visiting the Castelvecchio Museum on holiday, and deduced that the painting actually illustrated a mental handicap. He had treated three mentally disabled children who exhibited some of the features he discerned immediately in the painting, including the compulsive smile. He concluded that his patients in England had a syndrome which had been recognized since the time of that painting (presumably 1515), which is now known as "Angelman syndrome".[6] Thus, the painting may hold a far different meaning than was originally presumed by most viewers. A depiction of a mentally disabled, "different" boy – without prejudice – in the early 16th century is certainly far more surprising than a child's drawing.

The region of Verona was chosen for educational research by our group for a number of reasons. There are many indications that the region was an important setting for the first discussions of what was to be called "geology" in the 17th and 18th centuries, addressing the origins of mountains and their shapes, along with the connections with landscapes that could possibly be established with sacred books. In England, the systems of Burnet, Whiston, and Woodward can be taken as landmarks of an early stage of geology in which miracles were thought to be necessary elements for explaining landscapes. In Charles Lyell's own words, these thinkers worked "to call the Deity capriciously upon the stage, and make him work miracles, for the sake of confirming (…) preconceived hypotheses", as he wrote in his famous "Principles of Geology".

Many sources – including popular and widely used science textbooks – state that, during Darwin's era, there was no idea of geological time and a literal understanding of the Bible was dominant. This version reinforces many science curricula, justifying a straightforward approach to paleontology. However, this view finds no confirmation in Lyell's words regarding the "geologists of Italy who preceded (…) the naturalists of other countries in their investigations into the ancient history of the earth (…). They refuted and ridiculed the physico-theological systems of Burnet, Whiston and Woodward". He mentions the works of Vallisneri (1727), Spada (1737), Moro (1740), and Spallanzani (1758), along with the letters between Alberto Fortis and Domenico Testa (1793). These works, as original sources, contributed to the establishment of a scientific approach to questions related to the origins and ages of geological deposits.

The conclusions of the latter group were rooted in sound evidence, since they pointed to several places near Verona where fossils had undergone an extraordinary fossilization process and comprised animals and plants of the "torrid zone",[7] which were identified as "evidently contemporaneous" with volcanic eruptions, but did not belong to present-day European fauna and flora. This was considered a crucial point, since changes in climate were considered evidence of "land revolutions", which could not be explained by short-term processes.

Works published in the 18th century left the fact that so-called "scherzi di natura", as fossils were then known, could only be taken as remains of real living beings beyond all doubt. As early as 1721, Inquisition had approved Antonio Vallisneri's (1661–1730) book on the marine remains found in mountains, in which he put forward a new view of these fossils. In addition, Vallisneri proposed a new way of explaining mountain formation in terms of land elevation (a second edition was published in 1727). Soon after his death, Anton-Lazzaro Moro (1687–1764) would confirm Vallisneri's conclusions with different sorts of proofs, including a mathematical approach showing the impossibility of a universal deluge. He examined Burnet's and Woodward's systems in detail. The fact that locations hundreds of meters above sea level held perfectly preserved tropical fauna and flora, documented in extremely well preserved and abundant fossils, brought renown to Verona and the Veneto region. Marine fossils discovered high in the mountains couldn't be understood without the theoretical glasses of deep time, since marine deposits from a past tropical climate found 850 m above sea level were clearly incompatible with present-day conditions.

The sociocultural climate of the time was surprisingly supportive of this research, since several Catholic abbots were involved in the debates about the meaning of marine fossils found high in the mountains. Lazzaro Spallanzani (1729–1799) and Alberto Fortis (1741–1803) advocated for the ideas of deep time and great "land revolutions". Others, such as Giovanni Serafino Volta (1754–1842) and Domenico Testa (1746–1825), who soon became *Monsignori*, and the priest Ermenegildo Pini (1739–1825), firmly opposed their positions, arguing in favor of a literal interpretation of the Bible. Tropical fish found in the Italian mountains would prove the universality of the deluge, which would have brought fish from faraway places to temperate regions. Furthermore,

> during the last two decades of the eighteenth century, the Catholic Church adopted a clear position towards geological questions related to the history of the Earth, because of the need to re-establish the authority of the Bible, above all against Buffon's Epoques de la Nature (1778). Thus, especially from the 1780s onward, diluvialism was resumed in several states of Italy. (Candella, 2009, p. 90)

While Napoleon was setting fire to the old aristocratic structures of continental Europe, a new way of interpreting geological evidence was becoming widespread within the scientific community. Spallanzani and Fortis faced serious difficulties in their time, and others, such as Giovanni Arduino (1714–1795) and Déodat Dolomieu (1750–1801) were openly criticized in Catholic circles. Their explanation of geological events was essentially secular, and dismissed the need for miracles and divine interventions to explain present-day phenomena (Ciancio, 1995, 2010).

The so called "debate of the three abbots" (Gaudant, 1999) was emblematic of the clashes of ideas occurring at the end of 18th century. Huge fossil fish collections from Monte Bolca, near Verona – which belonged to rich Verona families – took center stage in discussions about the truth of the universal deluge. The debate featured scientific terms and approaches used in a range of attempts to answer the question of the origin of these fossils. Volta, a diluvialist, argued that they belonged to living species, now found far away from Italy, in America, Africa, and Asia. In fact, he was the author of the masterpiece, "Ictiolithologia Veronese", which includes detailed descriptions and drawings of the Bolca petrified fish. He used Linnean terms and included the alleged popular names of the species, which might therefore be assumed to not be extinct. Fortis firmly opposed this view, arguing that the fossils were comprised of extinct species, which would have disappeared long ago. Testa, in turn, argued that the fish must have died some hundreds of years ago, in the time of the Greeks, and that they would have then been found everywhere in the near sea, since warmer salt water would have been provided by the volcanic activity of that time (see Del Prete, 2008, pp. 389–393).

The published letters of these three abbots were part of the personal library of the Italian naturalist Giambattista Brocchi (1772–1826). He is considered an important forerunner of later theorists of biological evolution (Domini & Eldredge, 2010). Despite the fact that he considered species to be stable entities, he admitted – unlike Lamarck – that species, including the living organisms of his day, could eventually perish; this view is called "Brocchi's analogy". He left his handwritten personal perspective on the debate about living versus extinct fish species from Verona in his own reprint of Domenico Testa's (1793) work. He wrote:

> The author is Ab. Testa of Rome now Monsignor, even though little versed in Natural History felt inclined to write this petty pamphlet to flatter the papacy, fearing that the fish from Bolca could contradict Moses when speaking of the deluge. He longed for a Cardinal's Hat. (Brocchi, date uncertain[8])

This is a valuable observation, since Brocchi was a close friend of diluvialists like Ermenegildo Pini, despite the fact that he kept in close contact with non-diluvialists, such as Fortis. This note was certainly not intended to become public, as it would have brought him huge problems; however, it bears witness to his own interpretation of the interconnections in the debate about the deluge at the end of that troubled century. The debate was about the nature of fossil fish, and, particularly, whether they belonged to the local fauna. Brocchi himself had discussed the nature of the mollusk fossils of the Apennines at length, in his well-known work of 1814, noting that mollusks cannot move as fast as fish, which he deemed a significant factor regarding the question of the deluge. He concluded that many species of such fossils did not belong to the local Italian marine fauna, and might only be found far away. Behind the scenes, however, another game was being played, one closely connected with political power. In any case, the key endeavor was to determine if the fossils belonged to extinct species or to living ones; in other words, whether living and petrified animals and plants belong to the same species.

We call places where we find fossils which differ from present day living beings "paleoheteromorphic". Places where great similarities between petrified and living forms can be found are called "paleoisomorphic". The extent to which petrified and living forms differ may have led to the "habituation" of different cultural perspectives.

Our research intentionally involved listening to students from these two different sorts of places. In northeastern Brazil, fossils are very similar to living species, while, in the province of Verona, local fossils are very different from living species; students in both places are able to observe these phenomena. In both Brazil and Italy, we encountered no mention of fossils as "scherzi di natura". Some Brazilian students from Evangelical families mentioned the deluge in explaining the presence of fossil fish in places far away from the sea.

However, the idea of geological time seemed to be an equally difficult concept for young students' minds in both places.

8 History of Science and Classroom Practice: A Bottom-up Approach

We suggest taking these three levels of scrutiny into other contexts in order to provide hypotheses for prioritizing certain aspects of historical research. Instead of simply waiting for historians to produce work that is useful for classroom practices, we could offer them feedback regarding the need to more deeply research specific topics in certain ways. In other words, this 'bottom-up' proposal offers methods for determining and giving precedence to the particular sorts of historical research which will be most useful to science education. Furthermore, it may give rise to hypotheses for historiographical reappraisal. Therefore, using first-hand discourses and anthropological findings in key areas of educational concern may generate interesting indications for historical research, as the following example shows.

In listening to students discussing their ideas about evolution after being taught about the subject, we were able to identify the repeated use of heuristic procedures. One of these was the recurrent reference to humans, as related to evolution. Apparently, students centered and organized their knowledge about biological evolution around ideas about human evolution (Bizzo, 1994).[9] This was surprising, since the scientific subject was taught – according to the prevailing version in the educational field at the time – in such a way as to deliberately avoid the issue altogether. Despite the fact that there was a public debate about the place of humankind in the evolutionary scenario after the publication of Darwin's main work in 1859, human evolution was not seen as part of the constitutive structure of the development of the conceptual structure of the theory. Two main ideas prevailed in the culture of the time, one of them maintaining that Darwin actually had encompassed humankind in his theorizing (Bajema, 1988). A slightly different view held that he had decided not to be explicit about human evolution (Cooke, 1990). Another perspective, on the contrary, argued that Darwin's references to humans, especially with regard to cultural evolution, did not mean that he was giving an evolutionary account of human origins; from this angle, human evolution was framed as an issue he would have specifically decided to avoid when writing *The Origin of Species* (Bowler, 1989).

Students' conceptual structures showed how important the figure of the human species was in their minds. History of science scholars had no consensus on how central the human species was in the so-called Darwinian revolution, at least when *The Origin of Species* was first published (1859). Again, there were curricular consequences to this version of history, since humans were not addressed at the beginning of traditional expositions on the development of the theory of evolution. Instead, HS is invoked and the voyage of the *Beagle*, especially the Galapagos stopover, is explored in detail. It is therefore not surprising to find factually wrong information – such as the notion that the concept of natural selection was coined in the Galapagos islands, etc. – in school textbooks. Furthermore, despite the fact that there was, as shown above, lively debate about the inclusion of humankind in *The Origin of Species*, the version which has prevailed in the educational context is clearly linked to Galapagos finches, and not to humans.

This contradiction was of paramount importance in the proposed heuristic tool, generating a hypothesis for the reappraisal of the position of humankind in the bases of evolutionary theorizing. It led to a research proposal which suggested the "mining" of primary historiographical sources in several places in the U.K., including the Cambridge University Library, the British Library, and Down House, all of which was carried out in 1990–1991.

Our results showed that there was much fruitful discussion among historians about what was in Darwin's mind when writing the *Origin of Species*. The debates generally concluded with strong support for the view that the book was primarily intended to deal with humans. The well-known quote on the origin of humankind by means of sexual selection and the famous phrase "*light will be thrown on the origin of man and his history*" could be taken as indications of his will to include humankind in his evolutionary views. The reasons why his ideas on natural and sexual selection were to be applied to human beings were unclear in his 1859 book, an omission which was seemingly due to chance factors and to a dramatic irony[10] (Bizzo, 1992b). Our "mining" showed a broad contrast between historical documents and debates and HS as it is actually taught, demonstrating how difficult it is to separate "real history" from "pseudohistory". Furthermore, our findings from this study, as well as from the various research projects mentioned above, suggest that students' academic performances and discourses, along with key sociocultural factors, can be effectively analyzed in order to devise priority areas for research in HS.

Acknowledgements

The author wishes to express gratitude to the following people and institutions: Ann Nauman, Charbel El-Hani, Eduardo Mortimer, Elisabetta Albrighi, the Cerato Family (Bolca), the Vaccari family (Roncà), the Zardini family

(Montecchia di Crosara), Gian Paolo Romagnani, Giuseppe Pellegrini, Luca Ciancio, Massimo Cerato, Renata Del Sal, Roberto Zorzin; Biblioteca Bertoliana (Vicenza), Biblioteca Civica di Bassano del Grappa, Biblioteca Civica di Verona, Biblioteca di Storia (Università Degli Studi di Verona), Biblioteca Marciana (Venezia), Museo dei Fossili di Bolca, Museo Civico di Storia Naturale (Verona), Universidade de São Paulo, Università Degli Studi di Verona.

This work was made possible by grants from CNPq (300786/93, 304243/2005-1, and 300652/2007-0) and FAPESP (10320/2010).

Notes

1 In 1936, in the USSR, Boris Hessen was secretly judged and executed for 'political reasons'. Twenty years later, after the Stalin era, he was posthumously rehabilitated.

2 This does not mean that Darwin would have a strikingly different view in this regard, but the argument is an *a priori* one. It does not follow that different organs are equally efficient in different groups of animals, for instance, the lungs: an avian lung is far more efficient than a mammalian one. From a Lamarckian perspective, this would be impossible.

3 In discussing how science does not produce absolute truths, but, at the same time, does yield outcomes which cannot really be in doubt, Ziman used the example of genetics: "Who would doubt the credibility of Mendelian genetics, now completely confirmed ate the molecular level by the deciphering of the genetic code?" (Ziman, 1978, p. 9).

4 "Pelorism" was considered a deviant form. In the case of snapdragon flowers, the deviant (recessive character, in Mendelian terms) and common (dominant character) are known nowadays as zygomorph and radial flowers.

5 See, for instance, how misconceptions about insects are spread by literature and cinema at: http://www.umass.edu/ent/BugNetMAP/r_misconcept.html

6 In the 1980s, this syndrome was studied intensively, as it was recognized as an example of non-Mendelian inheritance. Angelman Syndrome is caused by the loss of the normal maternal contribution to a region of chromosome 15, most commonly by deletion of a segment of that chromosome. When such a loss occurs in the paternal chromosome, a different syndrome appears (Prader-Willi Syndrome). This demonstrates sex-related epigenetic imprinting in both cases.

7 The area between the Tropic of Cancer 23.5° N and the Tropic of Capricorn 23.5° S. It covers 39.78% of Earth's surface.

8 The handwriting belongs, without a doubt, to Giambattista Brocchi, and can be found in ink upon his personal reprint of Testa's 1793 work. The note could have been written anytime between 1793 and 1821. I am indebted to Biblioteca Civica di Bassano del Grappa and Dssa. Renata del Sal for access and technical support.

9 A conference on evolution education research was held in the US concurrently to our research, in December of 1992, at Louisiana State University in Baton Rouge, Louisiana, as an initial effort to shift the discussion from religious concerns to epistemological and cognitive aspects of learning. The conference eventually led to a special issue of the journal issued by NARST. As is well known, religion and science are still a matter of concern when the teaching of evolution is addressed. For instance, a paper from that seminal conference concluded: "(...) classroom teachers are to be able to teach evolutionary theory well and defend it against all attacks (...) and diligently respect the boundary between science and religion" (Smith, 1992, p. 194).

10 This 'dramatic irony' involved the death of Darwin's son (Charles Waring Darwin), who may have had Down's Syndrome, and who would have been taken as a case of "reversion" in Darwin's theoretical framework for heredity. This supposition was supported by more recent evidence (Keynes, 2001, p. 207).

References

Allchin, D. (2004). Pseudohistory and pseudoscience. *Science & Education, 13*, 179–195.

Bajema, C. (1988). Charles Darwin on man in the first edition of the "origin of species". *Journal of the History of Biology, 21*, 403–410.

Bateson, W. (1902). *Mendel's principles of heredity: A defense*. London: Cambridge University Press.

Berger, P. L., & Luckmann, T. (1966). *The social construction of reality: A treatise on the sociology of knowledge*. New York, NY: Anchor Books.

Bishop, B. E. (1996). Mendel's opposition to evolution and to Darwin. *Journal of Heredity, 87*, 205–213.

Bizzo, N. M. V. (1992a). História da Ciência e Ensino: onde terminam os paralelos possíveis? *EM ABERTO, 11*(55), 29–35. Retrieved from http://www.emaberto.inep.gov.br/index.php/emaberto/article/view/815/733

Bizzo, N. M. V. (1992b). Darwin on man in the "Origin of Species": Further factors considered. *Journal of thc History of Biology, 25*(1), 137–147.

Bizzo, N. M. V. (1993). Historia de la ciencia y enseñanza de la ciencia ¿Qué paralelismos cabe establecer? *Comunicación, Lenguaje y Educación, 18*, 5–14.

Bizzo, N. M. V. (1994). From down house landlord to Brazilian high school students: What happened to evolutionary knowledge on the way? *Journal of Research in Science Teaching, 31*(5), 537–556.

Bizzo, N. M. V. (1999). On the different interpretations of the historical and logical development of the scientific understanding of evolution. In *Toward scientific liter-*

acy: Proceedings of the IV HPSST Conference (pp. 99–112). Calgary: Faculty of Education, University of Calgary.

Bizzo, N. M. V., & El-Hani, C. N. (2009). Darwin and Mendel, evolution and genetics. *Journal of Biological Education, 43*(3), 108–114.

Bury, J. B. (1909). *Darwinism and history*. Retrieved from http://www.stephenjaygould.org/library/modern-science/chapter27.html

Bowler, P. (1989). Darwin on man in the "origin of species": A reply to Carl Bajema. *Journal of the History of Biology, 22*, 497–500.

Brush, S. G. (1989). History of science and science education. *Interchange, 20*(2), 60–70.

Bush, V. (1945, July). As we may think. *Atlantic Magazine*. Retrieved from http://www.theatlantic.com/magazine/archive/1969/12/as-we-may-think/3881/1

Callender, L. A. (1988). Gregor Mendel: an opponent of descent with modification. *History of Science, 26*, 41–75.

Candela, A. (2009). On the Earth's revolutions: Floods and extinct volcanoes in northern Italy at the end of the eighteenth century. In M. Kölbl-Ebert (Ed.), *Geology and religion: A history of harmony and hostility* (pp. 89–93). London: The Geological Society, Special Publications, 310.

Ciancio, L. (1995). *Autospie della terra*. Florence, Italy: L.S. Olgschki Ed.

Ciancio, L. (2010). *La Fucina segreta di vulcano*. Soave, Italy: Consorzio di Tutela Soave.

Corsi, P. (2003). The Italian geological survey: The early history of a divided community. In G. B. Vai & W. Cavazza (Eds.), *Four centuries of the word 'geology': Ulisse Aldrovandi 1603 in Bologna* (pp. 255–279). Bologna, Italy: Minerva.

Clough, E. E., & Wood-Robinson, C. (1985). Childrens' understanding of inheritance. *Journal of Biological Education, 19*, 304–310.

Cooke, K. (1990). Darwin on man in the "origin of species": An addendum to the Bajema-Bowler debate. *Journal of the History of Biology, 23*, 517–521.

Darwin, C. R. (1868). *The variation of animals and plants under domestication*. London: John Murray.

Dominici, S., & Eldredge, N. (2010). Brocchi, Darwin and transmutation: Phylogenetics and paleontology at the dawn of evolutionary biology. *Evolution: Education and Outreach, 3*, 576–584.

Dougherty, M. J. (2009). Closing the gap: Inverting the genetics curriculum to ensure an informed public. *American Journal of Human Genetics, 85*, 6–12.

Franzolin, F., & Bizzo, N. (2008). Biology concepts in basic education and in the academy: An analysis methodology. In *XIII IOSTE symposium proceedings* (pp. 287–293). Kusadasi, Turkey: Kusadasi University Press.

Gale, B. (1972). Darwin and the concept of struggle for existence: A study of the extrascientific origins of scientific ideas. *Isis, 63*, 321–44.

Gaudant, J. (1999). La querelle des trois abées (1793–1795): Le débat entre Domenico Testa, Alberto Fortis et Giovanni Serafino Volta sur la signification des poissons pétrifiés du Monte Bolca (Italie). In J. Tyler (Ed.), *Miscellanea Paleontologica, VIII* (pp. 159–206). Verona: Museo Civico di Storia Naturale.

Giordan, A. (Ed.). (1987). *Histoire de la biologie* (Vol. 2). Paris, France: Tec & Doc Lavoisier.

Greene, J. C. (1981). *Darwin and the modern world view*. Baton Rouge, LA: Louisiana State University Press.

Hessen, B. (1931). The social and economic roots of Newton's *Principia*. In G. Freudenthal & P. McLaughlin (Eds.), *The social and economic roots of the scientific revolution. Boston studies in the philosophy of science* (Vol. 278). Dordrecht: Springer. Retrieved from http://webfiles.mpiwg-berlin.mpg.de/rereadingClassics/Hessen.pdf/V1_Hessen.pdf

Heywood, D. (2002). The place of analogies in science education. *Cambridge Journal of Education, 32*(2), 233–247.

Hokayem, H., & BouJaoude, S. (2008). College students' perceptions of the theory of evolution. *Journal of Research in Science Teaching, 45*(4), 395–419.

Iltis, H. (1966). *Life of Mendel*. New York, NY: Hafner Publishing Co.

Jordão, M. P. (2006). *A estranha química do cinema e dos comerciais de* televisão (Unpublished Master's thesis). São Paulo, Brazil: FEUSP.

Kargdo, D. B., Hobbs, E. D., & Erickson, G. L. (1980). Children's beliefs about inherited characteristics. *Journal of Biological Education, 14*, 137–146.

Keynes, R. (2001). *Darwin, his daughter and human evolution*. London: Riverhead Books.

King, C. J. H. (2010). An analysis of misconceptions in science textbooks: Earth science in England and Wales. *International Journal of Science Education, 32*(5), 565–601.

Koyré, A. (1939). *Études galiléennes*. Paris, France: Hermann.

Lederman, N. G. (2004). Syntax of nature of science within inquiry and science instruction. In L. B. Flick & N. G. Lederman (Eds.), *Scientific inquiry and nature of science* (pp. 301–317). Dordrecht: Kluwer Academic Publishers.

Lederman, N. G., & Lederman, J. S. (2005). The nature of science and scientific inquiry. In G. Venville & V. Dawson (Eds.), *The art of teaching science* (pp. 2–17). Sydney: Allen & Unwin.

Limón, M., & Mason, L. (Eds.). (2009). *Reconsidering conceptual change: Issues in theory and practice*. Dordrecht, The Netherlands: Kluwer Academic Publishers.

Lorenzano, P. (1998). Acerca del 'redescubrimiento' de Mendel por Hugo de Vries. *Epistemología e Historia de la Ciencia, 4*(4), 219–229.

Lubbock, T. (2010, February 26). Great works: portrait of a young boy holding a child's drawing (circa 1515), Giovanni Francesco Caroto. Museo di Castelvecchio, Verona. *The Independent*. Retrieved from http://www.independent.co.uk/arts-entertainment/art/great-works/great-works-portrait-of-a-young-boy-holding-a-childs-drawing-circa-1515-giovanni-francesco-caroto-1910800.html

Martill, D., Bechly, G., & Loveridge, R. F. (2007). *The Crato fossil beds of Brazil: window into an ancient world*. Cambridge: Cambridge University Press.

Martins. L. A. C. (2007). *A teoria da progressão dos animais de Lamarck*. Rio de Janeiro, Brazil: Booklink; FAPESP; GHTC.

Martins, R. A. (2003). O *Comentariolus* na obra de Nicolau Copérnico. In N. Copernicus (Eds.), *Comentariolus* (2nd ed., pp. 25–93). São Paulo, Brazil: Ed. Livraria da Física.

Matthews, M. (1990). History, philosophy and science teaching: A rapprochement. *Studies in Science Education, 18*, 25–51.

Mayr, E. (1991). *One long argument: Charles Darwin and the genesis of modern evolutionary thought*. Cambridge, MA: Harvard University Press.

Nersessian, N. (1992). Constructing and instructing: The role of "abstraction techniques" in creating and learning physics. In R. A. Duschl & R. J. Hamilton (Eds.), *Philosophy of science, cognitive psychology, and educational theory and practice* (pp. 48–68). New York, NY: SUNY Press.

Osborne, J., Collins, S., Ratcliffe, M., Millar, R., & Duschl, R. (2003). What "ideas-about-science" should be taught in school science? A Delphi study of the expert community. *Journal of Research in Science Teaching, 40*, 692–720.

Prete, I. D. (2008). *Scienza e società nel settecento veneto: Il caso veronese (1680–1796)*. Milan, Italy: Franco Angeli.

Rossi, P. (1992). *A ciência e a filosofia dos modernos*. São Paulo, Brazil: Editora Unesp.

Rogoff, B., & Lave, J. (1999). *Everyday cognition*. Cambridge, MA: Harvard University Press.

Rose, M. R. (2000). *O espectro de darwin: A teoria da evolução e suas implicações no mundo moderno*. Rio de Janeiro, Brazil: Jorge Zahar. [Originally published in 1998 as *Darwin's Spectre* (*Evolutionary Biology in the Modern World*). Princeton, NJ: Princeton University Press]

Rudwick, M. J. S. (1976). *The meaning of fossils: Episodes in the history of palaeontology* (2nd ed.). New York, NY: Neale Watson Academic Publications.

Rudwick, M. J. S. (1985). *The great Devonian controversy: The shaping of science among gentlemanly specialists*. Chicago, IL: The University of Chicago Press.

Rudwick, M. J. S. (2010). *Worlds before Adam: The reconstruction of geohistory in the age of reform* (Vol. II). Chicago, IL: Chicago University Press.

Secker, C. (2001). *Feasability evaluation of science in the cinema* (Final report). Retrieved from dpcpsi.nih.gov/eo/documents/00-102-OD-OSE.pdf

Smith, M. U. (1992). Semantics, epistemology and the philosophy of science in evolution education. In R. Good, J. E. Trowbridge, S. S. Demastes, J. H. Wandersee, M. S. Hafner, & C. L. Cummings (Eds.), *Proceedings of the 1992 evolution education research conference* (pp. 187–196). unpublished.

Toulmin, S. (1972). *Human understanding: An inquiry into the aims of science*. Princeton, NJ: Princeton University Press

Tudge, C. (2000). *In Mendel's footnotes: An introduction to the science and technologies of genes and genetics from the 19th century to the 22nd*. London: J. Cape.

Ziman, J. (1978). *Reliable knowledge*. Cambridge: Cambridge University Press.

Ziman, J. (2000). *Real science*. Cambridge: Cambridge University Press.

CHAPTER 19

The History and Philosophy of Chemistry (HPC) in Teaching and in the Professional Development of Teachers: Contributions to the Debate from Science Education Research

Mario Quintanilla Gatica

Abstract

Scientific knowledge, as a deeply human activity, has been constructed through the centuries in a transcultural way, on foundations that have been determined (and many times conditioned) by the coexistence and simultaneousness of multiple beliefs about the structure of matter, conflicts of power derived from the need of "controlling the knowledge", mythological traces, religious influences, political crises and value disputes generating the most incredible controversies and, in many cases, complex controversies. So, in the last decades the idea of historicity and philosophy of chemistry admits polar opposite interpretations in the meta scientific analyses: some of them try to explain the evolution of knowledge of our discipline from a reductionist view, valuing the objectivity of the datum itself. This is known as anachronistic view. Other interpretations would like to generate explanatory models of science theories, rising from the valuation of the period and context in which such knowledge was "socialized" and influencing the cultural development of a certain scientific community. This is known as diachronic view. This last way of understanding the history of chemistry, valuable in our opinion for natural science teachers, generates approaches distinguished in a substantial way among the so called facts of the past and historical facts, as they have been approached by several researchers. Collecting objective data and interpreting it without understanding the social expectations, the political conflicts and the economic controversies of a period contribute to a restrictive interpretation of the development of knowledge and the scientific activity. This is even more so if one ignores the values in dispute conditioning the discoveries and scientific inventions, all elements that are also part of the so called "historical datum". Such restrictive interpretation would not be adequate, according to many authors, for scientific education in general and the teaching of chemistry

 | DOI: 10.1163/9789004409088_019

in particular. In the same way one cannot ignore the difference sources of information, the scientific instruments, the characteristic languages and strategies of communication that influence scientific work and science teaching. These elements have been very diverse in causes, attributes and consequences at moments and periods also very different in the human history. What sense does incorporating the history of chemistry as a relevant meta theoretical component in the professional development have? That is the question we would like to invite you to reflect upon in this chapter.

Keywords

history of chemistry – teaching and in professional development of teachers

1 Introduction

Scientific knowledge, a deeply human endeavor, has been constructed through the centuries in a transcultural way, on foundations that have been determined (and, in many cases, conditioned) by coexisting, simultaneous, multiple beliefs about the structure of matter, conflicts of power derived from the needs of various actors to 'control knowledge', mythological traces, religious influences, political crises, and value disputes, thereby generating the most incredible and, in many cases, complex controversies. In the last decades, ideas about the historicity and philosophy of chemistry have produced polar opposite interpretations in metascientific analyses. Some have tried to explain the evolution of knowledge in our discipline from a reductionist viewpoint, valuing the objectivity1of the datum itself. This is known as the anachronistic view (Kragh, 1987). Other interpretations attempt to develop explanatory models of science theories derived from the values of the period and context in which a certain facet of knowledge was 'socialized' and influenced the cultural development of a given scientific community. This is known as the diachronic view.

In our opinion, this latter way of understanding the history of chemistry is valuable for natural science teachers, since it facilitates approaches which distinguish between the so-called facts of the past and historical facts as they have been approached by several researchers (Hildgatner, 1990). Collecting and interpreting objective data without understanding the period in which the data were situated – the social expectations, the political conflicts, the economic controversies, the values in dispute, etc., which conditioned the discoveries and scientific inventions of the time, in short, the many crucial elements that comprise 'historical data' – contributes to a restrictive understanding of the development of knowledge and scientific activity that, according to many authors, is inadequate for scientific education in general and for the teaching of chemistry in particular (Echeverría, 2002; Matthews, 1994; Izquierdo et al., 2007). In the same way, the different sources of information, scientific instruments, characteristic languages, and strategies of communication which have oriented science teaching are also very diverse as to causes, attributes, and consequences at different moments in human history. Given these two (or more) intricate fields, what might be some of the benefits of incorporating the history and philosophy of chemistry (HPC) – a branch of the history and philosophy of science (HPS) – as a relevant metatheoretical component in the professional development of teachers? That is the question I would like to invite you to reflect upon in this chapter.

2 The History and Philosophy of Chemistry: What Can They Contribute to Teaching?

The history and philosophy of chemistry promote a better understanding of the nature of science, its theories, instruments, and methods. Such historical-philosophical approaches connect and evaluate the development of the thoughts and ideas of individuals with the development of scientific ideas in a particular moment, or with the consolidation of the metatheoretical agreement of the community. Thus, the history and philosophy of chemistry are necessary and useful for understanding the "culture of scientific knowledge construction"; they question the excessively scientistic and dogmatic information commonly found in chemistry classrooms and textbooks. In examining the lives and eras of different individual scientists, HPC humanizes their ideas, rendering them both less formal and closer to student interests, as well as directly relevant to the specific topics and subjects of the discipline. HPC also connects chemistry with other types of learning and knowledge, integrating the natural interdependence of human knowledge in a complex, heuristic way, and thereby adding value for teachers and learners.

For Nieto (2007), one of the more obvious weaknesses of the history of science has been its notable dependence, throughout the 20th century, on the words and interests of professional scientists. In that respect, this Catalan researcher points out:

> From its role in validating the professionalization of science, inherited from 20th-century positivism, to its more or less soothing role in the

> academic world up to World War II, the history of science has searched among the scientists of the hard sciences for its potential core audience. The disciplinary histories have thus proliferated: physics, chemistry, mathematics, biology, medicine, technology and even subjects such as geology, astronomy, veterinary science, nursing, etc. These, frequently, provide views of the past designed to serve the interests of a predetermined professional group and reinforce what some authors have called dominant or hegemonic scientific communication processes. (p. 86)

Over the last few decades, some studies have pointed out that perspectives from historical and philosophical analysis are absent from chemistry teaching, as well as from science teaching and teacher training in general. On the other hand, scholars from validated and socially legitimated "academic fields"[1] (historians, philosophers, sociologists, etc.) have progressively blurred the classical borders between the so-called *science of the learned,* as such, and science as encountered in the larger society, or *secular science,* so that for many people the habitual strategies scientists use to get new audiences and thus legitimate their prestige and power are unequivocal factors in scientific practice itself. So, the "new theories of chemistry", with their experiments, machines, and instruments, are validated though complex social networks constituted by many experts in basic research, in much the same way that innumerable "anonymous visitors" to science museums ingest and legitimate the science depicted in those institutions. In this way, an almost-inevitable spotlight is turned, first by chemistry teachers and others in educational institutions, and, second, by massive networks of communication media, toward a reductionist and restrictive image of science which is very separate from the cultural, social, and political elements that chemists have contributed to the systematic, permanent, and continuous development of knowledge in different periods and contexts[2] (Echeverría, 2002; Shapin & Barnes, 1977).

As a consequence, the different scientific publics – students, teachers, experts, and laypeople – each tend to have a distorted and incomplete view of the nature of chemistry, its history, its philosophy, its scientific goals and research methods, as well as of how scientific knowledge is built and evolves. These various publics seem to intentionally ignore chemistry's diverse and valuable social impacts, which stance, in some if not most cases, produces an attitude of rejection toward the scientific matters typical of chemistry, making understanding and learning chemistry both less compelling and more difficult. In addition, the "culture of signs or symbols" required for learning about chemistry points to a complex process of valuation, interpretation, and contextualization of theories, where instruments and phenomena are studied with certain purposes and intentions (García & Quintanilla, 2005). If, as a scientific community, we agree on the fact that chemistry is a process of learned knowledge with many dimensions – not only theoretical, but also social, political, and cultural[3] – the incorporation of the history and philosophy of chemistry turns out to be highly relevant to the initial and ongoing processes of training science teachers, students, and scientists. It enables all such parties to engage in the systematic construction of the theoretical-concept scaffolding constituted by a "particular moment in human history" along with the "scientific problem" intended to be solved with the theories and instruments available in that particular moment in that scientific community. Adds Nieto (2007):

> From this point of view, it is professional scientists, highly specialized and with a weak education in the humanities – we return to Snow – who finally decide which contents of their own discipline (including historical contents) must be transmitted to new generations through education and spread to the general public (hence, the title of Stephen Hilgartner's famous article, "The dominant view of popularization" [1990]). In this context, reduced élites of experts will decide the contents of a textbook, for example, in physics, and those same organic intellectuals (in Gramsci's terms) will decide the image we give to students and to the general public about the scientific past. The result is to a great extent conceivable: we return to wrapping paper, to that history of decorative science at the service of certain determined professional interests, as well as, very frequently, putting political, social, or economic history at the service of the determined and unmentionable interests of those in power at the time. (p. 86)

3 Dalton's Atoms "Make History" in 18th-Century Chemistry

As an example – although the descriptions and analyses in such efforts are surely always incomplete and unfinished – I will explore the life of John Dalton as related to the history of atomic theory.[4] The first conference on topics about chemistry given by Dalton took place, according to historians, in Kendal, England, in 1791. By that time, the famous and arrogant Lavoisier had already presented his thoughts on *the nature of water and the impossibility of turning it into soil* (1770) before his colleagues in the Paris Science Academy; had rejected, in a convincing way, the theory of phlogiston (1781–1783); and had disseminated the *law of mass conservation* (1789) to almost all of France.

In turn, English chemists, far from the continent, had made huge strides in pneumatic chemistry, in this way also advancing the quantitative chemistry to which Dalton would be devoted for most of his life. In Germany, chemists guided by Wenzel (1740–1793) analyzed the composition of several salts, from which the data were tabulated and organized to describe the amounts of acid, base, and water in the formation of each. These data, however, were not entirely exact, mainly due to the lack of precision instruments. Even so, the chemist Richter (1762–1807) proposed a mathematical interpretation of chemistry in light of the regularities among the proportions in the combinations of different substances, which would be the base of modern stoichiometry.

In the autumn of 1792, as the new century approached, the ideas of the French Revolution started expanding throughout Europe. In England, with its enduring monarchy, Dalton continued to investigate the behavior of gases, developing the virtuoso idea that the tendency of elastic fluids moving through others is basically produced due to a supposed repulsion among particles which stems from there being a different caloric atmosphere around each particle's fluid. Dalton's interest in the solubility of gases, according to Pellón (2003), lay not "in the chemical process" but in the *mechanisms* by which gases are dissolved, and, in this vein, his idea calculates the relative weights of the *last particles* in the different gases.

In 1803, after multiple tests, successful decisions, mistakes, and many notes, all the while slowly correcting, refining, and completing the development of his ideas in his note book,[5] Dalton compiled, among others, the following ideas:

1. Matter is formed by small last particles or atoms.[6]
2. Atoms are indivisible and they can't be created or destroyed.
3. The atoms of different elements have different masses.
4. All atoms of a given element are identical and have the same invariable mass.
5. The mass of the particle of a compound is the sum of its constituent atoms.
6. The particle of a compound is formed by a fixed number of atoms.

The above statements were finalized by Dalton in September and October of 1803, and would not be modified in his later tests and conferences. That being said, let's investigate whether the atomic theory initially proposed by John Dalton in 1803, and slowly transmitted and redefined through his various public conferences and personal essays, can be considered a natural development of the studies about gases and work in quantitative chemical analysis being carried out in the period of transition between the 18th and 19th centuries. As set out by Aragón (2004),[7] this theory was the only possible way to explain relations among chemical reactions with the available instruments and experimental techniques, but it led to problems of interpretation in the scientific community regarding the determination of atomic weights, since the formulae in use were generally incompletely determined, as can be noted in the case of water. Gay-Lussac's (1778–1850) ideas about comparing densities and law of combining volumes were very useful, but were frequently badly interpreted and, as a consequence, rejected by various *audiences*. During the same period, far from Dalton's influence, Amadeo Avogadro (1776–1856) and Andrés María Ámpere (1775–1836) correctly showed how the relations of gas volumes could be interpreted, but failed to convince the leaders of the growing modern chemistry community, such as Berzelius (1779–1848). Berzelius, who, like Davy (1778–1829), was inspired by the work of Alejandro Volta (1745–1827), recognized the importance of Dalton's atomic theory more than any other chemist.

Berzelius' analytical studies, which were of great scientific rigor for the period, concerned how elements could be combined to form stable compounds; from these, he had some success in obtaining the atomic weights of different elements.[8] His system of symbol systematization was a powerful one, and constitutes the basis for the symbols of modern chemistry. Even as Dalton became a great authority on the chemistry of his time, he was frequently ambiguous and confused his "audiences". Over several decades, Dalton remained unable to sufficiently clarify his atomic theory so as to reach a unified concept of chemistry. In that time, another distinguished British scientist, William Proust (1785–1850), basing his work on that of Dalton, suggested that the atomic weights of all elements were exact multiples of the atomic weight of hydrogen, which implied that elements heavier than hydrogen could be formed by that element. But this is a topic for a separate historical analysis.

4 What Does This Brief Reminder of the Life and Times of the English Chemist Dalton Tell Us about HPC?

As we have argued in other articles (Izquierdo et al., 2007), it is important that chemistry teachers know about the history and philosophy of the sciences, and that they are inspired to communicate HPC developments to their students in the best possible ways. We will have to be increasingly cautious, addressing our attention and goals to the "history of science" itself, so as not to confuse HPC with the chemistry taught in classrooms. The question we have been asking ourselves (Izquierdo et al., 2007) – and will continue to ask – is: What are the limitations that scientific rigor imposes on didacticians or chemistry teachers when they venture into a discipline which has not been fully formed but which should be used to their benefit and, moreover, to that of their students?

5 History or Histories of Chemistry from Which Philosophy of Chemistry?

Admit that a positivist history of chemistry is not possible and neither is the strict diachronic history, without valuing enough the theory of history, which could lead us to a "*present minded*" view of the same, according to which history must *necessarily be compromised* with the present so as to have sense and value.

When supposing that the history of chemistry (HC) in particular is justified only if it contributes to present interpretations of chemistry, one can fall either into *idealism* or an extreme *pragmatism*. In either case, it is assumed that HC in itself is of little interest and only its reconstruction (that will subjectively "give life" to past situations while trying to revive them) makes it interesting and valuable (Izquierdo et al., 2007).

We have pointed out in other studies that Bachelard (1993; quoted in Izquierdo et al., 2007) proposed the term "recurrent history" or "sanctioned history" to refer to a past history that is evaluated according to the values of current science. If we apply this notion to the history of chemistry, it becomes *a history that is constantly being written*, but one in which we cannot pretend that the "scientific past" developed in a continuous way until arriving at the present moment.

With that in mind, Bachelard explains that a historian of science is not a historiographer of "facts" but a historiographer *of the truth*. Nevertheless, this can lead to avoiding ideas from science in general (and from chemistry in particular) that eventually turned out to be false when viewed "with today's and yesterday's eyes". By linking scientific endeavor exclusively to success, such an approach can significantly distort the meaning of scientific activity in classrooms or in professional preparation; or, even more seriously, may frame science as something that advances incessantly, leaving unacknowledged the supposed deviations from this triumphant journey (Izquierdo et al., 2007). There is no "one truth"; instead, there are "truths". Although Bohr's atomic model was a very important historical link in the development of quantum mechanics, it started showing theoretical inconsistencies when applied, between 1913 and 1925, to multielectronic atoms and molecules. By the end of the 1920s, the definitive formulation of quantum mechanics – with Schrödinger, Heisenberg, Born, and von Neumann, among others, working hand in hand – had substantially modified the initial conception of electrons in the atom: electrons were now set out around the core in three dimensions. This was impossible to understand! In addition, according to Heisenberg's uncertainty principle, electrons stopped being conceived of as orbiting around the core in defined paths (Quintanilla, in press; Quintanilla, Labarca, & Aristizabal, 2010). Several important questions arose from these new formulations: How are the chemical and physical worlds related? Is chemistry an autonomous scientific discipline or, on the contrary, a mere sub-section of physics? Are chemistry and physics interdependent –and, if so, to what extent? To approach these questions, philosophers of contemporary chemistry look for adequate scientific, philosophical, and historical answers that constitute satisfactory responses to such problems. The case of the concept of atomic orbits is an interesting example for illustrating how historical and philosophical issues have relevant repercussions not only in the fundamentals of chemistry but also in the ways that chemistry is taught and learned. In this sense, the philosophy of contemporary chemistry – a new sub-discipline of the philosophy of science (see Labarca, 2005) – is an essential resource for chemistry education, as it gives us analytical-conceptual tools not only for approaching the proposed subject matter, but also for a better understanding of the nature of chemistry and its place in the current context of the natural sciences.

6 The History and Philosophy of Chemistry from a Naturalized View of Scientific Activity

Anyway, to give a rational and reasonable answer questions such as those raised above, coherent in this sense with a "complex, heuristic and interesting" HPC, I contend that we have to set out to unearth the historical, controversial, and polemic origins of the main theories of chemistry, showing the processes of creation and disputation, along with the development of key scientific ideas and methodologies, framing all of these as the results of collective work and human construction, involving intrigues, tensions, and even relaxed atmospheres. In this way, we analyze the very real complexity of the relations between science, technology, society, and communication (STSC) along the course of human history, integrating the diverse implications of the political, social, and coexistence processes that these interrelations have generated for the scientific community and for the community of chemists (Christie, 1990; Izquierdo et al., 2006).

The idea of approaching professional development in chemistry and the teaching of chemistry from an orientation which incorporates citizenship and values allows us to permanently reread and reevaluate a range of frameworks for interpreting scientific phenomena which are now explained by means of current theories (for example, quantum theory, as mentioned above). These theories continue to evolve vertiginously, but we may nonetheless discern the relations between the science and culture of a specific period, analyzing in this way their influences on the development and consolidation of a determined society, one which shares values that are systematically redefined (Barona, 1994; Quintanilla, 2007; Solsona, 1997). This is known as a pragmatic realist or naturalized view of science and constitutes an interesting metatheoretical mechanism for approaching the history and philosophy of chemistry.

From this perspective, chemistry would be defined as *one more human activity* contributing to the genesis, development, disclosure, consolidation, and transformation of scientific knowledge, as well as to the educational processes contributing to its understanding, teaching, and learning. This orientation, having a cognitive nature, bases its *theoretical model* on the dynamic and static character of scientific knowledge. Thus, the interpretation and explanation of chemistry as an eminently experimental science will contribute to the (non-naïve) analytical and intellectual convergence of the concepts that have historically been in "tension": *the discovery* and *justification of scientific* facts, as I noted a decade ago in another publication (Quintanilla, 1999). In this way, cognitive scientists question classical philosophy by configuring ideals about science with which the "real chemists" do not work and with which they, as a consequence, do not identify. They thereby establish a distance from the philosophy and sociology of traditional science, trying to establish a discipline without any other goal than to explain science as a human construction in permanent change (Izquierdo, 2000).

From this epistemological view, the "non-naïve" pragmatic realist or naturalist conceptions of science hold that scientific theories "reasonably" describe how the world really is. This means that "there are things or entities" in the world that can generally be characterized by their specific properties, structures, and functions: substances, elements, animals, vegetables, types of energy, etc. That is to say, according to this line of reasoning, that the theories found in chemistry are real descriptions of what happens in the real world. As I will insist on more than one occasion, it seems to me that this is a stimulating standpoint for the analysis and interpretation of facts and scientific theories. Its main difficulty is found in the fact that two theories in chemistry can give equivalent explanations or interpretations of a phenomenon (for example, the different acid-base theories put forth by Bronsted and Lowry versus Arrhenius), provoking doubt as to which of the two interpretations is "more real". To address this issue, Chalmers (1993) proposes what he calls *non-representative realism,* which assumes that scientific theories have determined goals and represent certain particular aspects of the world and not others.

In his famous article, *Explaining Science*, Ronald Giere (1988) states that the selection of scientific theories is normally carried out by means of a complex process of intellectual elaboration which includes the social and cultural interactions, as well as the personal judgment, of the scientist him/herself. From this perspective, the laws "in chemistry", would not be well-confirmed empirical generalizations so much as ideas which depend on the judgments of the scientists involved and the cultural context in which they are analyzed. This conception is thus opposed to the categorical rationality of logical positivism, and does not privilege the value of scientific judgment in decision making. Consequently, in a *naturalist conception of science*, HPC's efforts vis-à-vis the constant approximation of truth that is part of the essence of scientific activity, along with its examinations of methodological agreements as a human activity, may illuminate the "natural" historical evolution of science. This entails that the relation between a theoretical chemistry model and the real world is complex, since the theoretical model is a formal object and not, in itself, a discovery or declaration; the connection between the model and the world to which it refers cannot be a "real relationship". The important thing is then to determine whether the model adjusts to the corresponding systems in the real world and up to what point it is *similar* to them (Giere, 1992). Such adjustments would not be global, but only relative to those aspects of the world that a given model initially tried to capture. The relation between what was discovered/declared and the world is indirect, and must necessarily be specified through the associated theoretical model (Giere, 1994).

In this way, we realize that over the course of human history chemistry formulae (as formed and as taught) have been justified by complex epistemological arguments that involve trajectories of schools of thought, prevailing ways of understanding the world, and even the experiences of individuals whose learning is situated in their own cultures, studies, and educational institutions. In general, chemistry is traditionally "taught" without incorporating history, promulgating in this way a stereotype which privileges the supposed objectivity, rationality, accuracy, precision, and mathematical formalization of knowledge. This makes students think that concepts and scientific phenomena are invariably generated over time, that is, in ways which lack the argumentation and complexity typical of their actual genesis, construction, and evolution (Quintanilla, 2006). Everybody knows that our techno-scientific culture, to use Echeverría's (1995) term, changed in spectacular ways over the last decades of the 20th century. The disciplines discussed in this chapter – the history and philosophy of science, chemistry, and so on – are today increasingly important to educational, cultural, and other dissemination projects (museums of science, magazines, television programs like the *History Channel, etc.*) and constitute an irreducible element of the political agendas of the different organisms and institutions wielding power over citizens. So, assuming the history and philosophy of science will be increasingly incorporated in education, in teacher training, and scientific preparation (both in the initial stages and on an ongoing basis in all cases), we may foresee that teachers and scientists (and chemists) will have more liberty to make decisions about, apply, and evaluate their curricula, activities, strategies, and means of transmission and understanding of knowledge. This is because HPS/HPC will allow them to explore, in a non-regimented way, the validity, complexity, and contexts of the relations between theoretical models and phenomena, as well as between spoken and written scientific language, as used in scientific communities over time (Izquierdo et al., 2006; Quintanilla, Labarca, & Aristizabal, 2009).

The cognitive models used in the sciences may be framed as *models of moderate rationality* (also known as *moderate rationalism or contextualized or hypothetical rationalism*); this approach lays a foundation for teaching of chemistry (Izquierdo, 2000), yet constructivism contends that knowledge is always constructed and determined in contexts, because it is the result of interactions, experiences, and the resulting human ideas and reactions to ideas. At the other extreme, working primarily with the theoretical components of the history of chemistry, from an axiomatized science approach which is focused on signs, formulae, and symbols, separate from the real world and exclusively instrumental, does not offer a complete picture of chemistry either: the history of science itself, with all its weaknesses and strengths, is effectively overlooked.

7 The Dangers of a "Hagiographic" History of Chemistry (HC) and of a Naïve Philosophy of Chemistry (PC)

The HC and PC to which we are referring, do not, therefore, consist of a history of dates or facts or of superficial reflection which disregards key elements, that is, they are neither *hagiographic history* nor *naïve philosophy*. That is why the huge connotative and denotative value of language and the many ways of communicating and spreading the history of chemistry have emerged as crucial factors. At the same time, it is necessary to note that *writing well* in chemistry is not an automatic consequence of having successfully performed experiments or having correctly understood what a science teacher "says or writes". Learning chemistry in school or at the professional level has to do with the evolution and differentiation of ideas and points of view, from very simple conceptions (Bohr's atom, for example) to more complex ones (say, from quantum mechanics), which becomes possible through social and educational interactions. It includes language in its various forms, whether via rhetorical matters, strategic expressions on the parts of individual chemists, communications between members of the "chemistry community" (and with the "other sciences"), through promoting modifications in certain ideas .... Given the added dimension of language, the incorporation of HC in key science learning processes (preparation, teaching, learning, and dissemination) requires a theoretically coherent and expertly modeled form of argumentation.

The chemistry that is taught and learned in schools and universities takes place via specific, duly identified and characterized modes of communication, ensuring that teachers, students, and the larger society share certain understandings. Nevertheless, if one of the functions of schools is educating students about science in such a way as to prepare them for a culture of citizenship, we have to think about the need for creating adequate cognitive and linguistic (verbal, gestural, or written) environments, favoring the communication of scientific ideas which will enable our students to learn to speak "real chemistry". Different scientific fields use language differently; utilizing terms, linguistic expressions, and algorithms which have meaning in the context of creating and testing hypotheses, interpreting results, or coming to conclusions, with each linguistic mode used primarily – and sometimes only – within a given scientific community. It is expected that chemistry students, teachers, and scientists learn to use the language of chemistry as it is found in articles, papers, and text books, so as to translate the questions, explanations, and doubts arising from their daily lives into terms which have shared meanings in that context. These are not necessarily the same doubts confronting scientists in the HPC in particular, or in the HPS in general, but rather the public opinion determinant in these representations (Besaud-Vincent, 2000).

8 Which HC Is Most Suitable for the "New Purposes" of Scientific Education?

In a more complex sense, it seems to me that HC's more naturalized perspective on scientific formation requires a complementary analysis grounded in both epistemological and methodological bases. In his book, *Human Understanding*, Toulmin (1977) sets up an interesting discussion about *conceptual change* and *scientific change*, with the goal of permanently evaluating the categories of analysis from which knowledge is mobilized, taking into account the logic of the object as well as the scientific individual. If we assume the "dynamic and changing" character of chemistry (new theories, new problems, new instruments and languages), it is evident that reflection on, and understanding of, these changes should be integrated into the initial and ongoing professional development of teachers and of the scientific community, as we have anticipated above. We assume a stance of moderate rationality regarding the facts, phenomena, methods, and contexts involved in the building of such knowledge and consequently, how it is spread and taught. In that respect, and even though it cannot be totally justified, we must accept certain aspects of the principle of induction in order to understand the meaning of chemistry as it is reconstructed and redefined through the HPC, configuring a closer-up image of the chemistry we teach at school, or "school chemistry". Currently, the cognitive sciences solve this problem in a certain way, by trying to explain *how scientific knowledge is generated and works in the minds of people*. From this epistemological point of view, as I have insisted throughout this chapter, the sciences are seen as deeply human activities: their purpose is interpreting the world in a "rational and reasonable" way, using the capacity we have for expressing judgments to do so.

This approach leads to a blurring of the boundaries between scientific and daily thought, giving way to new models of science to be taught, learned, and valued (Izquierdo, 2001; Copello, 1995).

9 What Are the Purposes of Disseminating and Teaching HC and of Understanding PC?

As I have illustrated, certainly in a very general and incomplete way, via the "controversies between old quantum theory and quantum mechanics" (the former understood as spanning the era from Plank in 1900 up to Broglie's (1924) thesis, without including the latter) or in describing part of John Dalton's scientific life, it seems important to me to insist on the fact that chemistry is a process of scientific constitution with not only historical but also social and cultural dimensions, all of which may derive from specific tensions in a particular moment of human history or in the history of ideas. In consequence, in the interest of fostering these theoretical guidelines, the inclusion of HC and PC in the initial and continuous professional development of scientists, science teachers, and divulgators (scientific journalists, monitors of science museums, researchers) will allow them to relate the conceptual frameworks formed in a "particular moment of history" to the "scientific problems" they were intended to solve using contemporary theories and instruments (Quintanilla, García, & Izquierdo, 2009). This idea of professional development and teaching enables us to understand several theoretical-conceptual frameworks by which we may also interpret the scientific phenomena that *we understand well today* as explained by means of current chemistry theories. It also allows us to explore the relations between chemistry and the culture of a given period and to analyze the influences on the development, consolidation, or abandonment of scientific theories. For instance, we may investigate the ways in which aspects of Bohr and Heisenberg's era "influenced" both "new answers" and "new questions" about knowledge.

10 How to Teach and Learn HC and PC Considering This "New Teaching Culture"

In general, chemists and chemistry teachers adopt a realist position when attributing a defined existence to the entities, for example, orbitals, that they teach and study. But, when computing, physicists and chemists adopt an instrumental perspective, essentially taking orbitals, etc., as mere fictions, useful for calculation. This existing tension between chemical descriptions and quantum mechanical descriptions usually goes unnoticed by chemists and teachers, not to mention the students learning elements of science "by heart" without even questioning them. To add even more confusion, a group of scientists announced the first "sighting" of atomic orbitals in 1999. Nevertheless, some theoretical non-quantum chemists and philosophers of chemistry quickly objected to the conclusions obtained by the researchers responsible for this experiment, pointing out the conceptual mistakes committed in their interpretation of what they "saw". From this debate, it is possible to formulate a group of questions about orbitals: Are they "real" entities? Do orbital entities exist in the chemical world but not in the physical world? Should we only use a mechanic-quantum model, leaving aside the traditional chemical model, along with its descriptions? Or, on the contrary, if we decide to continue using a chemical model, how might we argue in favor of its use, given this controversy?

Until now, we have been considering the relevance of HC and PC to the construction of scientific, educational, and pedagogical knowledge, as well as in the professional development of teachers and scientists, all of which we framed as grounded in ideas around teaching and disseminating chemistry. Chemistry's epistemological nature fits very well with the pragmatic realism or naturalism mentioned above, and has some attributes which render it distinct from other epistemological views, which tend more toward the traditional or dogmatic – or may be openly committed to certain political, economic, or social processes (Shapin & Barnes, 1977). In the same way – and due to the enormous complexity of the material and the scientific rigor required for any analysis or point of view of this nature – science historians and philosophers of chemistry have developed their work according to a double perspective on facts, periods, and ways of dealing with them, drawing from both so-called *vertical* and *horizontal historiography*. Vertical or *synchronic* historiography is mainly characterized by the *delimitation* of a problematic scientific notion and by following its evolution and transformation over the history of science. On the other hand, horizontal or *diachronic* historiography is interested in the "science of the past" itself, for its theories, methods and instruments, and organization, and for the highly complex networks of intellectual influences located in every historical situation, coinciding with the value frameworks and metatheoretical and methodological agreements shared in the scientific communities of that period (Barona, 1994; Izquierdo, 2001). In this sense, we concur with Crombie's (2000) idea, which delineates and connects the questions, answers, and contexts which represent the goals to which scientific historians, ideally, aspire. As he explains,

> ... the goal of the science historian ... is finding which problems worried the scientists before being solved, which were their hypothesis and expectations and what they considered as answers and explanations .... (p. 19)

We have noted that several schools and researchers in fields of a metascientific nature, such as epistemology, history of science, philosophy of chemistry, and science education, conceive of chemistry as a human activity involving the production, evaluation, application, and dissemination of learned knowledge. This activity is immersed in historical, social, and cultural contexts, which give meaning to scientific activity and specify the scientific goals which are privileged or devalued, as well as the values which are supported, disregarded, or at risk, in communities and scientific institutions, the conditions and actions of which are in turn determined by multiple factors and processes. Hence, we can discern the huge importance of engaging in reflection *about or upon the process of the historical construction* of scientific knowledge in new curricular projects for chemistry teachers and scientists. According to Izquierdo (2001), the experimental sciences can be characterized by four main dimensions: an *essential goal* (Why do we want to know, describe, and interpret the world through chemistry?); a *methodology* (How are the different experiments and chemistry theories related to each other?); *rationality* (How and why do theories "about chemistry" change along the course of human history?); and a stance on the *nature of scientific representations* (Does chemistry tell us something about the real world?). From the consolidation of these dimensions in the different curricular approaches *about* and *in the teaching of chemistry*, the essential outcome would then be teaching students how to *think, talk, and act* vis-à-vis different situations in which people systematically interact with the physical or material world.

11 Understanding Chemistry as a "Deeply Human Activity": A Pending Challenge

In any case, we are not discussing naïve approaches to the nature of chemistry and its teaching, since we started from a perspective grounded in the notion that these are constructed via the lives of persons and of social groups, both of which bring certain determined epistemological intentions to the process, but, as set out by Bourdieu (2003), also carry ideological and social objectives. For this French sociologist, science refers to a very defined range of problems, in which a particular *paradigm or disciplinary matrix* (see Bourdieu, 2003, p. 34) is accepted by an important faction of the scientists in that field, who then impose that paradigm or discipline on others in their field in a continuous and disciplined way – not only in order to validate the currently accepted knowledge constructed in that area, but also to legitimate the *authority* of their actions, procedures, and convictions in the scientific community. This undoubtedly makes many who are interested in science and who delve into its various outlets (conferences, books, etc.) think that its language is difficult to understand and construct, and, therefore, to teach and learn.

In summary, the non-linear development of a scientific notion in the HC may be seen as similar to the ways in which old meanings are abandoned as new ones are slowly incorporated in scientific language, or compared to the evolutions of the interconnections established in dynamic and deliberate ways throughout the history of science (Estany & Izquierdo, 1990). This "logic" of the construction of historical and philosophical knowledge determines new "theoretical models", strengthening some families of scientific theories to the detriment of others, as noted by Giere (1992), who contends that the concepts, theories, and specific procedures of science are part of a dynamic network of connections between everyday and learned knowledge (Figure 19.1). In an analogous way, as pointed out by Toulmin (1977), science, scientific communities, and the individuals involved are in *constant* evolution, undergoing changes that are sometimes imperceptible.

According to this researcher/philosopher of the "new science", concepts, procedures, and theories "work" in ways similar to those whereby individual biological entities function within a biological system, that is, they are subjected, as Darwin would say, to the laws of natural selection, to a dynamic of constant change and renewal. For example, the concept of "periodic law" we teach today in general chemistry lessons (as discussed in more detail in the next section), is not *exactly the same* as that taught and spread toward the end of 19th century (Camacho & Cuellar, 2007). The deciding factor of *scientific evolution* that would correspond in biological terms to *adaptation to new environments* is the practical use of scientific knowledge and its meanings (Toulmin, 2007). We can, likewise, offer a metatheoretical explanation for such phenomena, since the continuous and permanent evolution of scientific concepts in the history of chemistry is similar to the continuous and permanent changes in the ideas we have about the nature of science when we learn how to research, learn about, and discuss that vast subject. Toulmin recognizes that the progressive change of a scientific discipline like chemistry implies adjustments and mismatches, conceptual tensions, and uncertainties typical of its specific knowledge modes and of the logic with which it is built and modified over the course of the history of science.

As a consequence, teaching students to understand *how to learn to write and talk about and do chemistry* requires a permanent commitment from the chemistry teacher not only to the transmission of specific notions, but also to the identification and characterization of the ways or styles with which students are perceiving (and understanding) reality and conceiving of the world/their worlds. Teachers must then take into account and encourage the ways students become more skilled – through learning chemistry – in modeling and interpreting the world/their worlds in a discursive way, with ideas and words arising from personal experience working in tandem with the thinking tools gleaned from the chemistry they have learned.

12 Toulmin's Model Applied to the Scientific Notion of "Periodic Law": An Example of the Usefulness of HC and PC

Toulmin's (1977) original proposal established three alternative methodological ways of representing the historical process of conceptual change: the *transversal*, the *longitudinal*, and the *evolutionary* (or *combined*) way. The concept of periodic law may be approached from the evolutionary angle with the purposes of investigating the ways in which this scientific concept was developed and consolidated in the HC of the 19th century and of presenting arguments that enable us to demonstrate that such evolution occurs in a dual process involving innovation-conceptual variation and intellectual selection.

Regarding the conceptual variation of this specific scientific notion, Camacho and Cuellar (2007) contend that during the historical development of periodic law various proposals were verified as new concepts appeared that could account for particular problems. The main function of such shifts was not exactly in verifying the truth or falsity of an empirical proposal or measuring the required frequencies as measures of probability per se (Toulmin, 1977, p. 213; quoted in Camacho & Cuellar, 2007), instead, they corresponded to the possibility that the concepts in question could be rearranged to obtain better – that is, more exact, more detailed and, in general, more intelligible – charts of the objects, systems, and facts involved (Toulmin, 1977, p. 213).Besides, as was shown in the previous section, periodic law was, like most scientific endeavors, a community effort, and a fundamental requirement for recognition as a genuine conceptual variant is that "it takes something more than the personal reflections of open-minded individuals to create an effective pool ofconceptual variants in a science .... The individual's innovation, that is, must be seen as providing a possible way of dealing with the problems which are the sources of collective dissatisfaction" (Toulmin, 1977, pp. 213–214).

Periodic law has remained a vital aspect of HC since its formulation in 1869, as can be seen in the different publications that still focus on it as a topic, in scientific, philosophical, and, certainly, educational (see Christie & Christie, 2003; Schmidt & Baumgärtner, 2003; Vihalemm, 2003; quoted in Camacho & Cuellar, 2007) discussions, and in its pedagogical use as one way of tackling the problem of teaching the properties of chemical elements.

Camacho and Cuellar (2007), referring to Toulmin's notion of intellectual selection (which gives an account of the reasons and causes behind the evolution of scientific notions), found that it was possible to establish that conceptual changes in periodic law were produced via collective activity, in a concerted search for solutions to problems encountered in the organization of chemical elements. These problems were recognized as a key issue in the field by the community of chemists at the end of the 19th century. Camacho & Cuellar (2007) conclude that relevant changes in the concept of periodic law could emerge and were possible thanks to these sorts of intellectual considerations and that these, among other aspects, allowed Mendeleev's formulation to be accepted by the community of chemists and physicists in the 19th century. This in fact occurred in such a way that it – besides providing explicative aspects – also sustained his proposal in its predictive aspects, exemplifying Toulmin's point that collateral effects *in favor of conceptual innovation may be more powerful than their expected consequences* (1977, p. 233). Besides producing a "scientific model" for the field of chemistry, Mendeleev's work eventually also turned into a "model of teaching" the subject.

13 Challenges for a "New History of Chemistry" (NHC) and PC in Science Education

The approach I present in this chapter seems to be the most pertinent – as well as perhaps the most difficult and challenging – approach to scientific dissemination, chemistry teaching, and the professional development of teachers. We have studied it from several angles, using various sources (Camacho, Quintanilla, & Cuellar, 2007; Uribe & Quintanilla, 2005). This approach can orient us as to how we may best teach and think about chemistry, assuming that we conceive of chemistry not from a restrictive or reductionist perspective, but rather as something that is connected to its social and temporal contexts, including values, philosophical ideas, languages, instruments, and certain purposes. This gives us a sense of the interventions and transformations inherent to scientific processes and provides us with useful didactic-theoretical tools for both schools and professional training (Izquierdo et al., 2006).

14 Final Reflections

In this chapter, I have presented some theoretical ideas that are especially relevant when thinking about the ways in which we are fostering the language of chemistry, and, with that, ways to learn about a "new history and philosophy of chemistry" (NHPC). I have attempted both pedagogical and epistemological analyses, since these referents give consistency to decision-making and to the means by which teachers manage scientific knowledge in their classrooms, whether in schools or universities, and even in restrictive learning contexts. Nevertheless, we can easily see that in real life little or nothing is known of these elements of analysis. This is regrettable, since the ideas in this chapter might actually be helpful in overcoming the great doubts and contradictions that may eventually emerge in the minds of chemists and chemistry teachers, and even from the students themselves, who – often in spite of their teachers – may still learn chemistry. Perhaps the most important thing about chemistry, as taught in schools and in the professional development of science teachers and chemists, is that it is best taught and learned with students and teachers sharing goals, mirroring the process of the construction of chemistry itself. The doubts, contradictions, and agreements I have presented in this chapter arise from this very fact, which is often ignored. The purposes of this experimental science characteristically coincide with the goals of scientific knowledge, that is, interpreting phenomena by thinking and talking about them in a shared disciplinary environment in which *science is discussed, written, communicated, and spread.* This is how chemistry has been constructed over the course of history, and the study of aspects of that trajectory undoubtedly holds great merit in helping us to think about the world.

Acknowledgements

This chapter was written with the support of the FONDECYT 1110598 and AKA04 projects, which are under the direction of the author, and are financed by the National Commission on Scientific Investigation and Technology of the Republic of Chile (CONICYT) and the University of Helsinki. In addition, the author would like to express her gratitude to Dr. José Antonio Chamizo, professor and researcher at the Universidad Nacional Autónoma de México, for his generous and insistent invitations to share the always developing and partial ideas in this chapter.

Notes

1 In Bourdieu's (2003) terms, the value in terms of lack of legal standing made by the "scientific community".
2 Nuria Solsona's historical analysis and educational proposals, which may be found in his books *Mujeres científicas de todos los tiempos* and *El saber científico de las mujeres,* are of unequivocal interest in this respect.
3 In this way, the idea of a "symbol" or "culture of the formula" operating without a rhetorical view of the language of science in the context of the production of knowledge is unclear.
4 To delve more deeply into this example and find similar notes from this epistemological analysis, see M. Quintanilla's (2007) *La enseñanza del modelo atómico de John Dalton desde una visión naturalizada de la historia de la química.*
5 Many of Dalton's notebooks and instruments were lost during the bombing of Manchester by Nazi troops on the nights of December 23rd and 24th, 1940.
6 Many of Dalton's notebooks and instruments were lost during the bombing of Manchester by Nazi troops on the nights of December 23rd and 24th, 1940.
7 Quoted by Pellón (2003).
8 Berzelius carried out a series of experiments to measure the proportions by which elements were combined and, in 1816, managed to study up to 2000 different compounds (Gribbin, 2005, p. 306).

References

Barona, J. (1994). *Ciencia e Historia*. Valencia, Spain: Universidad de Valencia Editorial SEC.

Bensaude-Vincent, (2000). *L'opinion publique et la science: A chacun son ignorance.* Paris, France: Institut d'édition Scenofi-Synthélabo.

Bourdieu, P. (2003). *El oficio del científico.* Barcelona, Spain: Anagrama.

Camacho, J., & Cuellar, L. (2007). La ley periódica desde la propuesta de Toulmin: Aportes para la enseñanza de la historia de la química. In M. Quintanilla (Ed.), *Historia de la ciencia: Aportes para su divulgación y enseñanza* (Vol. 2). Santiago de Chile, Chile: Editorial Arrayán.

Camacho, J., & Quintanilla, M. (2006). *A methodological proposal to assume themselves to the resolution of problems and to promote linguistic cognitive strengths. Proceedings of the Ninth International Conference on History, Philosophy & Science Teaching,* Calgary, Canada.

Camacho, J., Quintanilla, M., & Cuellar, L. (2007). *Ley periódica ¿modelo de enseñanza o modelo químico? Reflexiones desde la historia de la ciencia. Actas del X Encuentro de Educación Química,* Universidad de Talca, Chile.

Chalmers, A. (1993). *¿Qué es esa cosa llamada ciencia?* Madrid, Spain: Siglo 21 Eds.

Christie, M. J. S., & Hodge, R. C. (Eds.). (1990). *Companion to the history of modern science.* London: Routledge.

Copello, M. (1995). *La interacción maestra-alumnado en el aula: Dilemas sobre acciones favorecedoras del acercamiento entre los significados en relación a contenidos de ciencias naturales* (Master's thesis). Barcelona, Spain: Departamento de Didáctica de las Sciencias, Universidad Autónoma de Barcelona.

Crombie, A. (2000). *Historia de la ciencia: De San Agustín a Galileo* (Vol. 1, Siglos V-XIII). Madrid, Spain: Alianza Editorial.

Echeverría, J. (1995). *Filosofía de la ciencia.* Madrid, Spain: Akal Ediciones.

Echeverría, J. (2002). *Ciencia y valores.* Barcelona, Spain: Destino.

Estany, A. E., & Izquierdo, M. (1990). La evolución del concepto de afinidad analizada desde el modelo de S. Toulmin. *Llull, 13,* 349–378.

García, A., & Quintanilla, M. (2005). *Historia de la ciencia y formación docente: Algunos elementos para el debate didáctico. Actas de las IV Jornadas Internacionales para la enseñanza preuniversitaria y universitaria de la Química,* Universidad Autónoma de México, México.

Giere, R. (1988). *Explaining science. A cognitive approach.* Chicago, IL: University of Chicago Press. (Trad. cast. La explicación de la ciencia: un acercamiento cognoscitivo. Consejo Nacional de Ciencia y Tecnología, México, 1992)

Giere, R. (1992). Cognitive models of science, XV–XXVIII. In R. Giere (Ed.), *Cognitive models of science.* Minneapolis, MN: University of Minnesota Press.

Giere, R. (1994). The cognitive structure of scientific theories. *Philosophy of Science, 61,* 276–296.

Hilgartner, S. (1990). The dominant view of popularisation: Conceptual problems, political issues. *Social Studies of Science, 20,* 519–539.

Izquierdo, M. (2000). *Relacions entre la história i la didáctica de les ciéncies. Actes de les V Trobades d'historia de la ciencia i de la técnica*, Barcelona, Spain.

Izquierdo, M. (2001). Fundamentos epistemológicos. In F. J. Perales & P. Cañal (Eds.), *Didáctica de las ciencias experimentales: Teoría y práctica*. Madrid, Spain: Editorial Marfil Alcoy.

Izquierdo, M., Quintanilla, M., Vallverdú, J., & Merino, C. (2007). Una nueva reflexión sobre la historia & filosofía de las ciencias y la enseñanza de las ciencias. In M. Quintanilla (Ed.), *Historia de la ciencia: Aportes para la formación del profesorado* (Vol. I). Santiago de Chile, Chile: Editorial Arrayán.

Izquierdo, M., Vallverdú, J., Quintanilla, M., & Merino, C. (2006). Relación entre la historia y la filosofía de las ciencias. *Alambique, 48*, 78–91.

Kragh, H. (1987). *Introducción a la historia de la ciencia*. Barcelona, Spain: Crítica.

Quintanilla, M (in press). Relaciones entre la Historia de la Química, la filosofía de la química y la didáctica de la química. In M. Labarca & J. Martínez (Eds.), *Introducción a la Filosofía de la Química*. Ediciones CONICET, Universidad de Buenos Aires, Argentina.

Mathews, M. (1994). Historia, filosofía y enseñanza de las ciencias: La aproximación actual. *Enseñanza de las Ciencias, 12*(2), 255–277.

Nieto, A. (2007). Las 'historias de la ciencia' y sus adaptaciones a la enseñanza: un debate abierto. In M. Quintanilla (Ed.), *Historia de la ciencia: Aportes para la formación del profesorado* (Vol. I). Santiago de Chile, Chile: Editorial Arrayán.

Pellón, I. (2003). *Dalton, el hombre que pesó los átomos*. Madrid, Spain: Editorial Nivola.

Quintanilla, M. (1999). El dilema epistemológico y didáctico del curriculum de la enseñanza de las ciencias: ¿Cómo abordarlo en un enfoque CTS? *Pensamiento Educativo, 25*, 299–334.

Quintanilla, M. (2006). Science, citoyenneté et valeurs – socle d'une approche réaliste et pragmatique de l'enseignement des sciences: mise en perspective historique. In UNESCO (Eds.), *International Science, Technology & Environmental Education Newsletter, 31*(3–4).

Quintanilla, M. (2007). La enseñanza del modelo atómico de John Dalton desde una visión naturalizada de la historia de la química. In M. Quintanilla (Ed.), *Historia de la ciencia: Aportes para la formación del profesorado* (Vol. I). Santiago de Chile, Chile: Editorial Arrayán.

Quintanilla, M., García, A., & Izquierdo, M. (2009). *Ideas del profesorado de química en torno al papel de los gases en la comprensión del cambio químico en el siglo XVIII y sus implicaciones didácticas. Actas del XI Encuentro de Educación Química*, Universidad de Concepción, Chile.

Quintanilla, M., Izquierdo, M., & Adúriz-Bravo, A. (2005, July 15–18). *Characteristics and methodological discussion about a theoretical model that introduces the history of science at an early stage of the experimental science teachers' professional formation. Science & Education* IHPST 8, University of Leeds, England.

Shapin, S., & Barnes, B. (1977). Science, nature and control: Interpreting mechanics institutes. *Social Studies of Science, 7*, 31–74.

Solsona, N. (1997). *Mujeres científicas de todos los tiempos*. Barcelona, Spain: Ed.Talasa.

Solsona, N. (2003). *El saber científico de las mujeres.* Barcelona, Spain: Ed.Talasa.

Toulmin, S. (1977). *La comprensión humana: El uso colectivo y la evolución de conceptos* (Vol. 1). Madrid, Spain: Alianza Editorial.

Uribe, M., & Quintanilla, M. (2005). *Aplicación del modelo de Toulmin a la evolución del concepto de sangre en la historia de la ciencia: Perspectivas didácticas. Actas del VII Congreso Internacional de Enseñanza de las Ciencias*, Granada, Spain.

CHAPTER 20

Contributions to Physics Education from the History and Philosophy of Science

Irene Arriassecq and Verónica Guridi

Abstract

This chapter analyzes the historical evolution of HPS in connection with science teaching and learning, the different lines of research that have developed over time, and some examples of teaching proposals – aimed at students, teachers and preservice science teachers – which were designed taking research results into account. Some special phases in this evolution are described: a first historic landmark that considered HPS as playing a limited role in the process of institutionalizing science education research in Brazil; the consolidation of the area during the 1980s; the application of HPS in the development of variations of the conceptual change model (Posner et al., 1982), giving special emphasis to other philosophers besides Kuhn and Lakatos; and a last phase of the institutionalization process of science education research in Brazil. A critical analysis of the present situation regarding the incorporation of HPS in science education is carried out, and some ideas are suggested in order to make progress along these lines.

Keywords

physics education – history and philosophy of science – research in physics teaching – historical evolution of research trends

1 Introduction

An analysis of the possible contributions to physics education from the history of science (and of physics, in particular) and the philosophy of science – HS and PS, respectively, and HPS, when referring to both – constitutes a formidable task. In order to understand the complexity of this task, we must first identify the multiple theoretical aspects that converge in these endeavors.

 | DOI: 10.1163/9789004409088_020

On the one hand, the scientific discipline of physics has among its main objectives the goals of explaining, understanding, and predicting natural phenomena. These objectives are achieved by, first, intervening in those phenomena using specific methodologies and, then, utilizing specific language in order to communicate the findings of those interventions.

On the other hand, HPS is considered to be a metascience, or second-order criteriology, which claims the full range of scientific disciplines as its objects of study (Losee, 1972; Klimovsky, 1994). It is a theoretical reflection on scientific knowledge and scientific activity which focuses on the study of the processes, conditions, and results of scientific innovation, justification, systematization, application, evaluation, and communication (Adúriz-Bravo et al., 2006).

At the same time, multiple contexts coincide in the area of science education: it exists as a discipline to be taught; in theoretical frameworks, teaching and learning processes, teaching proposals, and curricula; as conceptions of a given discipline; and in its own specific manifestations as taught and learned by both teachers and students, reflecting students' and teachers' backgrounds, environments, institutional cultures, etc. Research on these aspects of science education did not always consider the possibility that analyses of these topics could contribute effectively to science education, particularly in the area of teaching and learning physics. This is beginning to change as more HPS-related factors are being taken into account in research into science education. Such studies have emerged in the wake of original work by researchers like Salomon (1988) and Lakin and Wellington (1994), among others, who contend that teachers' views on science – whether explicit or not – affect what and how they teach. Nevertheless, it is important to remember that the level of influence of teachers' viewpoints on their actual teaching practices may vary widely; other researchers, like Lederman (1999, 2006), have pointed out that relations between epistemological conceptions and teaching practices are not so simple and straightforward, since several intervening factors may impede or enhance the various factors involved. Thus, the contributions of HPS to science education should perhaps be taken into account with a measure of prudence. At the same time, many other authors – such as Lantz and Kass (1997) and Duschl (1997) – hold that the required training of science teachers should involve not only a scientific education but also learning about the nature of science; teachers would thereby gain knowledge about science which would include its purposes, methods, and relationships with technology and society.

In a time when scientific literacy has become one of the major goals of science education in many countries, gaining a deeper understanding of the history and nature of science is of paramount importance. The scientifically literate individual is expected to be capable of distinguishing between scientific and non-scientific knowledge, as well as between science and pseudoscience; to be aware of the limits and scope of science (as well as of what science can and cannot explain); and to be able to identify scientific methodologies. As an individual and as a member of a particular society, he or she should also be able to discern what is relevant to the scope of science, taking positive as well as negative aspects into account.

However, the concept of the nature of science (NS), as Acevedo-Diaz et al. (2007) point out, is complex and dialectical and, therefore, difficult to define with accuracy and by general consent. Experts often discuss descriptions and representations of the NS which are as dynamic as scientific knowledge itself, and so it is impossible to support the idea of a single or unified NS capable of representing either scientific knowledge in general or all of the scientific disciplines. Therefore, any representation of the NS will be partial and will compete with other incomplete representations.

As regards science didactics, even though there is consensus about the importance of the NS in science education (Bell et al., 2001), the means of achieving its objectives are not clear (Bell, 2005).

Several international groups are currently studying the applications of the NS to science education at different levels. One of the most important is the International History, Philosophy and Science Teaching Group (IHPST), which has been holding conferences since 1989, and has encouraged international magazines with great prestige within the scientific community and among science education teachers – such as the *International Journal of Science Education*, among others – to dedicate special editions to the NS as related to science education. Another major landmark was the creation of the journal *Science & Education: Contributions from History, Philosophy, and Sociology of Science and Mathematics*, which has promoted the inclusion of history and philosophy of science and mathematics courses in science and math teacher education programs since 1992. Moreover, this journal promotes discussions around the philosophy and purposes of science and mathematics education. It also considers the contributions of HPS to the intellectual and ethical development of individuals and cultures. This journal, which is edited by Michael Matthews, is associated with the IHPST.

The First IHPST Latin American Conference was held in 2010. This conference focused on the presentation and discussion of papers about the use of history and philosophy of science in science education, in accordance with the guidelines drawn up by the IHPST for international conferences.

In Latin America, Brazil was the first country to consolidate science education research as a specific area of study, and then to incorporate the study of the contributions of HPS as a line of research within science education itself.

Science education researchers in Argentina have been establishing links with researchers in Brazil since beginning to develop work in this area. This has been occurring, on the one hand, via the supervision of Argentinian researchers' theses by prestigious Brazilian researchers, such as Marco Antonio Moreira, from the Federal University of Rio Grande do Sul (UFRGS). On the other hand, Argentinian researchers have traveled to Brazilian institutes and universities to study and collaborate with research groups and have published reports from their research projects in pioneering Brazilian journals.

As Villani et al. (2010) point out, the development of science education research in Brazil was very similar to that which took place in many other Western countries, at least until the early 1990s. Systematic research on science education first appeared 40 years ago, as a consequence of an overall renovation in the field of Brazilian science education. This process was also related to political events taking place in the country. Krasilchik (2000) points out that in the 1960s, during the Cold War, the U.S. invested considerable amounts of money in the production of large-scale science education projects, with the intention of positioning the U.S. as the leading country in space exploration. In this effort, science education was directed toward producing an elite group of students who would then be guaranteed scientific careers. During this period, several organizations were formed which were intended to organize and develop science education. These included the Physical Science Study Committee (PSSC), the Biological Science Curriculum Study (BSCS), the Chemical Bond Approach (CBA), and the Science Mathematics Study Group (SMSG). Brazil imported these projects for domestic use, with several universities collaborating in their expansion across the country.

During this period, Brazil was in the hands of a dictatorial military government. In 1964, that government unilaterally changed the political scenario of the country and, consequently, the expected role of the school. In this turbulent context, the U.S. directly intervened in Brazil's educational policies and research on education. This interference became clear from 1964 on, but was especially so after 1968. A 1966 MEC/USAID (Education and Culture Ministry[1]/ United States Agency for International Development) agreement decreed that professional and technical education would be ideal for Brazilian education.

In 1963, MEC created six science centers in some of the most important capital cities in the country. Some of these centers were related to universities, and some to bureaus of education. The first science education research groups in appeared in these centers, sharing a special interest in investigating the implementation of the projects mentioned above, all of which had, again, been imported from the U.S. As a consequence, research on science education in Brazil was strongly influenced by foreign sources.

After this period, however, Brazilian researchers became less dependent on those foreign sources, and original lines of research appeared, differentiating the development of this period from the preceding one. Eventually, the history and philosophy of science emerged as important intermediaries during the stabilization process, not only unifying a great number of the research projects carried out during that period, but also serving as important means for participation in the political and ideological arenas.

During the initial phase of science education research, the history and philosophy of science played only a limited role in the institutionalization of science education research in Brazil. Articles and books of that time were based on the simplistic idea that mere knowledge of the history of science would in itself stimulate students' motivation and facilitate their learning of scientific concepts. Some publications in the field barely addressed the subject of teaching, and there was very little progress in exploring how the history and philosophy of science might be used more efficiently and effectively in the classroom. However, some members of the community of researchers involved in educational projects at that time considered contributions from the history of science – including practical results, such as the production of teaching material to complement projects for teaching physics – to be very important.

During the 1970s, some researchers went on to further study the history of science, and to publish works that described the connections and commitments between scientific development and economic and political power. This task was considered a way to combat the military dictatorship then still in power in Brazil, and to thus implicitly denounce the pacts between universities and the government.

The contributions of the history and philosophy of science to the consolidation of science education as a research area were stronger during the 1980s. Journals founded during this period disseminated their ideas to others who were also interested in the history and philosophy of science. The *Revista de Ensino de Física* (the Journal of Physics Teaching), for instance, was launched in 1979, but, until 1993, had no specific section dedicated exclusively to the history and philosophy of science. Nonetheless, each edition contained individual articles dedicated to the theme. In contrast, right from its beginnings in 1984, the *Caderno Catarinense de Ensino de Física* (Santa Catarina Journal of Physics Teaching) included a section entitled "The History and Philosophy of Physics".

Other important contributions came from students traveling abroad to develop studies about the history and philosophy of science. This was also pointed out by one of the interviewees in Villani and colleagues' (2010) study.

During the 1990s, more theoretical contributions from the philosophy of science also fostered the development of variations of the conceptual change model (Posner et al., 1982), giving special emphasis to other philosophers besides Kuhn and Lakatos. More specifically, some researchers developed analogies about the teaching of science based on the ideas of Feyerabend, Laudan, Popper, and Bachelard. Villani (1992), for instance, explored the flexibility of Laudan's approach to scientific progress (Laudan, 1984) as applied to understanding the changes that take place in students' ideas in schools. Based on Bachelard's idea of an "epistemological profile" (Bachelard, 1978), Mortimer (1995) proposed a new version of the same concept, which became known as a conceptual profile. He posited that the process of cognitive evolution intrinsically combines old ideas with new ones and argued that teaching should promote changes in students' profiles by broadening their spectra of useful ideas. However, it is important to clarify that Mortimer's conceptual profile does not have the exact the same meaning as Bachelard's epistemological profile. Mortimer (1996) explains that the notion of the conceptual profile has some characteristics in common with Bachelard's epistemological profile, such as its hierarchy of different zones, in which each zone has greater explanatory power than its predecessors. Nevertheless, several elements were added to Bachelard's notion. First of all, in Mortimer's model, it is necessary to distinguish between the ontological and epistemological characteristics of each profile zone. In a conceptual profile, it is possible for each zone to be epistemologically and/or ontologically different from the others. A second difference is related to the fact that pre-scientific levels are not determined by philosophical currents, but by the epistemological and ontological commitments of individuals. Finally, it is important to point out that, as it has evolved, Mortimer's theory has moved away from Bachelard and towards sociocultural approaches.

During the last phase of the institutionalization process of science education research in Brazil, two trends developed in relation to the history and philosophy of science: towards reforming the cultural and scientific knowledge on which high school education was based, and towards training teachers to develop corresponding lines of research.

The first trend, largely carried out in schools, involved the reformation of scientific knowledge content to focus on the genesis of conflicts in science and the evolution of scientific theories, as well as on successes and failures. As an extension, this line of reasoning may also include reflection on the presuppositions, images, and basic intuitions at play in scientific advances. The training of science teachers should foster the acquisition of these forms of knowledge for two purposes: first, from the cultural standpoint, to enrich the quality of the content to be taught and, second, to adhere to the methodological requirements deemed most effective for science teaching

The second trend was embodied in a group of researchers devoted to the history and philosophy of science, with their own undergraduate and graduate courses, journals, and conferences.

As the work of these researchers became more technical (Martins, 2000; Pietrocola, 1992), their methodology became more precise, and case studies were included, addressing, for instance, Becquerel (Martins, 1997) or the alchemist Sendivogius (Porto, 2001). At the same time, efforts were made to maintain connections with science teaching through pedagogy-related studies devoted, for instance, to Newton's theory of colors, which is full of technicalities but nonetheless useful for teaching purposes (Martins & Silva, 2001, 2003).

Starting in the 1970s, the community of physics teachers started to organize regular events and meetings. The first National Symposium of Physics Teaching (SNEF) was held in the beginning of that decade, reflecting the expansion of Brazilian education, in which physics was included in the final phases of the education process. This symposium was organized at regular intervals (nowadays, it is held every two years) in different Brazilian cities and became the main scientific meeting for physics education researchers in the country (Nardi, 2005).

Around the same time, the University of São Paulo (USP) instituted one of the first research groups in this area, enlisting the participation of primary and secondary school teachers. During the 1980s, many more research groups and graduate studies courses were created across the country, contributing to the growth of academic research in the area of physics education. In the same decade, the idea for the EPEF (Encontro de Pesquisa em Ensino de Física – Meeting of Research in Physics Education) emerged during a scientific meeting in which participants noted the need for a specific forum for the discussion of *strictu sensu* research (Nardi, 2005).

In Argentina, the process was analogous to the Brazilian one described above. As Orlando et al. (2008) point out, the physics teaching community received great encouragement in the 1980s. One indicator of this encouragement was the organization of the Physics Education Meeting (REF), which attracts more than a thousand teachers of this subject. This meeting was first organized in 1983 in Cordoba, and is still periodically held. As a consequence of these regular meetings, the community of physics teachers became more cohesive and the Argentinian Association of Physics Teachers (APFA) was created. This association was in charge of organizing periodic meetings and events, among other things. The success of such events was reflected not only in the number of people present and topics discussed, but also in the increasing number of research projects presented. This necessitated a larger platform for such discussions, and led to the first "Symposium on Physics Education Research" (SIEF) in Tucumán in 1992.

APFA is responsible for gathering researchers at alternating REF and SIEF meetings every two years. Its research community has been growing in number and its tasks have become more specific. In the past, researchers had to go abroad to attend graduate courses; at the very least, their dissertations and theses had to be developed under the guidance of foreign supervisors. Nowadays, it is possible for researchers to complete degrees at all levels in Argentina.

The growing number of projects presented at the various symposia throughout the years has reflected researchers' increasing interest in working on topics related to physics education in the different areas of educational research.

The increase since the first SIEF in the number of articles that follow these criteria for research projects shows advances in the construction of knowledge in the field of physics education, as well as incremental developments in specific training for science education researchers.

Nevertheless, it should be noted that the research projects presented at these symposia deal with heterogeneous topics. Most projects investigate problems connected with teaching and learning, curricular aspects, contextual issues, teacher training at the college level, and educational knowledge transmission. Only a small percentage of these projects deals with topics related to theoretical or epistemological aspects of physics education or methodological developments in the field. This fact should be taken into account, since such limited reflection on the process of producing knowledge does not contribute much to the consolidation and validation of methods in the process of developing scientific theories.

Several key questions emerge from this situation: Did the curricula used in a given project – especially high school curricula – include HPS? Do the curricula of the teachers' colleges studied provide for training in HPS? What happens with practicing teachers who did not, as undergraduates, have the chance to become familiar with HPS' contributions to science education? Do the class textbooks in question (both student and teacher versions) incorporate such contributions? If so, how did this happen? Are there any resulting teaching proposals based on research results?

This chapter analyzes the historical evolution of HPS in conjunction with science teaching and learning, the different lines of research that have developed over time, and some examples of teaching proposals – aimed at students, teachers, and science teacher trainees – which were designed taking research results into account. We present a critical analysis of the present situation vis-à-vis the incorporation of HPS in science education, and suggest some ideas for making progress in this area.

2 The Origins and Evolution of the Incorporation of HPS in Science Education

A consensus has been reached, based on discussions over the last fifty years, regarding the need for solid science education in the formation of scientifically literate citizens. As early as 1982, the director of the National Science Foundation (NSF) in the U.S. drew attention to the widening gap between the scientific "elite" and the scientifically "illiterate". The ordinary citizen is unable to grasp new scientific knowledge because of its inherent complexity and the speed at which it is developed. The challenge at hand is to determine how teachers and curriculum developers can narrow that gap. Duschl (1997) considers the study of the nature of science and scientific inquiry in science classes to be crucial to that endeavor, since these subjects allow teachers to introduce new teaching models in their classrooms and provide them with different criteria by which to select and design teaching proposals, as well as with the capacity to sequence and determine the importance of their chosen topics. This would contribute accordingly to improvements in students' learning.

In the mid-20th century, the most ambitious modifications in science teaching and learning ever attempted were implemented on an international scale. The main objectives of these changes were to update teachers' and students' knowledge and arouse students' interest in science. Their central goal was to form a new generation of scientists. This approach was based on a philosophy of science that stresses the justification of knowledge. Accordingly, two strategies were developed that would dominate science education for more than two decades. One of these was an *emphasis on processes,* which gives priority to the generic skills and techniques that science employs to collect, manipulate, and interpret data. The other was an *emphasis on inquiry* that stresses the role of manipulation and research activities in which the student plays an active role as researcher trainee. The Physical Science Study Committee (PSSC) project, set up by MIT in 1956, was probably the best known such group in the world. In the 1950s and 1960s, the revised curricula – which was intended to train students to "think" as scientists, historians, and philosophers – challenged well-established models regarding the nature of science. The debates that took place among philosophers in those decades took research from the history of science into account and began to reject the idea that observations and theories could be treated as separate entities. This led to a new perspective that considered science as an activity in which

- the theoretical framework adopted determines the associated observation patterns;
- progress is not cumulative and change (of hypotheses, of theories) is one of the main characteristics of scientific activity.

That is to say, central aspects of logical positivism, such as the clear division between observations and theories, and the role of logic as applicable only to the verification and justification of theories, were beginning to be brought into question.

Unfortunately, the notions of science presented in most of these curricular changes were still influenced by empiricism. The processes of science were emphasized, but in a reductionist way, since the focus was on training students to observe, measure, establish relationships among variables, formulate hypotheses, draw graphs, control variables, and design experiments, while neglecting both explanation and evaluation. The latter two are fundamental processes of scientific activity, as they are the essential products of scientific research; moreover, evaluation criteria are required by the scientific community when deciding to accept any explanation.

Another aspect of the nature of science that was not expressed in these curricular reforms was the fact that, even though science is a rational activity, its products are always provisional. At the same time, as Duschl (1997) points out, a deep analysis of scientific activity reveals many false starts, uses of defective logic, and wrong assumptions. These, too, are science – and they should be presented in science classes if our aim truly is the formation of scientifically literate citizens who are able to appreciate a less-stereotyped view of science and those who develop it, and are therefore capable of making critical decisions about scientific and technological issues in democratic societies. There is, then, a consensus that it is necessary for the 21st-century citizen to not only know about science, but also to know how it is produced, changed, and validated, how it develops over the course of history, and how it relates to social and cultural environments. This is called the science, technology, and society (STS) approach, which supports the incorporation of HPS in science education as part of preparing students for citizenship (Adúriz-Bravo et al., 2002).

During and after this period of incorporating – on a worldwide scale – the most widespread curricular reforms in history, there were efforts to bridge the gap faced in classroom science education between these new curricular proposals and discussions that took place in the field of HPS.

Generally speaking, research related to the nature of science can be divided into several areas which, although closely related, are different:

- Investigations of students' conceptions of the nature of science.
- Studies of teachers' conceptions of the nature of science and their efforts to offer alternative interpretations.
- Identification of the relationships between teachers' conceptions, teaching practices, and students' conceptions.

- The development and implementation of curricula designed to improve students' conceptions of the nature of science.
- Analyses of the incorporation of HPS into textbooks.
- The design, implementation, and evaluation of teaching and learning sequences that incorporate contributions from HPS.
- The incorporation of HPS in teacher education.

3 Students', Pre-Service Teachers', and Teachers' Conceptions of Science

In an important and detailed analysis of research into students' and teachers' conceptions of the nature of science, Lederman (1992) states that the earliest suggestions for encouraging students toward better understandings of the nature of science could be glimpsed in the 1907 findings of the "Central Association of Science and Mathematics Teachers". Over the following decades, strong arguments for placing more emphasis on "scientific method" and "science processes" emerged, for instance, in the 1960s, when the central focus in science education was on the analysis of scientific and research processes.

More recently, the analysis of the nature of science has been incorporated as a vital component of scientific culture (AAAS-American Association for the Advancement of Science. (1990), among others).

The first formal instrument for detecting students' conceptions was developed by Wilson in 1954 (Wilson, 1977). At that time, according to students from 43 high schools in Georgia, scientists' main aim was to "discover true natural laws". Klopfer and Cooley, who were the first to use a qualitative methodology for the analysis of data in this area, obtained similar findings in 1961.

In turn, Aikenhead (1972, 1973) produced comparable results from research conducted in different countries. These studies revealed that students lacked sufficient knowledge about: (a) the role of creativity in science; (b) the role of scientific models; (c) the role of theories and their interrelations with research; (d) the differences between hypotheses, laws, and theories; (e) the relationships between experimentation, models, theories, and absolute truth; (f) the fact that science does not only involve the gathering of data and the classification of facts; (g) what a scientific explanation is; and (h) the interrelationships and interdependences among the different branches of science.

Two other research projects, carried out by Rubba et al. (1977) and Bady (1979), supplied yet more data. The former found that about 30% of high school students believed that "scientific research reveals absolute truth". Most of the individuals studied also thought that scientific theories may, after repeated tests and verifications, eventually become laws.

Bady researched the ways in which students understood the testing of hypotheses. Most of them, regardless of their school level, believed that hypotheses could be adequately tested through verification. He concluded from those results that the majority of the students in his study had a simplified and naïve view of the nature of hypotheses and scientific theories.

There is plenty of literature on science education research regarding teachers that points out the need for science teachers and curricula developers to take present ideas about the philosophy of science into account. The ways in which science classwork is approached may, at any level of education, be a determining factor in students' understandings of scientists' work, and in their attitudes towards science and scientists.

The last few decades have seen an increased interest in studying teachers' conceptions of the nature of science. Several studies place science teachers firmly within the camps of the various forms of positivism. Nevertheless, dissenting results suggest that this is a more complex problem. Studies such as those conducted by Acevedo (1994) and Lakin and Wellington (1994) point out that, although teachers exhibit empiricist characteristics, they should not be simplistically classified as holding naïve inductivist views. According to Koulaidis and Ogborn (1989), science teachers often take an approach more closely connected with a Kuhnian view than with empiricism – and a high percentage of them in fact have conceptions of the nature of science which cannot be linked to any particular philosophical approach.

Nowadays, the science education research community agrees on the need for science teachers, especially physics teachers, to understand – at least to a certain degree – philosophy of science issues and their relevance to science education. In this regard, Abell and Smith (1994) cite studies such as those carried out by Wolfe (1989), Zeider and Lederman (1987), Smith and Neale (1989), and Tobin and Fraser (1988), in which the authors note that teachers' conceptions of the nature of the discipline they teach may influence students' conceptions of science and may limit the kind of science that students produce in class. At the same time, Abell and Smith's (1994) research identified the following general characteristics of pre-service science teachers:

- Pre-service science teachers consider science to be the way we discover what exists in the world. This is a perspective based on naïve realism, in which the object of research is assumed to behave independently of the researcher.
- They share a positivist approach to the science process: universal natural laws have empirical bases.

- Science is perceived as free of imagination and human creativity; its main objective is seen as understanding the "whys" and "hows" of the natural world.
- Only a few pre-service science teachers used the term "theory" and none of them discussed how theories develop, become accepted, or help to organize knowledge and guide research.
- Neither the roles of logic, induction and deduction, nor even the role of invention were discussed.
- Science is regarded as unburdened by human elements: there are no flaws, guilty consciences, prejudices, or ethical dilemmas involved.

Despite the growing number of books and articles addressing the main questions in the philosophy of science, teachers are still misinformed. This is partly due to the fact that they have been trained in a system that favors scientific knowledge above all and pays little attention to the history and philosophy of science. This limited understanding of the nature of scientific knowledge is even shared by some scientists, who are thus at least 20 or 30 years behind in their awareness of developments in the philosophy of science.

4 The Incorporation of HPS into Textbooks

There are also investigations that focus on how to incorporate HPS into the physics textbooks used by both teachers and students. Arriassecq and Stipcich (2000) follow this line of scientific inquiry in their critical analysis of the incorporation of HPS into the high school physics textbooks written after the Argentina's 1993 educational reforms. One of their main conclusions is that HPS is not incorporated into school textbooks according to the consensus theoretical framework agreed upon in HPS, but is reduced, rather, to mere anecdotal vignettes or, at best, to single chapters on epistemological aspects which, unfortunately, do not bear any relationship to the structures and approaches in all of the other chapters. They also draw attention to the need for developing and incorporating concrete proposals for bringing HPS content into classrooms.

Later studies (Arriassecq & Greca, 2004, 2007) revealed that:

- The textbook appears to be the main resource used by teachers in preparing their classes, especially at the high school level. These same textbooks are used by students
- The ways in which topics are addressed may profoundly condition the learning results achieved by students.

In the same studies, which specifically refer to the ways textbooks present special relativity theory (SRT), the researchers contend that those teachers who face the task of approaching SRT for the first time will generally resort to using their textbooks as the sole guide for their classes. In many cases, teachers will not have had the opportunity to reflect on SRT and decide for themselves which concepts are most relevant for understanding this theory, they will thus tend to follow the plan offered by the textbook (or textbooks) they have selected for class preparation, without adapting the material to fit their own criteria.

In turn, the results presented in this study coincide with those obtained on the same topic in other countries. This demonstrates that the available teaching materials provide teachers with insufficient guidance for approaching topics such as SRT from epistemological and contextualized perspectives.

Based on that fact, and taking budget concerns into account, the production of teaching material for teachers and students which facilitates meaningful learning by appropriately and clearly introducing contents from conceptual and motivational perspectives seems an obvious necessity. Such materials should offer serious, research-based discussions about the various contextual aspects relevant to physics theories.

Kragh's contextualized or "anti-whig" approach analyzes historical events in light of the beliefs, theories, and methods of the time in which a given idea (or group of ideas) was conceived. This view offers a more realistic approach to history, one which integrates the obstacles and mistakes in scientific work. This view of science seems to be more accurate than the "whig" version, since it considers rejected hypotheses to be as important as the successful ones in the construction process of scientific knowledge. This view produces more reliable, comprehensive results, as it encompasses the successes, as well as the failures, experienced by scientists during their processes of developing, elaborating and experimenting on, and validating scientific theories.

5 The Design, Implementation, and Assessment of Teaching and Learning Sequences That Incorporate Contributions from HPS

Teixeira et al. (2009), in a thorough and methodologically rigorous work, investigate the implementation of teaching which incorporates HPS in physics classrooms, with the aim of obtaining critical and reliable information on this subject.

The vast majority of the studies selected for analysis support the idea that there are similarities between students' spontaneous understandings of scientific concepts and the historical development of those concepts. Such findings support the idea of conceptual change as a key tool for science education, despite the large amount of criticism found in the literature regarding this type of approach. While the authors examined a variety of teaching strategies based on HPS, comparatively few were found to provide the pedagogical references necessary to justify the use of these strategies, and even fewer were concerned with assessing students' prior knowledge of HPS.

The analyzed studies presented various ways of utilizing HPS in physics teaching: in relation to teaching objectives (learning concepts, nature of science [NOS], attitudes, argumentation, and metacognition); in relation to teaching strategies (as integrated with the subject of physics, integrated with another teaching strategy, and non-integrated); in relation to didactic materials (historical narratives, biographies, replicas of historical experiments, historically contextualized problems, and stories of scientists' lives).

The authors found positive effects resulting from the didactic use of HPS in the learning of physics concepts, despite the lack of consensus on this subject; they also indicated a lack of agreement about the occurrence of conceptual change. More research efforts are therefore needed to investigate these aspects, especially when the aforementioned limitations in research procedures are taken into account. In the same way, no consensus was found as to how HPS might promote improvements in students' attitudes towards science, which led Teixeira and his colleagues to conclude that this subject also requires further investigation.

When looking closely at the effects of the didactic classroom employment of HPS, favorable results were found in the areas of argumentation and metacognition, despite a relative dearth of studies dealing with these areas. This type of approach also appears to promote a more mature understanding of NOS among students, which should be taken into consideration when planning physics curricula and/or teaching strategies. This shows that potentially important areas are being explored, and suggests that HPS-based physics teaching be given a high position on the science education research agenda.

Arriassecq's (2008) doctoral dissertation addressed the problem of teaching special relativity theory (SRT) at the secondary school level in Argentina. This project involved several areas of research, focusing on: the epistemological difficulties presented by SRT itself; teachers' difficulties in dealing with SRT at that level; and the textbooks that both teachers and students used as a teaching-learning resource. The results of these studies showed that there is a wide gap between the teaching proposals presented in documents from Argentina's Ministry of Education, as well as in research reports, and their actual practice in classrooms. In order to narrow this gap, the author developed a teaching proposal in which SRT is approached via its historical and epistemological contexts.

This teaching proposal was designed within a framework that comprises epistemological, psychological, and didactical aspects, and consisted partly of written material (in textbook format) to be used by teachers and students, taking account of the deficiencies the adopted approach considers as such. Several studies conducted prior to the design of this teaching proposal (and the supplementary teaching material used in its implementation) adopted an approach that assigns the history of science and epistemology important roles in the design of concrete class proposals, sharing equal billing with psychological and didactical frameworks. This contextualized approach places great conceptual emphasis on the topics discussed, which is essential for rendering historical-epistemological discussions meaningful to students.

The use of history of science and epistemology is considered to facilitate, among other things, the determination of epistemological obstacles – knowledge that can be used to guide the selection of relevant teaching content – as well as discussions about the production of scientific knowledge, the role of the social and cultural contexts within which a given aspect of scientific knowledge was developed, and its impact within and outside of the scope of science. All of these activities should take place with an aim toward the eradication of science stereotypes and practices that discourage students from taking an interest in science.

The thesis draws on elements of Bachelard's epistemology (1991) for an epistemological analysis of SRT content (see also Arriassecq & Greca, 2012). Our analyses defined the central, most meaning-laden concepts students should learn as space, time, and those notions related to reference frames (system, observer, simultaneity, and measurement) which are essential for understanding the relativity-based perspective on space-time.

Our assessment of the implementations of the teaching proposal – which included use of the related written material – demonstrate that the acquisition of key concepts of SRT in that context was superior to the results obtained when SRT was approached in the "traditional" way, in which the standard textbook is the main teaching resource used by the teacher (Arriassecq, 2008; Arriassecq & Greca, 2012).

The supplemental text produced as part of the design of the teaching proposal addresses SRT – a topic that, despite its importance, has not been sufficiently investigated within the area of physics education in Argentina – in greater depth. In it, SRT is presented in a clear yet innovative way, within an original theoretical framework that comprises epistemological, psychological, and didactical elements.

6 The Incorporation of HPS in Teacher Training in Argentina

The general guidelines for epistemological studies that are followed by physics teachers are derived from the regulations developed by the institutions responsible for education policies at different educational levels. In her thesis, Islas (2010) compiles all such regulatory documents concerning the incorporation of HPS into the syllabi of physics teacher training colleges in Argentina.

The document *"2008-Science education year"*[2] extracts what the *Report of the National Commission for Improvement in Natural Science and Mathematics Education* points out when it stresses

> the need to overcome both the simplistic views on science and scientific work as well as those views of scientific work as something extremely difficult, which lead to school failure. [...] At the same time the program aims at arousing interest in those disciplines, which follows from understanding what producing science and producing mathematics mean, their usefulness and importance for citizenship; demystify the process of knowledge development for students and teachers at different education levels, encouraging them to value it as an activity for social construction; promote future scientific vocations.

In the recommendations section of the document, the above objectives are translated into suggestions for teaching practice. The author proposes that the different aspects of scientific knowledge – including, among other things, its empiricism and use of models, along with the ways in which results and their interpretations are discussed and debated – should be taken into account by teachers and those who plan curricula.

There are other points in the document regarding classwork that are worth mentioning. For one, the author contends that in order for students to build solid knowledge, teachers must integrate experimentation, encourage questions, foster Socratic dialogue, and use rigorous, logically sound, and simple reasoning in their classrooms. All of these are characteristic of 'proper thinking' in the science classroom. But they are also distinctive aspects of the ways scientists think when doing research.

In differentiating between the contexts where knowledge is developed – in the classroom or in the research community – the author (of *2008-Science education year*) claims that

> the most significant difference between both activities is that, whereas the scientific community generates new knowledge on the borderline

> between the known and the unknown, students in class build concepts that, despite being new to them, have already been validated by science.

Finally, it is worth pointing out that the *Report of the National Commission for Improvement in Natural Science and Mathematics Education* states that one of the obstacles detected via its diagnostic methods is the "stereotyped picture of science and scientists, which is also shared by teachers".

In Argentina, most teachers who are members of universities or state institutes have some curricular time at their disposal for the study of topics related to HPS. In addition to contents per se, Islas (2010) also reviewed other elements of syllabi, including bibliographies, objectives of the subjects covered, and requirements for passing courses.

A shared characteristic of the teacher training programs included in this analysis is the inclusion of non-standard epistemologies. The common syllabus subheading, "Tendencies among contemporary epistemologies" (or something along those lines), is considered an indicator of the presence of innovative explanations. Authors such as Kitcher, Giere, van Frassen, Habermas, and Gadamer appear in the bibliographies of some syllabi, in addition to those most frequently considered "contemporary", including Lakatos, Feyerabend, Laudan, Toulmin, and, to a lesser extent, Bachelard.

7 Final Remarks

Over the last twenty years, we have seen some significant rapprochements between the areas of science education and HPS. Even though advocates for the incorporation of aspects of HPS into science teaching are aware of the existence of differing opinions, they stress the importance of a contextualized approach. That is to say, they propose that science should be taught in a way that enables students to learn how to think critically, and to analyze the ways in which social, historical, philosophical, ethical, and technological contexts are closely linked to the development, validation, and application of scientific knowledge.

However, it is worth pointing out that, as Matthews (2000) states,

> It is unrealistic to expect students or preservice teachers to become competent historians, sociologists, or philosophers of science. We should have limited aims in introducing questions about the nature of science in the classroom: a more complex understanding of science, not a total or even a very complex, understanding.

Nevertheless, it is essential to more closely evaluate the reformulation of curricular projects at different educational levels – including science teacher training in teachers' colleges and other institutions – in part, by taking analyses of the results obtained into account. It is also necessary to consider supplemental training for practicing teachers who never had the opportunity (in college or graduate school) to approach science from this contextualized perspective.

The increasing volume of innovative written material for students and teachers which incorporates HPS should be beneficial for science education writ large. Over the last few years, research into science education has produced teaching units for different science topics which incorporate historical and epistemological contexts and take key aspects of the NOS into consideration; and such efforts are generally on the rise. However, very few of these positive developments actually reach in-service teachers, pre-service teachers, or students.

Notes

1 In Portuguese, Ministério da Educação e Cultura.

2 The magazine has been published twice a year since 1992.

References

AAAS-American Association for the Advancement of Science. (1990). *Science for all Americans*. New York, NY: Oxford University Press.

Abell, S., & Smith, D. (1994). What is science?: Preservice elementary teachers' conceptions of the nature of science. *International Journal of Science Education, 16*(4), 475–487.

Adúriz-Bravo, A., Couló, A., Kriner, A., Meinardi, E., Revel Chion, A., & Valli, R. (2002). *Three aspects when teaching the philosophy of science to science teachers*. Retrieved from http://www1.phys.uu.nl/esera2003/program/listofauthors2.htm

Acevedo-Díaz, J., Vázquez-Alonso, A., Manassero-Mas, M. A., & Acevedo-Romero, P. (2007). Consensos sobre la naturaleza de la ciencia: fundamentos de una investigación empírica. *Revista Eureka, 4*(1), 42–66.

Adúriz-Bravo, A., Salazar, I., Mena, N., & Badillo, E. (2006). La epistemología en la formación del profesorado de ciencias naturales: aportaciones del positivismo lógico. *Revista Electrónica de Investigación en Educación en Ciencias, 1*, 6–23.

Aikenhead, G. S. (1973). The measurement of high school students' knowledge about science and scientists. *Science Education, 51*, 539–549.

Aikenhead, G. S. (1974). Course evaluation I: A new methodology for test construction. *Journal of Research in Science Teaching, 11*, 23–30.

Aikenhead, G. S. (1974). Course evaluation II: Interpretation of student performance on evaluative tests. *Journal of Research in Science Teaching, 11*, 31–41.

Arriassecq, I. (2008). *La enseñanza y el aprendizaje de la teoría especial de la relatividad en el nivel medio/polimodal* (Doctoral dissertation). Universidad de Burgos, Spain.

Arriassecq, I., & Greca, I. (2004). Enseñanza de la teoría de la relatividad especial en el ciclo polimodal: Dificultades manifestadas por los docentes y textos de uso habitual. *Revista Electrónica de Enseñanza de las Ciencias, 3*(2), Article 7. Retrieved from http://www.saum.uvigo.es/reec

Arriassecq, I., & Greca, I. (2007). Approaches to special relativity theory in school and university textbooks in Argentina. *Science & Education, 16*(1), 65–86.

Arriassecq, I., & Greca, I. (2012). A teaching–learning sequence for the special relativity theory at high school level historically and epistemologically contextualized. *Science & Education, 21*(6), 827–851.

Arriassecq, I., & Stipcich, S. (2000). La visión de ciencia en los textos de física de nivel básico. *Memorias V Simposio de Investigadores en Educación en Física* (CD). Santa Fe, Argentina.

Bachelard, G. (1978). *A filosofia do não*. São Paulo, Brazil: Abril Cultural.

Bady, R. J. (1979). Students' understanding of the logic of hypothesis testing. *Journal of Research in Science Teaching, 16*(1), 1979.

Bell, R. L., Abd-El-Khalick, F., Lederman, N. G., McComas, W. F., & Matthews, M. R. (2001). The nature of science and science education: A bibliography. *Science & Education, 10*(1–2), 187–204.

Duschl, R. (1997). *Renovar la enseñanza de las ciencias*. Madrid, Spain: Narcea.

Islas, S. M. (2010). *La formación epistemológica de los profesores de física respecto de los debates producidos al interior de la comunidad de investigadores en la disciplina* (Doctoral dissertation). Universidad Nacional de Córdoba.

Klimovsky, G. (1994). *Las desventuras del conocimiento científico: Una introducción a la epistemología.* Buenos Aires, Argentina: AZ Editores.

Klopfer, L. E., & Cooley, W. W. (1963). The history of science cases for high schools in the development of student understanding of science and scientists: A report on the HOSG instruction project. *Journal of Research in Science Teaching, 1*, 33–47.

Koulaidis, V., & Ogborn, J. (1989). Philosophy of science: An empirical study of teachers' views. *International Journal of Science Education, 11*(2), 173–184.

Krasilchik, M. (2000). Reformas e realidade: o caso do ensino das ciências. *São Paulo em Perspectiva, 14*(1), 85–93.

Lakin, S., & Wellington, J. (1994). Who will teach the "nature of science"?: Teachers' views of science and their implications for science education. *International Journal of Science Education, 16*, 175–190.

Lantz, O., & Kass, H. (1987). Chemistry teachers' functional paradigms. *Science Education, 71*, 117–134.

Laudan, L. (1984). *Science and values*. Berkeley, CA: University of California Press.

Lederman, N. G. (1992). Students' and teachers' conceptions of the nature of science: A review of the research. *Journal of Research in Science Teaching, 29*(4), 331–359.

Lederman, N. G. (1999). Teachers' understanding of the nature of science and classroom practice: Factors that facilitate or impede the relationship. *Journal of Research in Science Teaching, 36*, 916–929.

Lederman, N. G. (2006). Nature of science: Past, present, and future. In S. K. Abell & N. G. Lederman (Eds.), *Handbook of research on science education*. New York, NY: Routledge.

Losee, J. (1972). *Introducción histórica a la filosofía de la ciencia*. Madrid, Spain: Alianza Editorial.

Martins, R. A. (1997). Becquerel and the choice of uranium compounds. *Archive for History of Exact Sciences, 51*(1), 67–81.

Martins, R. A. (2000). Que tipo de história da ciência esperamos ter nas próximas décadas? *Episteme: Filosofia e História das Ciências em Revista, 10*(1), 39–56.

Martins, R. A., & Silva, C. C. (2001). Newton and colour: The complex interplay of theory and experiment. In F. Bevilacqua, E. Giannetto, & M. R. Matthews (Eds.), *Science education and culture: The contribution of history and philosophy of science* (pp. 273–291). Dordrecht, The Netherlands: Kluwer Academic.

Matthews, M. (2000). *Time for science education: How teaching the history and philosophy of pendulum motion can contribute to science literacy*. New York, NY: Kluwer.

Mortimer, E. F. (1995). Conceptual change or conceptual profile change? *Science and Education, 4*(3), 267–285.

Mortimer, E. F. (1996). Construtivismo, mudança conceitual e ensino de ciências: para onde vamos? *Investigações em Ensino de Ciências, 1*(1), 20–39.

Nardi, R. (2005). Memórias da educação em ciências no Brasil: A pesquisa em ensino de física [Memories of science education in Brazil: Physics education research]. *Investigações em Ensino de Ciências, 10*(1), 63–101.

Orlando, S., Gangoso, Z., Lecumberry, G., & Ortiz, F. (2008). *Educación en física, ¿Qué?, y ¿Dónde investigamos?: Una mirada a la producción nacional.*

Pietrocola, M. P. O. (1992). *Élie Mascart et l'optique des corps en mouvement* [Élie Mascart and the optics of bodies in movement] (Doctoral dissertation). Université de Paris VII, Paris, France.

Porto, P. A. (2001). Michael Sendivogius on nitre and the preparation of the philosopher's stone. *Ambix, 48*(1), 1–16.

Posner, G. J., Strike, K. A., Hewson, P. W., & Gertzog, W. A. (1982). Accommodation of a scientific conception: Toward a theory of conceptual change. *Science Education, 66*, 211–227.

Rubba, P., Homer, J., & Smith, J. (1977). A study of two misconceptions about the nature of science among junior high school students. *School Science and Mathematics, 81*(3), 221–226.

Salomon, P. (1988). *Psychology for teachers: An alternative approach*. London: Hutchinson.

Teixeira, E., Greca, I., & Freire Jr., O. (2009). The history and philosophy of science in physics teaching: A research synthesis of didactic interventions. *Science & Education, 21*(6), 771–796.

Villani, A. (1992). Conceptual change in science and science education. *Science Education, 76*(2), 223–238.

Villani, A., Silva Dias, V., & Valadares, J. (2010). The development of science education research in brazil and contributions from the history and philosophy of science. *International Journal of Science Education, 32*(7), 907–937.

Wilson, S. (1977). The use of ethnographic techniques in educational research. *Review of Educational Research, 47*, 245–265.

PART 6

Science Education in Non-Formal Settings

CHAPTER 21

Non-Formal Education in South America: A Preliminary View

Francisco Julián Betancourt Mellizo

Abstract

From the middle of the last century, this essay describes, in a bird's eye manner and in a preliminary way, the development of processes that dynamized non-formal education in our region. Emphasis is placed on the information obtained in the congresses of the Popularization Network of Science and Technology in Latin America and the Caribbean, Red-POP. In addition, some processes, activities and conceptualizations developed by the Museum of Science and Game of the National University of Colombia are presented.

Keywords

non-formal education – informal education

1 A Brief Overview

The rationalist ideas of the Enlightenment had a great influence on our region. As was the case in most South American countries, such developments were originally brought to Colombia by European travelers. The Spanish physician and naturalist José Celestino Mutis introduced Colombians to the ideas of Linnaeus and Newton, as well as to many math concepts, all of which were disseminated via lectures on natural philosophy, conferences, and the hard work of local sympathizers (Mutis, 1975; Arboleda, 1987, 1995; Schumacher, 1988; Amaya, 2004).

Latin Americans of various ethnicities and nationalities were inspired by the liberating ideas of the French Revolution and by rationalist notions emerging from the Enlightenment. Francisco José de Caldas, the astronomer and botanist, was one such innovator, who – through various print media, including *El Papel Periódico de la Ciudad de Santafé de Bogotá* (the Newspaper of the City

 | DOI: 10.1163/9789004409088_021

of Santafé de Bogotá), *El Correo Curioso* (The Curious Mail), and *El Semanario del Nuevo Reino de Granada* (the Weekly Magazine of the New Kingdom of Granada), which he founded – published his scientific works on astronomy, geography, and botany (Shumacher, 1988; Arboleda, 1994; Bateman, 1998).

These early developments, and others which occurred at the end of the colonial era – when national independence was on the rise and the many young republics born in the 19th century were just being created – were aimed at the privileged, literate sectors of society.

A shift occurred in the second half of the last century, with the emergence and growth of processes involving the modernization and homogenization of different states in the region, the creation of the various National Councils of Science & Technology, and movements in civil society concerned with the popularization and dissemination of science. Various perspectives have inspired ideas for the development and material progress of Latin American countries, as well as for encouraging scientific literacy, the dissemination and popularization of knowledge, and ways to generate critical awareness in the region.

There has been a tradition of non-formal educational activities in Latin America for at least six decades (Sánchez-Mora, 2006; Franco, 2011). Different approaches to the field have ranged from literacy and adult education to job training and popular education. Such efforts have been initiatives of the state or critical sectors of civil society; in some cases, they have involved cooperation between both sectors.

Paulo Freire, who first developed proposals and programs for adult education in his native Recife, Brazil, stands out as a key figure in this field. Freire's writings have been highly influential in Latin America and around the world. His most famous book, *Pedagogía do Oprimido* (Pedagogy of the Oppressed; Freire, 1970), has been translated into at least 18 languages and his writings advocating for popular public education have been a source of inspiration for many groups in our region (Gadotti, 2004). Beginning in the 1960s, and flourishing in the 1970s and 1980s, these groups developed processes and dynamics in the field known as popular education that were not necessarily linked to the states in which they were created. Freire`s ideas have in fact inspired many people from many countries and backgrounds (Perez, 2010). His thought has also been influential in the vast field of non-formal education, which may also be referred to as civic education or citizenship education, and is based on universal human ethics (Freire, 2005).

In 2010, Independence Day or "cry of independence" (El Grito de la Independencia) was celebrated all over Latin America. It had been 200 years since 1810, when the process that led from the colonial era to the emerging republics – where supposedly all are equal before the law – had begun.

Two centuries later, our democracies are fragile. For many members of the Latin American population, the relationship between citizens and the state is diffuse and there is little understanding of rights and duties and their concrete expression in citizens' everyday lives. This also implies a weakness of the states in this region. It is almost impossible for the majority of individuals to internalize citizenship codes of all kinds, as expressions of strong intersubjectivity, in democracies that are socially and economically vulnerable.

In the various programs and projects undertaken by the Museum of Science and Games (MSG [in Spanish, Museo de la Ciencia y Juego]) at Colombia's National University, we have conducted many workshops for teachers from different regions of Colombia who work in public and private schools. In several of these programs, we have posed the following question: What roles should physics, chemistry, biology, or general science play in the formation of citizens? We have asked similar questions about art and philosophy. Teachers were surprised and puzzled by these questions, and their reactions provided us with ample materials for investigation. But, at the same time, it was painful to find that what is taught in schools is typically not contributing to the education of conscious citizens, that is, to the development of whole people who are able to engage in concrete ways in the processes of democratic society, as well as in their particular social and cultural contexts. This research led us to conclude that what was taught in schools often had almost nothing to do with culture – and that students probably felt like school was entirely apart from life.

If so, what function is the body of knowledge taught in school serving?

It has been suggested that there are serious problems with science teaching (Giordan, 1982; Zuleta, 1995; Segura, 2000). Science is generally taught in a decontextualized way which systematically ignores the social contexts of knowledge production, and, instead, presents a view of the scientist as a kind of genius, an aseptic superman with no selfish qualities. In this regard, the question is similar to the picture presented to us of the founding fathers of political history who, stripped of any self-interest, are framed as angels (Lechner, 2000). This is a "history of bronze", of statuary rather than human beings.

Nor do science classes tend to explore the effects and social implications of current scientific issues such as biotechnology, genomics, and cloning, despite the fact that information about these topics is widely covered in the media, albeit with varying comprehension of the given subject and from different angles, whether demonizing, exalting, or somewhere in between.

In turn, it is not common for teachers or their students to perform those processes of knowledge recontextualization or mediation vis-à-vis everyday life contexts which allow people to construct deeper understandings of what they learn.

Students rarely learn or internalize the values underlying the scientific process, including recognition of others, respect, argumentative power, teamwork, etc., many of which were instrumental in the construction of Western democracy (Bronowski, 1968).

These deficiencies are linked to the fact that the average amount of schooling in the region is 8.5 years. This points, first, towards the tremendous efforts to be undertaken at the level of formal education if we aim to build democratic processes on a daily basis in our effort to overcome the weaknesses of our societies. Second, it reveals the importance of the possibilities for the construction of citizenship that may be found in non-formal and informal education processes.

Even though there has been progress, many of the circumstances that generated Freire's thinking have persisted. Many of the values of the colonial era are still alive in our countries; in other words, the transformation from vassals of the Spanish Crown to strong democratic republics populated by knowledgeable and engaged citizens has not been fully realized. This is why we insist on the relevance of citizenship education incorporating both local and global knowledge.

The field of non-formal education has been permeated by endogenous processes, such as the impacts of Freire's thinking, and by exogenous processes, including the direct influence of international currents of thought, or the perspectives and interventions of world power centers regarding what some what some call the periphery. An example of the latter was the U.S.-driven Alliance for Progress' efforts toward modernization and homogenization in Latin American countries. In 1961, President John F. Kennedy proposed a 10-year plan, stating:

> ... We propose to complete the revolution of the Americas, to build a hemisphere where all men can hope for a suitable standard of living and all can live out their lives in dignity and in freedom. To achieve this goal political freedom must accompany material progress...Let us once again transform the American Continent into a vast crucible of revolutionary ideas and efforts, a tribute to the power of the creative energies of free men and women, an example to all the world that liberty and progress walk hand in hand. Let us once again awaken our American revolution until it guides the struggles of people everywhere-not with an imperialism of force or fear but the rule of courage and freedom and hope for the future of man. (Kennedy, 1961, n.p.)

In addition, the Russian's launch of the Sputnik 1 in 1957 surprised the United States and generated, as is well known, a series of educational reforms, especially in science teaching, the purpose of which was to beat the Soviet Union (as it was known at the time) in the space race. These educational reforms influenced Latin American countries not only in science education, but also in the popularization and dissemination of scientific ideas.

This confluence of endogenous and exogenous influences also resulted in a trend towards the creation of interactive museums and science and technology (S&T) centers. This process began in Latin America in the late 1970s with the creation of an interactive museum called the Centro Cultural Alfa, in Monterrey, México. It then proceeded rather gradually throughout the 1980s with eight additional museums, and peaked in the 1990s, with the creation of 93 institutions. Although the information we have is still incomplete, it seems to indicate that the trend over the first decade of this century was toward a decrease in new institutions of this type, since we estimate that from 20 to 30 interactive museums were created during that period (Betancourt, 2002).

Museums are institutions of informal and non-formal education; they also often provide support for formal education. In our region, interactive museums and S&T centers accomplish important functions in these areas. To clarify: *Formal* education leads to certifications or diplomas recognized by the state and, for that reason, must adhere to certain rules and formalities associated with those achievements. The sequence of formal education is well-known: from elementary school on to middle and high schools, followed, possibly, by undergraduate and, more rarely, graduate (specializations, master's degrees, and doctorates) degrees. *Non-formal* education shares the formalities and rituals of formal education, but does not lead to a diploma. Non-formal education is found, for example, in continuing education courses, hobby classes, work-skills workshops, etc. *Informal* education has no such formalities and occurs in multiple contexts: in homes, neighborhoods, and work environments; via the movies, the press, TV, the Internet, and museum exhibitions .... These distinctions are also used in talking about place, encompassing formal (for example, schools) versus non-formal or informal spaces of learning (for example, museum exhibitions). In the first context, learning is subject to formalities, obligatory, and subject to a curriculum or program of study; in the latter case, learning is informal: "It is voluntary (no one is mandated to learn). It is learner motivated and guided by learner interest. It is nonlinear and open ended. It can occur in a variety of settings ... it is ubiquitous and ongoing-it occurs in many places, at any time of the day, and at any time of one's life" (Diamond, Luke, & Uttal, 2009).

In some countries of the region, such as Brazil, the term informal education is not used, and so non-formal is also used to refer to what is elsewhere called informal. In such circumstances, non-formal education corresponds to a very broad spectrum of activities.

This can be seen, for instance, in the events of the Network for the Popularization of S&T in Latin America and the Caribbean (RedPOP) which organizes a meeting every two years convening the community that popularizes and disseminates S&T in the region. At these events, network members present experiences and reflections in five fields or areas of work: non-formal education (NFE), museum studies and museology (MM), scientific journalism (SJ), production of materials (PM), and professionalization of the field (PRO). A quick look at the papers presented at these meetings offers information relevant to this chapter.

Table 21.1 shows the number of papers accepted for the last eight meetings of RedPOP, beginning with the sixth meeting, held in Rio de Janeiro, Brazil, in 1999, and ending with the thirteenth, in Zacatecas, Mexico, in May of 2013. This period was chosen because the information from previous meetings of the RedPOP network was incomplete. The data for the meeting in Zacatecas were provided by a member of the organizing committee, Miguel Garcia.

Papers in the field of NFE are obviously the most numerous. This is because, in the region, this field involves not only museums, but also other organizations that focus on NFE. In 2001 alone, there were 19 institutions of non-formal education linked to RedPOP, including: Mundo Nuevo, Argentina; Grupo de Estudo de Pesquisa em Educação Não Formal e Divulgação da Ciência, GEENF, Brazil; Club de Ciencias, Chile; Centro de preparación para la Ciencia y la Tecnología, CEPRECyT, Perú; Programa Ondas-Colciencias, Colombia; Fundación para la Ciencia y la Tecnología, CIENTEC, Costa Rica; Programa de Promoción Educativa Comunitaria, ASPRODIC, Guatemala; and Café Científico de Baja California Sur, México.

The field of NFE itself includes organized courses, conferences, scientific expeditions, science fairs, workshops, science film clubs, amateur astronomical observation groups, training programs for science teachers, science clubs, scientific theater groups, and even science fiction-related activities, among other things, resulting in the inclusion of a very wide variety of educational pursuits in this category (Betancourt, 2008).

Not all of the RedPOP network activity categories were launched simultaneously with its creation in Rio de Janeiro, Brazil in 1990. For example, the field of professionalization was initiated at the eighth meeting, which was held in 2003 in Leon, Mexico, when our colleagues in Mexico noted the need to include those individuals working in the dissemination and popularization of S&T.

It is no coincidence that Brazil has hosted RedPOP meetings four times, while Mexico has hosted three: both countries have extensive, strong communities engaged in S&T and its popularization. Argentina, Chile, and Colombia also have established communities in the field. This is reflected in the Latin

TABLE 21.1 Selected papers as oral presentations (o) or posters (p) in the areas of non-formal education (NFE), museum studies and museology (MM), scientific journalism (SJ), production of materials (PM), and professionalization of the field (PRO), at meetings of the RedPOP 1999–2013. Percentages refer only to oral presentations

Meetings	No days	NFE	MM	SJ	PM	PRO	Total papers
1. Brazil 1999	4	0 13 (p) 31%	36 (o) 8 (p) 31%	0 2 (p) 16%	27 (o) 10 (p) 22%		118 (o) 33 (p) 151
2. Chile 2001	3	33 (o) 9 p 47%	18 (o) 2 (p) 22%	7 (o) 0 (p) 8%	19 (o) 2 (p) 23%		77 (o) 13 (p) 90
3. México 2003	4	61 (o) 6 (p) 31%	55 (o) 8 (p) 29%	15 (o) 5 (p) 9%	27 (o) 4 (p) 14%	35 (o) 2 (p) 17%	193 (o) 25 (p) 218
4. Brazil 2005	1	20 (o) 26%	26 (o) 34%	11 (o) 14%	11 (o) 14%	9 12%	77 (o) 25 (p) 102
5. Costa Rica 2007	3	40 (o) 22 (p) 48%	21 (o) 7 (p) 25%	9 (o) 3 (p) 10%	0 (o) 0 (p) 0%	14 (o) 1 (p) 17%	84 (o) 33 (p) 117
6. Uruguay 2009	3	74 (o) 45%	22 (o) 13,5%	25 (o) 15%	33 (o) 20%	9 (o) 6.5%	163
7. Brazil 2011	3	52 (o) 40%	20 (o) 15%	10 (o) 8%	27 (o) 21%	21 (o) 16%	149 130 (o) 130 (o)
8. Zacatecas 2013	4	75 (o) 16 (p) 46%	26 (o) 6 (p) 16%	22 (o) 16 (p) 13%	25 (o) 14 (p) 15%	15 (o) 2 (p) 10%	163 (o) 54 (p) 217

American Prize for the Popularization of S&T: Mexico has eight winners, while Colombia has six, Brazil has four, and Argentina and Chile have two winners each.

At the meeting in Brazil in 2011, 149 oral presentations were accepted, but the published proceedings included only 130.

Another way to visualize the data in Table 21.1 is presented in Figure 21.1 in a column graph that tracks the changes in the categories (NFE, MM, SJ, PM, PRO) over the course of the same series of network meetings.

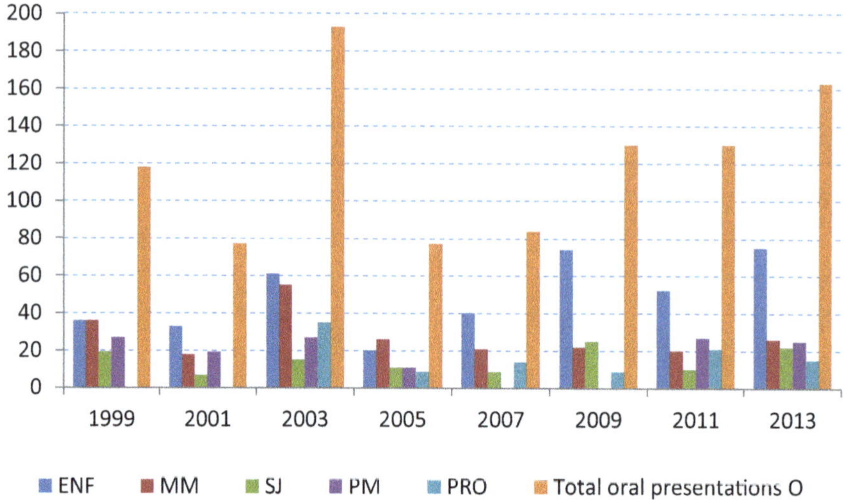

FIGURE 21.1 Comparisons between oral presentations (o) in the five fields

In the graph, the preponderance of the NFE field in oral presentations selected for the various meetings of the network is clear. The papers presented in the various fields describe final or ongoing research results; depict activities that bring something new to the field or might be inspiring for other groups; and/or reflect on various aspects of S&T popularization and dissemination activities in the region.

What NFE themes emerged as important in the first decade of the third millennium? To answer this question, we explored the records of the seventh RedPOP meeting, held in Santiago de Chile, Chile in 2001, and of the last event of this decade, the eleventh meeting, which was held in 2009 in Montevideo, Uruguay.

In Santiago, 33 oral presentations were selected for the NFE category, out of which we classified 18 as within the area of dissemination and popularization of S&T (radio programs on science aimed at school students, electronic school newspapers about S&T, S&T TV programs for schoolchildren, informative conference cycles). This also included activities, such as workshops on various S&T topics, school camps, extracurricular activities for youth, training processes for mediators (monitors and guides), open laboratories, science fairs, and health and science clubs.

In Montevideo, 74 papers were selected, out of which 31 – or almost half – involved the dissemination and popularization of S&T. In addition, there were seven papers on civic education and social inclusion, while had been only one such paper in Santiago. Thus, this issue gained momentum over the decade. Questions about the relationships between democracy, citizenship, and knowledge are emerging once again. Suffice it to say that the main presentation of the Montevideo conference (delivered by Dr. Rodrigo Arocena, rector of the Universidad de la República, the foremost public university in Uruguay), was called "The Democratization of Science: Major Problems and Opportunities for Twenty-first Century Democracy", and that the main theme of the event was "Building Identity and Citizenship". It is remarkable to note the presence of works that assess and analyze NFE activities from critical perspectives.

The twelfth RedPOP meeting occurred at the start of the second decade of this century in Campinas, Brazil, not long after the entire Latin American region celebrated Independence Day in 2010. There were, surprisingly, only two poster presentations on the subject, as if S&T centers, interactive museums, etc. were not interested in thinking about the causes behind our poor civic education, the reasons why knowledge is not one of the foundations of our nations, and why knowledge and civic education have not played an important role in building our democracies.

The topic of social networks in the areas of non-formal and informal education began to emerge in Campinas and is expected to grow enormously in the future.

Projects regarding clubs, fairs, and traveling exhibitions were again present, proclaiming upcoming events for the next decade; along with works involving science weeks and other events convened by state institutions presenting S&T activities across the country. Seven papers on learning processes in various non-school environments were also presented, while in Santiago the emphasis had been more on teaching than on learning. Health was a predominant theme. Topics that had been discussed since meetings in the 1990s – involving such subjects as physics, biology, mathematics, and environmental education – were included, along with emerging or trending issues like cloning, genomics, robotics, and mechatronics.

As mentioned before, works on science fairs have been frequently presented at network meetings. These fairs began in the 1960s, and have since advanced in many ways. Most South American countries organize school science fairs at the national level. With increasing globalization, there may be a tendency to standardize the requirements and fields of knowledge for local science fairs, so that winners can participate in major international fairs that take place, for example, in the U.S.

2 A Particular View

In our case, the Museum of Science and Games (MSG) has developed small fairs in Bogota, the capital of Colombia (Arango, 2003). The city is divided into areas, each of which is operated by a local mayor, who answers to the mayor of the city as a whole. We have also had the support of two local municipalities, Engativá and Teusaquillo, for these fairs.

These fairs are structured by varied and numerous meeting processes with teachers from participating institutions. The two core activities developed with teachers are workshops on the proposed theme for the exhibition and educational meetings held in each institution to refine the project which is going to be presented at the events. The subjects addressed in the fairs concern large areas of knowledge about the environment and the city. The workshops are structured around three axes: pedagogical, ecological/environmental, and social/heritage (Avila & Bautista, 2008).

In the pedagogical axis, educational aspects of the disciplines (biology, physics, and chemistry) are developed within the context of the topic chosen for the event, and with reference to everyday life. In the social and heritage axis, work resolves around the methodology of social cartography that allows us to map the environmental heritage, both historical and urban, in which the participating schools are immersed. Since "social mapping is useful in constructing more comprehensive and reasonably accurate representations of social and cultural phenomena" (Paulson, 1996), it allows better visualization of all the social landscapes of a given environment. In the ecological and environmental axis, students investigate local environmental conditions, the impact of human activities on that environment, and the need to build global and local awareness of our responsibilities towards the environment.

Each workshop also emphasizes what we call "basic cultural competences". These competences relate to the fact that many actions, even commonplace human actions, entail both observation and exploration, comparison and relationship, inference and argument. These are paired skills that humans have been using since ancient times, which are subject to the worldviews prevailing at any a given time. We use these skills unconsciously and spontaneously in daily life, but they are used consciously and methodically in scientific endeavors. These skills are inherent to humans and are also deeply connected to rational knowledge; hence, it is essential that students learn about them and that teachers grasp their importance, regardless of the subject they teach (whether natural or social sciences, art or philosophy ...).

Another aspect of the workshops developed with the teachers is the integration of objects and elements of everyday life, an approach that is especially emphasized in the workshops of the pedagogy axis. One of the messages we intend to spread thereby is simple: we want people to take heart in the fact that they can do these activities in their schools or with their families, using readily available resources. This is linked to another important purpose: using everyday objects in science experiments and demonstrations is a way of bringing the field of science into everyday life and the family realm. This is a process of recontextualizing science into everyday contexts, as mediated by common objects.

Another message is that one does not necessarily need sophisticated tools to explore natural and social environments. We should instead cultivate an open and questioning attitude that leads us to explore, to connect, to infer; students also learn that these can occur at the collective as well as individual levels. In the latter case, students' reasoning and communication skills are likely to be refined and something wonderful may appear: collective construction.

In our activities, everyday objects are generally employed in ways that differ from their common use. Disrupting functionality is an important step towards the construction of creative processes. Our workshops are also based on game dynamics. We develop our own games, some of which are based on board games that are re-contextualized and mediated to the realities and purposes of a particular subject matter. Hence, the proposed workshop activities are open and flexible. We have found that teachers in our workshops are able to replicate these activities and, even better, to change and adapt them to their educational needs.

In all of the above processes, we are not exactly working on knowledge ownership, that is, on what and how students learned vis-à-vis a given piece of information. Our aim, instead, regards discovering how to provide conditions that encourage individuals and collectives to learn in enriching and exciting ways. This means that, for us, learning is a contextual and idiosyncratic process which flourishes under particular conditions.

In the processes leading up to the actual fair, our work with the students is indirect, that is, it is mediated by the teachers, who must submit a project they have produced with their students that somehow addresses the theme selected for the event. These rarely involve new projects, since, typically, school projects that are already under development are adapted to fit with the theme of the fair. This process is based on our educational meetings, where these projects are initially refined for exhibition at the fair. In these meetings, we also work with the teachers on another, non-mandatory requirement: student-run mini-workshops integrating everyday objects or games.

At the last fair, which was held in the locality of Teusaquillo in September of 2010, all projects were accompanied by activities in which visitors could participate, even though projects were presented in traditional exhibition formats, such as posters, objects, and students giving explanations. This allowed us to highlight students' efforts in the collaborative work that was required to execute the various activities. These collaborative efforts were indispensable in generating a dynamic in which students felt what it was like to perform socially useful work for different audiences.

Institutions participating in the process that leads up to the fair have the option of bringing exhibitions of our program, called "Museum Suitcases", into their schools. The Museum Suitcase program consists of exhibitions on a given topic which are packed into suitcases; they include small format, hand-held posters and games (Avila, 2009). Two of these institutions presented suitcases at the last fair that had to do with the projects they had worked on and presented at the event, which reaffirms the benefits of our proposals and our design.

Similar suitcases are an important part of another program, the Mobile Astronomy Classroom, an initiative made possible with the help of other institutions, including the National Astronomical Observatory of the University at Bogota, the National Learning Service (SENA), and Colciencias, the state agency for the promotion of science. The Mobile Astronomy Classroom travels the country visiting the many towns and villages far from the capital which would otherwise have no access to such experiences.

Workshops with teachers are also essential to the development of other museum programs. Bogotá's Secretary of Education offers a program called School-City-School, through which students from different localities in the city take "pedagogical expeditions" to visit sites of interest. These visits offer great opportunities for the development of a range of activities for children and young people which vary depending on grade level.

Prior to the expeditions with students, teachers from the school groups that visit the MSG participate in a workshop which offers related materials and activities. They are encouraged to conduct similar activities with their student groups. The teachers retain the material they have worked with during the workshop and are also given a flexible guide that allows them to carry out activities according to their interests.

Later, in what we call the socialization stage of the pedagogical expedition experience, the teachers present the work they have done with their students. In addition to the experiences acquired during the pedagogical expeditions themselves, this work is influenced and inspired by visits to the MSG's interactive room, as well as by the above-mentioned teacher workshops and workshops for students held during their time at the MSG.

Since the institutions participating in the expeditions have access to our museum suitcases, their contents, including posters and games, serve as models for school activities, such as the development of similar games adapted to different contexts; this influence is also seen in the socialization stage and beyond. In several cases, these MSG activities were the source of inspiration for school science fairs, not only in choices of subjects, but also in the use of everyday objects and games.

During the socialization stage, teachers were required to choose students to do a presentation about how they viewed the process. Experiential aspects were apparent in these presentations, involving emotional, group, and learning features which have been very important to us since the birth of MSG 28 years ago.

In the mid-1990s, we discovered an article written by Sheldon Annis (1984) called "The Museum as a Staging Ground for Symbolic Action", which became most influential in our work. Based on his ideas, we constructed a model (Betancourt, 2000, 2007, 2012) that addresses some of the issues arising in our activities.

Annis stated that the museum visit is an encounter between two worlds (his word), or contexts (our word): that of the exhibition and that of the visitor. To try to understand this meeting "space" between the two worlds, Annis considered it as composed of at least three planes: the first one, which he calls *oneiric*, is non-rational; the next, the social plane, he calls *pragmatic*; and he designates the third plane, the *cognitive*, as rational in nature.

One interesting thing about Annis's ideas is that in this discussion he somehow captures "the human being": we are non-rational beings, we are social beings, and we are rational beings. Our model was constructed based on a process of re-contextualization: we changed Annis's *worlds* to *contexts* and selected the general characteristics of Annis's planes, allowing us to take into account aspects he did not foresee or include in his essay. For example, our non-rational plane unfolds on other planes containing emotional, cultic, play, intuitive, and evocative elements, among others. The pragmatic plane involves social factors and the rational plane involves learning. For this reason, we call the latter the plane of learning. This plane in turn splits into implicit learning, social learning, and meaningful learning. If we represent the contexts of our model by spheres or circles, we can visualize the meeting space of the contexts of the exhibition and the public as the place where these spheres overlap; we locate the staging ground for symbolic action in that overlapping area, with Annis's planes represented by lines.

The physical classification for the exhibition context and the intersubjective status of the public or visitor context has been taken from Falk and Dierking's (1992) contextual model. Their construction allows for the investigation of the influence of each context on the planes proposed by Annis and also raises questions for evaluating exhibitions, as well as in the production thereof. The two-sphere model (Figure 21.2) can be expanded to three spheres when considering that the public or visitor context is generic, and can thus be unpacked

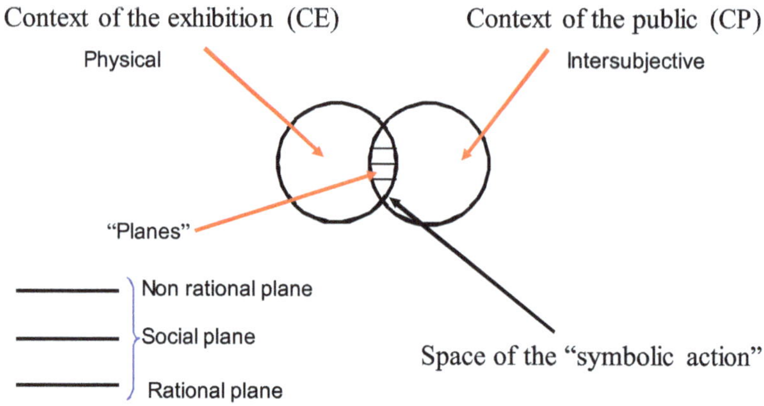

FIGURE 21.2 Representation of the two-sphere model based on the ideas of Sheldon Annis

into the "self" and "others", thus generating two contexts: the personal and the group (Figure 21.3). This provides a three-sphere model similar to the one proposed by Falk and Dierking (1992). The intersection of these three spheres generates a more complex surface than does that of two spheres, resulting in four distinct sites of overlap rather than one. Three of these represent interactions arising between pairs of contexts, and the fourth represents the interaction between all three contexts, thereby unfolding and elaborating the overlap area of the two-sphere model. Moreover, this more-complex model enables us to generate additional questions.

As we design and develop an exhibition, we must consider our evaluation processes. The two-sphere model helps us to ask questions that can be addressed in the design process. For example: how can we design an exhibition that thrills audiences? In the case of the three-sphere model, the resulting questions can be split into individual and group questions. Since the public context is split into two subcategories, the interactions between individuals and groups and their influences on Annis's planes can generate further, more precise questions. The two-sphere model allows us to ask only general questions, which cannot necessarily be implemented, while the three-sphere model enables more specific, operationalizable questions about individuals and groups.

The model that we have developed based on Annis's ideas can be applied to any type of exhibition. One might argue that human beings are actually immersed in exhibitions, since in their day-to-day lives they essentially go from exhibition to exhibition, from their homes, to their streets, work, classrooms, buses, shopping centers, to the many different spaces devoted to leisure in a given society, and beyond. These are all EC in which the model may be used. This is very important to us, because humans are three-dimensional beings

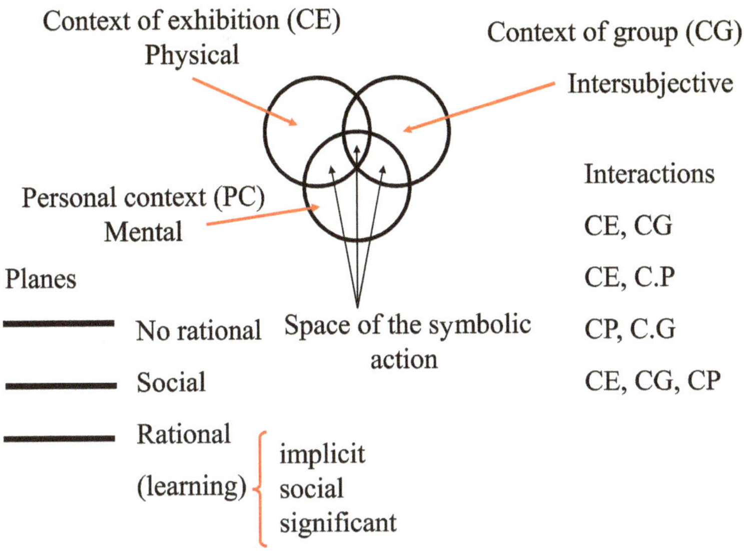

FIGURE 21.3 Three-sphere model. Each sphere represents a context: exhibition, personal, or group

that inhabit these three-dimensional settings which may be deemed exhibitions and are inherent to human culture. They are in fact the only three-dimensional form of human communication.

We therefore seek, in our educational activities, to invigorate participants on all of the planes proposed by Annis. For example, the non-rational plane is activated via game activities. We also design activities that encourage social interaction and incorporate everyday objects in order to generate a broader, more-inclusive learning environment. These are ways to cultivate the processes of intersubjectivity that are fundamental to the development of participatory, multicultural, polyphonic democracies. Annis's contextual model, by its very nature, allows us to design instruments for the evaluation of activities in terms of their interrelated emotional, social, and learning aspects.

In addition, it engenders new research questions and perspectives on work. For instance: The exhibition can be viewed as a re-creation of a physical space, in which the language of objects becomes central to that re-creation. Every language has a syntax, semantics, and pragmatics. The syntax of objects is in their juxtaposition (Barthes, 2009): How do the juxtapositions of objects influence the non-rational, social, and learning planes? What about the semantics of objects? The pragmatics? (Betancourt, 2013).

We wanted to give an overview of developments in non-formal education in Latin America, an endeavor which will always of necessity be partial given a field as extensive and complex as NFE. To that end, and as an example of such activities in the region, we have outlined several projects initiated and run by the MSG in that field, while also exploring related conceptual assumptions and lines of research.

References

Amaya, J. (2004). *Mutis, apóstol de Linneo* [Mutis, apostle of Linnaeus]. Bogotá: Instituto Colombiano de Antropología e Historia.

Annis, S. (1984). The museum as a staging ground for symbolic action. *Museum, 143*, 168–171.

Arango, M. (2003). De cómo una feria de ciencia se volvió un paseo de exploración [How a science fair can become a journey of exploration]. *Museolúdica, 9*, 38–49.

Arboleda, L. C. (1987). Acerca del problema de la difusión científica en la periferia: El caso de la física newtoniana en la Nueva Granada (1740–1820) [About the problem of scientific dissemination in the periphery: The case of Newtonian physics in New Granada [1740–1820]]. *Revista latinoamericana de historia de las Ciencias y la tecnología, 4*, 7–30.

Arboleda, L. C. (1995). Introducción de una cultura newtoniana en las universidades del virreinato de la Nueva Granada [The introduction of Newtonian culture into the universities of the viceroyalty of New Granada]. In *Newton en América* (pp. 29–66). Buenos Aires: Ed. Fepai.

Arboleda, L. C., Díaz, S., & Molinos, R. (1994). *Francisco José de Caldas*. Bogotá: Molinos Velásquez Editores.

Ávila, C. (2009). Las exposiciones enmaletadas [Exhibitions in suitcases]. In *Memorias de la XI Reunión de la RedPOP [Proceedings of the XI Meeting of RedPOP]*. Montevideo, Uruguay: Ciencia Viva-LATU.

Ávila, F., & Bautista, S. (2008). Los talleres en el museo [The workshops in the museum]. *Museolúdica, 20–21*, 96–101.

Barthes, R. (2009). *La aventura semiológica*. Barcelona: Paidós.

Bateman, A. (1998). *Francisco José de Caldas: El hombre y el sabio* [Francisco José de Caldas: The man and the wisdom]. Bogotá: Editorial Planeta.

Betancourt, J. (2000). Sobre recreación y museos [On recreation and museums]. *Museolúdica, 5*, 55–60.

Betancourt, J. (2002). Popularización de la ciencia y la tecnología: datos latinoamericanos de museos interactivos y centros de ciencia y tecnología en 2001 [Popularization of science and technology: Latin American data on interactive museums and S&T centers, 2001]. *Museolúdica, 8*, 26–31.

Betancourt, J. (2007). El modelo de Sheldon Annis. *Museolúdica, 18–19*, 54–63.

Betancourt, J. (2008). La RedPOP a través de sus reuniones [RedPOP through its meetings]. In N. Botinelli & R. Giamello (Eds.), *Ciencia, tecnología y vida cotidiana: Reflexiones y propuestas del nodo sur de la red pop* (pp. 13–23). Montevideo, Uruguay: Ciencia Viva, Red-Pop, Unesco.

Betancourt, J. (2012). *El modelo de Sheldon Annis* [Sheldon Annis's model]. Curso sobre evaluación. Seminario de Investigación Museológica, Dirección General de Divulgación Científica, DGDC, Universidad Nacional Autónoma de México, UNAM.

Betancourt, J. (2013). La re-creación del espacio [The re-creation of space]. In *Memorias de la XIII Reunión de la RedPOP* [Proceedings of the XIII Meeting of RedPOP]. Zacatecas, México: Consejo Zacatecano de Ciencias.

Betancourt, J., Bautista, S., Martínez, A., & Moreno, N. (2011). De evaluación, públicos y juego [On evaluation, publics and games]. *Museolúdica, 26–27*, 23–41.

Bronowski, J. (1968). *Ciencia y valores humanos* [Science and human values]. Barcelona: Ed. Lumen.

Diamond, J., Luke, J., & Uttal, D. (2009). *Practical evaluation guide: Tools for museums and other informal educational settings*. Lanham, MD: AltaMira Press.

Falk, J., & Dierking, L. (1992). *The museum experience*. Washington, DC: Whalesback Books.

Franco, M., & Linsingen, I. (2011). Popularizaciones de la ciencia y la tecnología en América latina: mirando la política científica en clave educativa [Popularization of science and technology in Latin America: watching science policy in key educational settings]. *Revista Mexicana de Investigación Educativa, 51*, 1253–1272.

Freire, P. (1970). *Pedagogía del oprimido* [Pedagogy of the oppressed]. México D.F.: Siglo XXI Editores.

Freire, P. (2005). *Pedagogía de la autonomía* [Pedagogy of autonomy]. México D.F.: Siglo XXI Editores.

Gadotti, M. (2004). *Historia de las ideas pedagógicas* [History of pedagogical ideas]. México D.F.: Siglo XXI Editores.

Giordan, A. (1982). La enseñanza de las ciencias [Teaching the sciences]. Madrid, Spain: Siglo XXI Editores.

Kennedy, J. F. (1961). *Address at a White House reception for members of congress and for The Diplomatic Corps of the Latin American Republics, MARCH 13*. Papers of John F. Kennedy, Presidential Papers, National Archives, Washington, DC (Digital Identifier: JFKPOF-034-013-p0001). Retrieved from https://www.jfklibrary.org/archives/other-resources/john-f-kennedy-speeches/latin-american-diplomats-washington-dc-19610313

Lechner, N. (2000). Orden y memoria [Order and memory]. In G. Sanchez & M. Wills (Eds.), M*useo, memoria y nación: Memorias del Simposio Internacional y IV Cátedra Anual de Historia "Ernesto Restrepo Tirado"*. Bogotá, Colombia: Ministerio de Cultura.

Mutis, J. C. (1983). *Archivo epistolar del sabio naturalista Don José Celestino Mutis* [Epistolary archive of the wise naturalist José Celestino Mutis]. Bogotá, Colombia: Instituto Colombiano de Cultura Hispánica.

Paulston, R. (1996). *Social cartography: Mapping ways of seeing social and educational change*. Levittown, PA: Taylor & Francis.

Pérez, T. (2010). Aportes feministas a la educación popular: Entradas para repensar pedagógicamente la popularización de la ciencia y la tecnología [Feminist contributions to popular education: Rethinking the popularization of science and technology from pedagogical perspectives]. *Educação e Pesquisa, 36*, 243–260.

RedPOP. (2001). *Memorias de la VII Reunión de la RedPOP* [Proceedings of the VII Meeting of RedPOP]. Santiago de Chile, Chile: CONICYT.

RedPOP. (2009). *Memorias de la XI Reunión de la RedPOP* [Proceedings of the XI Meeting of RedPOP]. Montevideo, Uruguay: Ciencia Viva-LATU.

Sánchez-Mora, C. (2006). *Una metodología para evaluar el aprendizaje informal a partir de exhibiciones de museo* [An evaluation methodology for informal learning in museum exhibitions]. Internal document, Dirección General de Divulgación Científica, DGDC, Universidad Nacional Autónoma de México, UNAM.

Segura, D. (2000). *Constructivismo: ¿construir qué?* [Constructivism: Constructing what?]. Bogotá, Colombia.: Escuela Pedagógica Experimental.

Shumacher, H. (1988). *Caldas: Un forjador de la cultura* [Caldas: a shaper of culture]. Bogotá, Colombia: Empresa Colombiana de Petróleos, ECOPETROL.

Shumacher, H. (1988). *Mutis: Un forjador de la cultura* [Mutis: a shaper of culture]. Bogotá, Colombia: Empresa Colombiana de Petróleos, ECOPETROL.

Zuleta, E. (1995). *Educación y democracia: Un campo de combate* [Education and democracy: A field of combat]. Bogotá, Colombia: Corporación Tercer Milenio.

CHAPTER 22

Science Education Research in Science and Technology Museums in Brazil

Martha Marandino and Guaracira Gouvêa

Abstract

The chapter provides an overview of the research on museum education. It seeks to characterize some of the theoretical and methodological perspectives on the subject developed by science education researchers in Brazil, concerned with studying science museums as educational spaces. It is emphasized the importance that this subject acquired in national and international investigations, due to the historical relevance and political importance of the museums throughout their existence. Considering those aspects, it initially introduces some of the historical aspects of science museums, both in the national and international context, and then discusses the meaning of museum education. In the sequence, the research on education in museums of science and technology is addressed and, at the end, investigations developed by the groups of research coordinated by the authors are discussed, presenting theoretical and methodological frameworks and some of the main results. It is understood that the research on education in science museums, in its articulation with the field of science education, makes an enormous contribution to the understanding of the role of these institutions in society and, consequently, to the expansion and deepening of educational research.

Keywords

science museum education – museum education research

1 Introduction

The educational character of museums has been established since their inception in the 16th century, even as understandings of science, education, and audiences have changed over the intervening decades and centuries. Today,

 | DOI: 10.1163/9789004409088_022

the educational objectives of these institutions are, to a certain extent, indisputable. However, considerations of the social, cultural and educational roles played by museums in modern society have produced diverse reflections from different fields of knowledge. In education research especially, countless papers focus on the educational roles of science museums.

Below, in our overview of research into museum education, we will refer to studies developed by groups working in a range of countries. Knowledge production in the museum realm occurs on a global scale, since museums are institutions of the modern world and, as such, are connected to other such institutions – and to those working with them – worldwide. Events organized by the International Council of Museums (ICOM), especially the meetings of the International Committee for Museums and Collections of Science and Technology (CIMUSET), as well as those organized by the International Committee of Museum Educators (CECA), are geared towards the exchange of cultural, educational, and research experience among museums.

In the United States, the National Science Teachers Association (NSTA) expressed its clear appreciation of science education situated in informal spaces through the publication of a political statement on informal or free-choice learning in 1998. An ad-hoc committee was also formed by the National Association for Research in Science Teaching (NARST) for the promotion of science education in informal spaces (Dierking, 2005). Both initiatives are based on the need to expand science teaching processes in informal education spaces.

American and European organizations associated with science museums and science centers, such as the Association of Science and Technology Centers (ASTC) and the European Network of Science Centers and Museums (ECSITE), reinforce the growing role that these spaces have been playing around the world, as places of public communication and informal education about science.

In Latin America, it is important to highlight the participation of museums of science and technology and science centers in the conferences organized by the Network for the Popularization of Science and Technology in Latin America and the Caribbean (RedPOP), an agency that brings together Latin American institutions which foster informal programs for the promotion of science and non-formal education.

According to Valente (2008), beginning in the 1950s and through the 1970s and beyond, conditions in Brazil were increasingly conducive to the development of science and technology museums; many different public and private museums are currently open to the general public. These institutions organize research projects and offer science education activities on their premises.[1] In

1998, the Brazilian Association of Science Museums and Centers (ABCMC) was formed, reaffirming the country's movement toward organizing and structuring these institutions.

Given this context, the importance of studies on the subject of education in science and technology museums and centers[2] is clear. This chapter initially introduces some key historical aspects of these museums, in both national and international contexts. Next, we evaluate the meaning and purpose of education in museums. Research on science education in science and technology museums will also be addressed. We finish by presenting the research results of a group that we coordinated, exposing theoretical and methodological assumptions, and offering our final reflections on the theme.

2 Historical Aspects of Science Museums

There have been many studies characterizing the evolution of science and technology museums over the centuries. Paulette McManus' (1992) illuminating work helps us to understand some of the main changes through which these institutions have gone. McManus differentiates three generations of science museums from their emergence not long after that of art museums to the present; each generation is based on the particular themes motivating museum creation. While each generation is loosely associated with particular inspirations and eras, there are overlaps and commonalities between all three in both time and space. The first generation originated in cabinets of curiosities. These 17th-century ancestors of science museums were characterized by the accumulation of particular sorts of objects (fossils, taxidermied animals, coins, scientific instruments, paintings, etc.), and were accessible only to a small, privileged group. The end of the 17th century brought a trend toward the sorting and structuring of these collections so that they could be used for demonstrations, studies, and the general dispersion of knowledge. Natural history museums emerged in this period; the first one to open to the general public was the Ashmolean Museum, at Oxford University in England, in 1683.

According to McManus (1992), first-generation science museums were and are, in a sense, sanctuaries of objects in open storage, where each and every item is displayed based on a classification system and in a repetitive fashion. The development of the natural sciences in the 19th century was characterized by the display of accumulated collections based on a particular area of research and on knowledge gleaned from and presented in museums. The display of such collections reflected the research developed in different scientific disciplines which, in turn, also started to become restricted to their areas.

Nineteenth-century museums were intended to work as educational spaces of popularization, as institutions following the democratic ideals of the previous century and included in general efforts toward societal modernization (Köptcke, 2001, 2002). On the one hand, these democratic ideals – which had been largely inspired by the French Revolution – spurred the opening of many more museums in Europe and America. On the other hand, they raised concerns about the educational trends in these institutions.

The first generation of museums extended throughout the 19th and 20th centuries. According to McManus (1992), a second stage of this first generation emerged in a movement started in the late 1960s, in which exhibitions were rated on how much they attracted and stimulated their audiences, with special consideration given to educational dimensions. In these cases, scientific research is no longer necessarily displayed for the public and exhibitions are strongly influenced by educational theories (Cazelli, Marandino, & Studart, 2003; McManus, 1992).

Influences from industrial technology and concerns about having science focus more on public interests characterize, for McManus (1992), the second generation of science museums; as well, the second generation placed increasing emphasis on the teaching functions of museums. The Conservatoire des Arts et Métiers (France, 1794) and the Franklin Institute (USA, 1824) are examples of this generation, which shared goals around informing people about the industrial production of the period, offering technical training, promoting scientific production, and educating the ordinary citizen.

In Europe, government projects of formal instruction started to include – sometimes as a mandatory component – visits to museums. In the context of the recently discovered educational benefits of school field trips to museums, these institutions created a range of so-called educational services. Reflections on the educational roles of these institutions vis-à-vis school curricula grew, together with proposals for finding better ways to work with this audience (García Blanco, 1999).

McManus's (1992) second generation of museums emerges in the early 20th century, as initiatives that involve more intense participation by audiences began to be proposed and implemented. The Deutsches Museum (Munich, Germany, 1903) is considered an important landmark in the realm of science and technology museums, since, in an attempt to foster dialogue and interactivity, and to leave behind stand-alone, static exhibitions, it presented devices to be operated by museum visitors side by side with its historical collection (Cazelli, Marandino, & Studart, 2003). This 'push-button' type of interaction, as well as, later on, other forms of visitor participation, became more and more widely disseminated (Cazelli et al., 1999).

McManus (1992) states that the greatest difference between the third generation of science museums and the previous ones is the role of historical collections. Contemporary exhibitions are more focused on scientific ideas and conceptions and less centered around objects. Device interactivity is highly appreciated for the ways in which it enhances connections between visitors and science. Examples of pioneering third-generation science museums include the Palais de La Découverte (Paris, France, 1937) and the New York Hall of Science (New York, USA, 1964).

Allard and Boucher (1991) also highlight the growing educational role played by museums over the course of the 20th century, which led to the introduction of strategies designed to facilitate communication with the public within exhibitions. During the first half of the 20th century, research was conducted with museum visitors in various countries which indicated the need to respect the characteristics and interests of various audiences, whether specialized, lay, or student. Given these sorts of findings, old-style exhibitions wherein the museum collection was simply displayed as such were gradually replaced by representative selections from specific themes. American museums became famous for their use of media devices and ambient reconstructions (dioramas) which facilitated visitors' understandings of themes and made exhibitions both more intelligible and educational (García Blanco, 1999; Marandino et al., 2008).

These changes in science museums are closely related to historical developments in the science teaching field. In the 1960s, following the enormous impact of the 1957 Sputnik launch, new approaches to teaching the sciences which aimed to minimize the prevailing scientific and technological illiteracy of the time were proposed. The first science center – the Exploratorium (San Francisco, USA, 1969) – was created in this context. The success of its approach and interactive devices inspired the creation of similar spaces all over the world, thus creating the whole industry of interactive science museums (Beetlestone et al., 1998). However, interactivity alone as a guarantee of learning in science museums is challenged by education research conducted in the 1980s and 1990s which was based on new theoretical perspectives and on debates about the role of museums taking science, technology, and society into consideration (Falcão, 1999).

Brazil's first museums were built in the 19th century. Created in the format of the great European and American museums, these Brazilian institutions were also concerned with collecting, cataloguing, and studying elements of the country's natural and cultural worlds. The very first was the National Museum (Rio de Janeiro, 1808), which established its collection based on the natural sciences. This model later inspired the founding of the Emílio Goeldi Museum, in Pará (Belém, 1866), the Museum of Paraná (Curitiba, 1883), and the Paulista Museum (São Paulo, 1895; Lopes, 1997).

According to Cazelli, Marandino, and Studart (2003), the movement to create science museums in Brazil may also be seen as an amplification of the promotion and teaching of science. The 1960s in Brazil were characterized by the mobilization of the country's scientific and educational communities, which became more organized as well as increasingly aware of the problems in the scientific field, and, especially, in science teaching.

According to Gaspar (1993), most Brazilian science education initiatives – formal and non-formal – relate back to the creation, in 1950, of the Brazilian Institute of Education, Science and Culture (IBECC), which is associated with the University of São Paulo (USP) and UNESCO. From the 1960s on, the teaching of traditional scientific disciplines such as mathematics, physics, chemistry, and biology required a greater variety of materials and equipment in order to assemble the educational labs which were increasingly seen as vital for learning such subjects. This led to the creation of a new type of organization, Centers of Science (CECIS), which were designed to centralize the production, implementation, and review of such materials; however, their most important function was to ensure greater accord with the realities of local schools and to provide teacher training.[3]

This was how, in the 1950s and 1960s, a range of institutional initiatives structured the development of science teaching in Brazil, thereby advancing educational practices in diverse formal and non-formal contexts (Krasilchik, 1987; Fracalanza & Megib Neto, 2006; Megib Neto, 2007). During that period, Brazil's experiences in some ways mirrored what was taking place in the rest of the world, as it was strongly influenced by the United States. Systematic actions were also taken to improve science teaching in the 1980s, the products of which included the creation of science teaching research groups, the implementation of graduate programs in the field, and the founding of research magazines on the theme of teaching science. In 1997, the Brazilian Association for Research in Science Education (ABRAPEC) was founded, indicating the consolidation of research on science teaching in Brazil. Research groups were also established under the auspices of science and technology museums; these also took part in national and international forums on science education.

The 1980s were an important period in the history of Brazilian science museums, since a number of government-funded science museums were founded in Rio de Janeiro and São Paulo during this period. Many of the museological institutions founded in this period were centered on (or at least promoted) audience-centric activities, with particular foci on introducing school-aged students to interactive museographic elements and/or developing educational activities for different grade levels (Cazelli, Marandino, & Studart, 2003).

The 1990s were also characterized by the founding of new science museums, as well as by the consolidation of previously founded institutions. Public bids were issued with the goal of encouraging science-promoting activities and contributing to museum funding. The national policy for the promotion of science was consolidated when the Science and Technology Popularization and Diffusion Department was formed under the Secretary of Science and Technology for Social Inclusion (SECIS) of the Ministry of Science and Technology. This department had, among its many goals, the explicit aim of contributing to the promotion of science education. This policy has ensured regular support for educational activities in science and technology museums, as well as research on science education in museums (Navas, 2008).

More museological institutions and initiatives in the field of the sciences emerged in the late 1990s as a product of such government funding, as well as donations from private entities, such as the Vitae Foundation. In addition, as previously mentioned, the Brazilian Association of Science Museums and Centers was founded in 1998, bringing a new organizational phase for such institutions.

If, on the one hand, financial resources dedicated to Brazilian museums have always been scarce (Cazelli, 2005), it is also possible to affirm that a variety of successful initiatives have recently been established. Around 2003, the Brazilian Ministry of Culture's Department of Museums and Cultural Centers instituted a foundation to discuss the development of public policies for Brazilian museums (Marandino et al., 2008), which later allowed for the creation of establishments focused on these spaces, such as the National Observatory of Museums and Cultural Centers (2006). In 2006, in a partnership between the Federal University of the State of Rio de Janeiro (UNIRIO) and the Museum of Astronomy and Related Sciences (MAST/MCT), a Graduate Program on Museology and Historical Heritage was conceived, with the goal of educating researchers to work in different knowledge areas specific to museums.

This context has fostered the formation of research groups in graduate programs on education, science education, science history, and science promotion, whose goal is to study different aspects of science and technology museums. Both researchers and institutions confront a wide range of factors in the fields of science education and science and technology museums. They have been influenced by its international context, including external policies and funding, as well as by national forces based on initiatives coming from communities of scientists, educators, promoters, and museologists, along with government policies and private programs and foundations.

Valente (2008) points out that the movement towards increasing the number of science and technology museums that occurred in the 1980s (with complementary growth in the 1990s), originated in the historical conditions around science and technology in the 1950s, '60 and '70s, which were supported by a development-oriented perspective linked to the growth of national sciences and technology. This meant that the scientific community was essentially consolidated by means of investment policies. Under these conditions, access to – and demands for – formal education grew, thereby stimulating the public desire and need for museums that could contribute to science education for the population.

Museums became partners in formal education in that they broadened access to education and culture for diverse segments of society. They have been urged to emphasize this role by prioritizing educational activities, as well as by encouraging, contributing to, and conducting research on education in and about these spaces.

Contemporary educational activities in the science museums of Brazil tend to be highly structured. In that context, increased inclusion has been one of the major demands placed on museums in the 21st century, since – beyond the concerns with teaching, learning, and entertainment which were consistently present in the previous century – there is now a strong emphasis on social commitment to the expansion of audiences as well as on institutional diversity and accessibility. As indicated by Tojal (2007) the main hurdles faced in the enjoyment of different cultural manifestations are symbolic in nature and therefore require a certain degree of literacy to enable code recognition. Museums, thus, bear an important role in broadening the cultural repertoire of citizens by promoting respect and the "recognition of cultural and social diversity which works toward not only the communication of cultural objects from a multicultural point of view, but also contributing to cultural democratization through processes of social inclusion" (Tojal, 2007, p. 81).

Contemporary pressures on these institutions result in some placing blame on the education sector. To Tojal (2007), however, changes cannot be restricted to addressing certain issues merely in order to expand the participation of specific types of audiences. It is necessary to work towards tearing down barriers to museum access – despite these being physical, sensory, financial, attitudinal, and intellectual – and the promotion of public involvement of audiences with such institutions.

Recent research outlined the demographics of a subsection of Brazilian museum visitors. According to a study conducted by the Institute of National Heritage and Art (IPHAN) regarding visitors to 11 museums in two Brazilian cities (Rio de Janeiro and Niterói) in 2005, museums tend to draw highly compensated working people. Among those visitors who stated that they did not have a job, more than half (53.4%) were students. The educational level of museum visitors was comparatively high, with 45.7% having a college degree (the average number of school years completed by those living in the Rio de Janeiro metropolitan region of is 8.3 years, i.e., up to 8th grade). The majority of both adult and youth visitors were European descent. In Rio de Janeiro, the museums which had the largest percentage of non-Caucasian visitors were the two natural science museums: the Museum of Astronomy and Related Sciences (48%) and the Museum of Life (46%).

On the other hand, Cazelli (2005) indicated that while Brazil's social and economic contrasts are also present in inequalities vis-à-vis visiting such public cultural spaces, Brazilian youngsters living in Rio de Janeiro nonetheless do benefit from opportunities to visit museums with their families and/or schools. Her investigation reveals that the social capital gained via school-museum activities contributes to the expansion of youngsters' cultural experiences, particularly for students in public schools: "In other words, public schools run by the city carry an equalizing and active role, especially relevant to those youngsters from families whose cultural capital volume is lower" (Cazelli, 2005, p. 206). The data, according to the researcher, reinforce the relevance and importance of having more effective, active policies around museum collection enhancement and educational programs.

The above-referenced investigations, although they may not represent the overall profile of museum visitors in Brazil, do present significant findings that must be carefully analyzed when considering goals such as the expansion of programs and better visitor preparation in Brazil's museums. Hence, such considerations must guide how research into science teaching in Brazil may contribute to the ways science museums promote visitor empowerment in terms of various aspects of scientific knowledge and its relations with society.

3 Aspects of Education Research in Science and Technology Museums in Brazil

Groups interested in researching science education in science and technology museums focus on different scientific topics depending on the theme being developed (e.g., education, communication, sociology and history of science, museology, and so on). The expansion of science education research as a theoretical and methodological perspective for understanding educational phenomena in museums has occurred both nationally and internationally – a fact that may be corroborated via recent initiatives from science education specialists attempting to add value to non-formal and informal education. Articles in renowned publications in related fields have also increasingly referred to the theme[4] over the last few years.

Brazilian science education groups who study museums of science and technology have thus been following a path drawn by countless prior contributions. Huge efforts are being made – by means of both academic work and educational activities – to understand the museum as an educational institution, using its own historical and educational specificities, along with its own pedagogic procedures, as references.

The museum has had, since its original constitution, an educational and cultural character. What has changed over time, especially since museums were opened to the general public in the 19th century, is the populations toward whom educational and cultural activities are directed (Valente, 2003). In opening their doors to a diversified audience, museums have had to redirect their educational actions.

In reflecting on educational research and activities in museums, we must – in addition to locating them in the history of museums as institutions – also situate them in regard to related areas of knowledge, which means we must consider epistemological, social, and political issues. Therefore, if we are to evaluate science education in a particular museum of science and technology, it is important to be clear as to the social conditions of science and technology production at each moment of that museum's history, not to mention over the historical trajectories of the subjects presented therein.

A quick overview of educational research into science museums in Brazil reveals recurrent themes and methodologies, but also emerging ones. Studies related to audience surveys, particularly concerning visitor profiles (individual, social, cultural) and agendas, as well as the characteristics of those who do not tend to visit museums, etc., have been more and more common. These studies have predominantly been developed under the auspices of the Observatorium of Museums, which convenes researchers from different Brazilian museums[5] in order to gain insights into museum attendance, among other things. Such surveys provide important contributions to the development of educational activities in museums.

Many other survey themes are more directly related to educational issues. With that in mind, studies of science museums that address formal and non-formal education aspects of scientific literacy have been undertaken in Brazil (Cazelli, 1992; Gaspar, 1993; Moreira, 2013; Cerati, 2014), along with others which investigate museum-school relations, characterizing the specificities of museums as institutions – such as their educational activities and relationships with school audiences – as related to student and/or teacher education

(Lopes, 1991; Cazelli, 2005; Martins, 2006; Pereira et al., 2007; Köptcke et al., 2008; Jacobucci, 2006). Another important research group approaches issues around learning and the production of meaning in science museums based, variously, on a theory of mental models and concepts of interaction and interactivity (Falcão, 1999; Falcão et al., 2004); the socio-historical theoretical point of view (Gaspar, 1993; Garcia, 2006; Campos, 2013); and activity theory (Bizerra, 2009). Studies published regarding human mediation in science museums are increasingly generating concerns related to the training of museum monitors (Nascimento et al., 2002; Queiróz et al., 2002; Massarani & Almeida, 2008; Marandino et al., 2008; Oliveira, 2008; Ovigli, 2010; Caffagni, 2010).

Other investigations focus on understanding the scientific paradigms and conceptions expressed/perceived in exhibitions, the processes of museographic transposition, and the re-contextualization of scientific discourse (Gouvêa & Marandino, 2000; Marandino, 2001; Gouvêa et al., 2002; Contier, 2009; Oliveira, 2010; Salgado, 2011). In addition to the above, aspects of the history of science and technology museums in Brazil and of the constitution of their public character are important themes of study for understanding education in these spaces (Valente, 2003, 2008; Lopes, 2009). These are just some examples of the ways academic production focused on museum education in Brazil has grown over the last twenty years.

In the following sections, we present examples from our recent research in order to more closely illustrate some aspects of museum education being explored by research groups in Brazil. The first section, "Science and technology museums, education, and media", includes examples from research conducted by Dr. Guaracira Gouvêa of the Federal University of the State of Rio de Janeiro). The second, "Museum education: studying the pedagogic specificities of science museums", contains the results of investigations conducted by Dr. Martha Marandino of the Department of Teaching Science and Mathematics in the Faculty of Education of the University of São Paulo, in her role as the coordinator of the Group for Study and Research in Non-Formal Education and Science Communication (GEENF).

4 Science and Technology Museums, Education, and Media

The research groups under our supervision have been conducting studies which address the production, distribution, and reading of discourses of science promotion recorded in different media; during these efforts, we have undertaken in-depth explorations of the concepts of language, discourse, genre, image, reading, and mediation. Deeper understandings of these concepts in turn support our considerations of discourses of science promotion through lenses grounded in the fields of science, education, communication, and information.

Within the context of this chapter, we will limit our focus to research regarding science and technology museum exhibitions. These research studies were based on three concepts which formed the foundation of our understanding of exhibitions: language, discourse, and genre. In this approach, we have used ideas about image, reading, and mediation to analyze museum exhibitions. In the following pages, we offer examples of studies developed with our guidance.

4.1 *Image-oriented research*

The theoretical perspective we assumed frames language, following Bakhtin (1986), as a category encompassing the different ways of producing utterances. Bakhtin defines language as a human production enacted via the various instances of enunciation that make up the realm of utterances. As we explore the discursive aspects of exhibitions in science museums, it is important to emphasize that we are referring to certain spheres of human communication and their respective discursive genres from a Bakhtinian perspective. Genre, for Bakhtin (2000), reveals the ways in which social norms, purposes, and actions – including utterances – are socially constructed. Discourse production is thus a constitutively social practice wherein the author, speaker, and reader each express themselves in a particular social place.

For the exhibitions surveyed, the interlocutors are the author (the creator of the exhibition) and the reader (visitor) and the discourses are produced and experienced in the spheres of science communication and science promotion. Thus, the exhibits constitute a discursive genre (Zamboni, 1997) consisting of hybrid semiotic texts. One of the components characterizing such hybridity is the inclusion of images.

The term "image" has different meanings. It can be defined as a representation of a thing, including, by extension, a mental representation of something that can be perceived by the senses. Images may be associated with shadows, figures, figurative imitations, or representations related to an object, its referent, whether in analogy or perceptual similarity. There may thus be aural, tactile, olfactory, visual, and even oral images.

Relations between texts, particularly between verbal written texts and texts consisting of imagery, are set in different ways depending on the type of communication desired. A graphic representation of an idea can be faithful to the text or the text can clarify it or be clarified thereby; it can go beyond the text or simply decorate the text. In the case of scientific texts, as with many other contemporary texts, the typical printed text is a semiotic hybrid (Lemke, 1998), i.e., incorporating verbal written language, pictorial elements, and graphics in the same page spaces. In exhibitions, verbal texts coexist with still images, animated images, historical objects, and interactive devices that – when articulated together – comprise a discourse the intention of which is expressed through that articulation.

In our research (Gouvêa, 2009), the relationships between texts and images in science and technology museum exhibitions were studied by considering each exhibition itself as a language phenomenon, a recontextualized discourse consisting of narratives with certain rhetorical structures structured to explain specific science models (Gouvêa, Izquierdo, & Martins, 2006).

The two museums chosen as research settings have three-dimensional collections of historical artifacts, present interactive exhibits, and use different media resources in their exhibitions. Two exhibitions, one from each museum, were analyzed: one (museum 1[6]), comprised guidelines for drafting an interactive exploration of the seasons of the year, and the other (museum 2), consisted of an interactive exhibit which also displayed technological objects from the history of communications in Brazil, as well as objects representing the evolution of forms of communication from the 19th century on.

In analyzing the relations between verbal written texts and images in both exhibitions, fixation emerges from polysemy, as pointed out by Barthes (1990), as particular means are chosen to explain phenomena. The image takes on the role of complementing that which the written text presents in each component of the exhibit. Taking these exhibitions as a whole, we could frame museum 1's exhibit in terms of communication that is based on a scientific-educational model of asking and answering questions, be they real or rhetorical, since the drafted discourse asks visitors to build an explanatory model for the seasons of the year. Museum 2's exhibit communicates an affirmative and masterful conception of science – science as linked to progress and as the sole source of explanatory knowledge for phenomena.

Both exhibitions presented facts and artifacts of everyday life (leisure, work, and home), experimental artifacts, and few abstract, tactical representations designed to bring science and technology closer to visitors' everyday life.

In performing this research, we considered images as something to be read, a process that is in this case intended to lead to certain conceptions of science. Images require a learning experience via that reading process, i.e., the production of meaning. Then again, what does reading mean? What is reading?

4.2 *Reading-oriented research*

The idea of exposure supposes a form of reading by the individual who is exposed, be it to works of art, ethnographic objects, scientific texts, a machine, a forest .... Exposure, then, is not limited to display, it also entails ways of looking (Marandino et al., 2003). The arrangement of certain objects in a particular space forms a set of semiotic markers that embody the meaning-producing intentions of those who place those objects in that space; an exhibition is therefore essentially an utterance, a communicative act.

Language is expressed by contemporary museums via different texts, providing unique articulations involving places, objects, and times (Van-Präet, 2003). Museum discourses are capable of communicating new conceptions of material or immaterial objects; tangible and intangible assets; built or natural spaces; memories; and new ways of thinking about time, thereby expanding our concepts of both museums and exhibitions.

The exhibition, a materialization of a particular articulation between place, objects, and time indicates a certain circuit of consumption/participation, a desired/planned way of being read. However, visitors can potentially circumvent the control mechanisms contained in this circuit as designed by its makers, and may themselves build hyper-texts via their own readings; this may result in multiple temporalities, multiple sequences, and the sharing of different cultural objects, whether at the time of the visit (with fellow visitors) or later, at another time (alone or with another audience).

Exhibition readings entail the production of meanings by the visitor and embody author-reader interactions. However, in any museum, authors bear a certain identity, as constituted by the collections housed by that museum and by the exhibition designs conceived and implemented therein. In this way, authors are themselves situated – named, given an address, and dated historically – even as they exercise their control over polysemy via ubiquitously placed explanatory display markers in their exhibitions. How such exhibitions are then read depends on the social conditions of the setting (Eco, 1993; Barthes & Compagnon, 1987), meaning, in part, that visitors do not forget the fact that they are in a science and technology museum and go about reading just any text. But what and how do visitors actually read?

In order to present examples of research addressing how visitors read exhibitions, we chose two studies that took place under our supervision. The first, conducted by Andrea Costa (2009), examined the narratives produced by visitors to a science and technology museum. In her work, Costa sought to learn about visitors' production of meaning about/readings of the historical value of scientific instruments during an exhibit consisting of a guided tour which was presented as part of an educational sky watching program. The second research study, conducted by Renata Monteiro (2011), analyzed spontaneous audience perceptions of the relations between science, technology, and society after listening to informative presentations and reading exhibitions. This research was undertaken in two institutions: (i) Museum 1, a Science Centers-style institution, which prioritizes the display of scientific results, concepts, and phenomena, with a focus on manual and mental interaction, and (ii) Museum 2, which is based on the exhibition of generativepre-industrial Brazilian techniques, arts, and crafts, with a focus on mental and cultural interactivity.[7]

In the first study, analyses of the narratives produced during interviews highlighted the presentation of historical objects in science museums as highly beneficial for visitors, since these can act as bridges between scientific content and history, further humanizing science and connecting it with the ethical, cultural, and political interests of individuals. In addition, such displays were perceived as making topics more exciting, thus inspiring reflection, and, in turn, increasing visitors' critical thinking skills.

In the second study, with reference to the work of Boaventura de Sousa Santos and to studies of science, technology, society, and the environment (in the Portuguese acronym, CTSA), Monteiro sought to determine the extent to which audiences' CTSA perceptions at the selected museums leaned more towards one of two analytical categories: monocultures or their opposites, ecologies. The identities, local scale, and memory aspects associated with the generative s displayed contributed to Museum 2 visitors expressing the highest number of central ideas concerning ecologies. Museum 1 visitors showed a higher occurrence of central ideas related to monocultures, which reinforce neutral readings of science and technology as displaced from the social, political, and economic contexts of their production. These results affirm the importance of using semiophores in science exhibitions in order to help visitors gain the ability to think critically about science and technology.

The results of these two studies also reiterate the key role of a material culture of science and technology in relation to other cultures on the agendas of the various educational activities offered at science and technology museums.

These studies of visitor-exhibition interactions seek to understand how visitors read and, further, to determine what, precisely, was read by them in that reading. What were the meanings produced via reading what was displayed? How was the reading of the intended science discourse mediated by the exhibition? Science center discourses are fragmented into many devices and objects; in order to be read by the visitor, many of these fragments necessitate further mediation, often in the form of in-person assistance with turning on a given device, enabling the observation of certain phenomena, or explanation of what was observed, thus engendering another form of mediation. So, what is mediation?

4.3 *Mediation-oriented research*

Every social practice involves communication processes wherein human mediation is embedded in the production, articulation, and use of information.

Contemporary studies point to the need to study information from a variety of angles including its production, use, and social appropriation of knowledge in networks, to examine the ways in which information is embedded in social movements, as well as its modes of communication and promotion of knowledge (Almeida, 2008). This view interrogates social interactions, emphasizing the symbolic nature of culture and the circulation of various cultural indicators among different social groups.

In this way, social interactions are understood as a form of cultural mediation, a convergence between two cultural worlds, aiming not toward the replacement or shrouding of one cultural medium by another, "but the changing of both man and the world" (Coelho, 1989, p. 8). The mediator is responsible for bringing these worlds closer together. Mediation is not, however, the mere transmission of knowledge from one individual who 'knows' to another who 'does not know'; if it were framed as such, we would run the risk of turning the mediation process into a one-way street, one lacking dialogue and ignoring the peculiarities, unique qualities, and responses of the receiver. Similarly, in educational practice, mediations involving teaching and learning do not rely only on the teaching sector, as manifested in the role of mediator/teacher, but regard the interactions between all parties concerned, so that relationships are established between the two cultural worlds of faculty and students.

Some authors (Nascimento, 2008; Marandino, 2005) point out that the concept of mediation allows for different theoretical interpretations and may be applied in various contexts. Thus, with Nascimento and Almeida (2009, p. 4), we assume the three functions of mediation to be: "1. static connections between subjects and objects; 2. negotiations of the meanings attributed by subjects to objects at various levels and 3. transformations of meaning via the actions of socio-historical subjects on objects".

We also hold that the act of mediation is a constitutive social practice of museums, wherein mediation is performed by different actors (curators, designers, museum experts, and educators) who may or may not be present during visitors' readings.

Mediation is conducted in the preparation of exhibitions prior to their availability to audiences. In this process, different semiotic resources are organized to make sense of included materials and subjects, which, in the case of museums, encompass combinations of place, time, and themes Davallon (1999), we can distinguish three languages/logics in the exhibition production process which articulate place, object, and time: the logic of discourse, the logic of space, and the logic of gesture. These correspond, respectively, to the moments spanning the preparation of the exhibition, through its execution, and nearing the time of visitation. For visitors, these operations are read as marks of intentionality and figure into their reconstruction of the content of the exhibition, allowing for easier processes of (re)attaching meaning to the data it contains. This entire cycle generates and encourages encounters between cultural worlds.

Research conducted by Simone Pinheiro Pinto (2014) aimed to examine the construction of discourse in the face-to-face human mediation found in traveling science promotion exhibitions. The corpus of this research consisted of two trunk exhibitions (linked to museums 1 and 2[8]), the mediator training courses for these programs, and the performances of the participating mediators. These travelling exhibitions aim to bring science to underserved regions lacking in cultural facilities and are composed of interactive devices that are handled by visitors (school and non-school audiences). Observations and recordings of mediator trainings at the institutions involved and of their performances during exhibition activities were conducted with the aid of the mediators themselves.

Our analyses pointed to a still-developing discourse of mediation, which was, in effect, outlined by other discourses; in other words, the prepared discourses did not – as presented – fit with the genre of science promotion, since there were difficulties in the mediators' recontextualizations of scientific and institutional discourses. Since both institutions aim to disseminate scientific culture, it was expected that mediators would, through their discourses, fulfill this goal. However, due to various elements of the social conditions of production of these discourses, the intended scientific discourses were not properly recontextualized during the actual mediations that occurred.

4.4 *A Few Considerations*

We have proceeded from an initial focus on the logical development of exhibitions to studying the interactions of audiences with exhibitions, which we refer to as reading or gestural logic. Since many museums are training mediators to facilitate their educational activities, we now turn to the study of face-to-face human mediations. We use this phrase because we understand mediation as always human. When we elaborate upon information in order to communicate, we, then, perform a first mediation; for museums, such materiality is expressed through exhibitions. The second mediation occurs in whoever reads what is communicated; in museums, this unfolds via visitor's readings; in all cases this entails the meeting of two cultural worlds. The third mediation – which is not always necessary and does not always occur – is performed by the mediator in person.

The studies presented here consider exhibitions as forms of language and as communicative acts. We analyze both the modes of production of exhibitions and the productions of meaning that transpire in visitor interactions. In order to more deeply understand forms of meaning production, our next research study will be focused on new ways for visitors to participate in exhibitions at science and technology museums.

5 Museum Education: Studying the Pedagogic Specificities of Science Museums

Given the relevance of studying the pedagogic aspects of science education, especially those involving exhibitions, the Group for Study and Research in Non-Formal Education and Science Communication (GEENF) (www.fe.usp.br) conducted a set of investigations into the pedagogical approaches used in science exhibits. Most of our research seeks to understand the educational processes developed by museums and to contribute to those processes in such a way as to help museums increase their audiences and reach their different audiences in more efficient, effective ways.

Much of our investigation analyzes museum education in two important dimensions. The epistemological dimension relates to understanding the teaching and learning process for scientific knowledge as present in museums. In our case, we sought to study how the transformation of scientific ideas is reflected in the ways science is taught and disseminated through exhibitions and other educational activities, as well as how visitors learn during visits to these places. In developing these studies, we based our analyses on Yves Chevallard's didactic transposition theory, along with his anthropological theory of the didactic.

The second dimension, the sociological, refers to the ways in which we understand the construction of museums as places of education. In this case, our analysis focuses on the influence of the political, economic, cultural, and social fields in determining the educational actions, including the production of exhibitions, of these institutions. We mainly based this dimension of our studies on Basil Bernstein's (1996) concepts of pedagogic discourse and recontextualization.

Further, we examine the contributions of science museums to the scientific literacy of their visitors as regards the articulation of theoretical frameworks from the fields of science education and public communication of science.

In the following sections, we present some of the research we developed in these epistemological and sociological dimensions; along with findings from our analyses which could enhance the development of educational activities in museums.

5.1 *Studying the Epistemological Dimension of Museum Education*

There are a number of studies investigating the deconstruction and reconstruction of science as it is transformed into an exhibition (Gouvêa et al., 2002; Marandino, 2004; Mortensen, 2010; Simonneaux & Jacobi, 1997). The science that goes into an exhibition undergoes a transformation process as it is appropriated from its origin in an academic science discipline, adapted to a museum context, and embodied in an exhibition. This process is similar to the one at the root of math and science didactics, where math or science knowledge as it exists in the scientific discipline is not considered to be a form knowledge that can be directly taught in schools (Achiam & Marandino, 2014).

The process of didactic transposition (Chevallard, 1991) holds that science and math content must be appropriated from its original context, usually a mathematical or scientific research discipline, and adaptively transformed in order to become teachable and learnable. Didactic transposition deals with the trajectory of knowledge in the process of being rendered teachable, and includes questions such as: Where does knowledge come from? How, and by whom, is it shaped? What is its degree of effectiveness in promoting learning? (Chevallard, 2007).

Questions such as these are also relevant in non-formal science education contexts such as science museums. Such places, especially the exhibitions therein, are teaching environments, presenting certain bodies of scientific knowledge which they wish to mediate for their visitors. Accordingly, evaluation of the didactical transpositions that take place in museums is a necessary and worthwhile undertaking if we, as researchers, are to produce results than can improve teaching practices – including the practices used in developing exhibitions – as well as school use of exhibitions, in ways that can potentially affect the learning outcomes of these environments (Achiam & Marandino, 2014).

The educational activities that take place in the museum are characterized by their relationship to three elements: *space, time,* and *objects* (Van-Praët & Poucet, 1993). The museographic transposition process involves the transformation of scientific knowledge into knowledge taught in an exhibition (Simonneaux & Jacobi, 1997). In such contexts, the scientific concept being presented is naturally a central research element. Studies focus on what happens to the central concept, along with related concepts from the reference knowledge field, and determining the differences that emerge as both concept and associated reference knowledge become exhibit knowledge. We designate this approach as epistemological, since it interrogates the differences between the network of concepts in the scientific knowledge realm and the new relationships that are created – another network – in the exhibit environment (Achiam & Marandino, 2014).

Hence, some of our studies of museums' museographic transposition processes review, among other topics, the transformation of the concept of biodiversity through the production of exhibitions, and, particularly, of dioramas. Dioramas are scenarios originally conceived of as a means for visitors to explore and discover 'authentic' environments much in the same way as a naturalist explores and discovers unknown natural areas. Our main questions are: "From which perspective on biodiversity do science museums speak?" and "What does the audience understand about biodiversity after visiting museum exhibitions?".

Understanding exhibitions and dioramas as devices imbued with educational intentions allows us to consider them as products of museographic transposition, a transformation process in which content from the domain of scientific research is transformed and adapted to become embodied in the final installation of a physical exhibit in a museum (Mortensen, 2010). The resulting exhibit, with its objects, texts, specimens, and other items, is thus an expression of the adaptations and selections that scientific knowledge and practices undergo, and comprises an expository discourse (Marandino, 2001). This discourse is a 'version' of ecology which may be substantially different from the ecology found in scientific discourse per se (Mortensen, 2010).

While studying exhibitions in two prominent Brazilian museums – the National Museum of Rio de Janeiro's Federal University and São Paulo University's Zoology Museum – we collected evidence which shows that the conservationist dimension of biodiversity is not treated as a central element in such exhibitions (Marandino & Monaco, 2009), although this does not prevent the theme being addressed in other areas of the museums. In the exhibitions studied, we found a strong emphasis on approaches to biodiversity as related to levels of biological, taxonomic, and evolutionary organization.

Another study, carried out in two other Brazilian museums – the Natural History Museum of Capão da Imbuia in the city of Curitiba in Paraná state and the Science and Technology Museum of the Catholic University in the city of Porto Alegre in the state of Rio Grande do Sul – shows that the dioramas in the exhibitions studied depict biodiversity in similar ways (Oliveira, 2010). Both focused on the identification of levels of species and ecosystem diversity, confirming the typical accord between the standard 20th-century conception of the natural sciences and its representation in natural history museums. In this period, dioramas gained popularity as an effective way to display ecology concepts in museum exhibitions. Thus, the eco-centric exhibit genre that dioramas represent (Fortin-Debart, 2003) continues to be emphasized in museum exhibitions and even in science centers, such as the Science and Technology Museum of the Catholic University in Rio Grande do Sul. Generally speaking, these dioramas use a variety of means to express the intricate interrelations among living organisms and between them and the environment.

Genetic diversity is rarely mentioned in exhibit texts or explicitly represented by objects in the dioramas studied (Oliveira, 2010). Since ecology emerged as a macroscopic domain of research in the 19th century, the concomitant development of the diorama had no way of depicting – or even being aware of – the sub-microscopic aspects of biodiversity. Presumably, this cultural link between the diorama and the perception of ecology as a macro-level science still exists in important, if subtle, ways; thus, today, dioramas still seem to favor macro levels of biodiversity – such as species and ecosystem – over micro levels, such as the genetic (Marandino, Achiam, & Oliveira, in press).

Results differed, however, in our analyses of recent immersion exhibitions wherein biomes are reproduced on a real scale and conservationist discourses are emphasized. The two immersion exhibitions studied – the Zoobotanic Foundation in the Brazilian city of Belo Horizonte in Minas Gerais and the Biodome in the Canadian city of Montreal (Marandino & Diaz Rocha, 2011) – underlined the centrality of human impact on ecosystems in their constructed narratives. Levels of biodiversity were extensively represented, with an emphasis on the species and ecosystem levels. Since the expository discourses of these exhibitions took a macroscopic approach, using texts as well as live and taxidermied objects to create narratives, the genetic aspect was rarely present. An evolutionary perspective on biodiversity was presented solely in the Zoobotanic Foundation, in a space that recreated the evolutionary history of plants.

Some GEENF projects have explored the learning potential of exhibitions in order to understand visitors' meaning-making regarding biodiversity (Garcia, 2006; Bizerra, 2009; Campos, 2013). These research studies explored the potential for exhibits to promote understanding about biodiversity and aspects of conservation through contact with exhibitions in museums. In a partnership between the University of São Paulo in Brazil and the University of Copenhagen in Denmark, our initial project studied adult audiences' perceptions of biodiversity after visiting dioramas in two museums – the Zoology Museums of the University of São Paulo and the University of Copenhagen. We are now analyzing the data and some of the findings have been published (Marandino et al., 2012; Achiam & Marandino, 2014).

Considering the existence of dioramas in many natural history museums around the world and the improvements in immersion exhibitions over the last few years and, at the same time, assuming that those elements are educational objects derived from a process of museographic transposition, such exhibits have strong research potential for museum education studies. We contend that the processes of biodiversity dissemination and learning in such exhibits merit further research; this will, in turn, assist museum educators and schoolteachers in improving teaching and learning processes in museums. In addition, these studies could help museum designers, curators, and other staff to better communicate and teach about biodiversity through exhibitions. We also believe that research into the processes of learning about biodiversity in museums could help promote scientific literacy in a more efficient manner.

5.2 *Studying the Sociological Dimension of Museum Education*

Some of our research assessed the construction of expository discourses in biology exhibitions in science museums based on Basil Bernstein's concepts of pedagogic discourse and recontextualization (Marandino, 2001, 2014). This framework allowed us to study the processes, stakeholders, and knowledge involved in the production of the discourses expressed in exhibitions in science museums. What for us constitutes an *expository discourse* results from selections gleaned from scientific culture, as mediated by the different knowledge backgrounds of the various stakeholders involved in the production of the exhibition. In this research, the recontextualization process was analyzed using ideas of power, classification, and framework, among others, to explain the relationships between the discourses of the scientific discipline of biology, the history of science, museology, education, and communication which transpire in the production of an expository discourse. We also make explicit the ways in which agencies in charge of museums' official recontextualization fields along with those from domains geared toward pedagogic recontextualization determine the final expository discourse of an exhibition.

Bernstein's pedagogic discourse framework was useful for our examination of how the constraints imposed by objects and texts in exhibitions help to create a specific manner of visitor interaction with these elements. Those aspects enabled us to explore the power tensions created in the production of expository discourses, by showing how distributive and evaluative rules, along with the rules shaping recontextualization, work in the context of exhibitions. Studying the dynamics forming expository discourses using Bernstein's framework reveals the individuals and institutions, the selection criteria, the negotiations, and the power relations involved.

Luciana Martins (2011; under my supervision), investigated the agents involved in the production of educational activities in natural science, humanities, and art museums in Brazil. Her work emphasizes public and private institutions, government agencies, and stakeholders that fund and regulate the educational practices of museums. In general terms, Martins (2011) indicates that the official recontextualizing field (ORF) of museums is made up of different agencies than those that regulate formal education. For example, in Brazil, the official policies which influence policies related to museums are created and overseen by the Federal Government's Ministry of Culture, and, to a lesser degree, by its state and municipal counterparts. A second governmental agency, the Ministry of Science and Technology of Brazil, develops public policies for museums, especially those related to the natural sciences. Finally, universities participate in the orf as well, mainly through its agencies of culture and extension, by providing funds to educational programs and defining the politics of museum education.

Martins calls attention to an interesting aspect of those agents directly related to the pedagogic recontextualization field (PRF) of museums: in the case of the three museums she studied, some of the agents in the ORF of museums are the same ones as in the PRF of museums. This happens because museum educators also research and publish in the museum education field and participate in defining policies for museums. Bernstein (1996) believes that such situations are exceptions and that they should be disclosed, considering that most often those who create official discourses are not the same individuals who recontextualize them. Given that museum educators may constitute all or part of the PRF, but also, possibly, participate in the ORF, those professionals playing a part in both fields have an extraordinarily high degree of autonomy in the production of museums' pedagogic discourses in the three types of museums studied by Martins (2011).

Martins (2011) provides examples of how the ORF and PRF of museums operate and influence the activities of educational sectors and their actors. Specifically, her examples explain how power works in these contexts, by defining the agencies that influence the selections of what can be promoted for museum education activities. Her work indicates the ways in which Brazil's Ministries of Culture and Science and Technology, along with its universities, influence exhibition projects by giving or withholding financial support. Her research also reveals how agents – such as museum educators – can make vital decisions in order to influence and define programs and, sometimes, policies leading to improvements and increased funding in the area of museum education.

5.3 *A Few Considerations*

On the one hand, the examples below are used to analyze the transformation process as scientific knowledge is rendered accessible through exhibitions, as well as how visitors understand such knowledge. On the other hand, Martin's work is dedicated to studying power relations, to revealing the agents and institutions involved in defining the educational activities of museums, including exhibitions. This dual approach allows for a deeper understanding of how content and idea selection are performed in museum education processes. Both aim to improve the awareness and functioning of the many actors involved in the educational processes of museums, by improving their comprehension, offering ideas, and helping them to conceive of and prioritize activities. Hence, her work has potential use for both educators and researchers in the museum education field, as well as for designers, curators, etc., as it may help them to better understand the teaching and learning processes that occur during museum visits, as well as to establish relevant criteria for evaluating quality in order to produce improved exhibitions in science museums.

6 Final Considerations

In this chapter, we have presented an overview of research on museum education, a sub-category of science education research. This overview, far from being conclusive, is aimed at identifying theoretical perspectives being used by researchers involved in the area of science education in Brazil and concerned with the study of science museums as educational spaces. We also sought to highlight the importance that these subjects have acquired in national and international research, especially due to increasing recognition of the historic relevance and political importance of museums over the centuries.

We contend that research on education in science museums, as connected with the field of science education, can offer enormous contributions to understanding the roles of these institutions in society and, as a consequence, enlarge and deepen educational research – all of which will in turn benefit students.

The chapter pointed, briefly, to the theoretical approaches used by author-affiliated research groups and gave examples of their research. A range of perspectives on education in science museums, including analyses of the production of exhibitions, the processes of mediation, and audience studies, provided useful contributions for better understanding the educational roles of museums. In these studies, we found that appropriating concepts from the linguistic field and basing analyses on the epistemological and sociological dimensions of education have great potential for deepening the field of museum education.

Taking into account the realities of life in Brazil, where much of the population has very low cultural capital, the need to expand access to museums becomes even more urgent. In partnership with schools, science museums can effectively enhance the cultural experiences of children, youngsters, and adults. To do so, however, museums need to develop activities that can promote effective educational processes, even as they find ways to increase access. Research has an important role to play in improving education as promoted in these spaces.

Notes

1 According to the ABCMC there are around 190 science popularization spaces spread throughout the country, including museums, zoos, aquariums, planetariums, observatories, and botanical gardens, offering a variety of programming for all age groups (www.abcmc.org.br).

2 In this chapter, the term science museum will be used to encompass museums of science and technology, natural history museums, science centers, zoos, and botanical gardens, among others.

3 It is worth mentioning that, in Brazil, the development and purpose of Centers of Science were different than those of American science centers. The Centers of Science in Brazil mostly oversee teacher education and the production of didactic materials. American science centers are dedicated to science communication and informal education for the general public.

4 In 1998, *Science Education* initiated a dedicated section covering informal science education. Since 2011, the *International Journal of Science Education* has pubished a Part B, which is dedicated to science communication and informal education.

5 The Observatorium of Museums and Cultural Centers (OMCC) is a research and service program for museums and related institutions which has proposed the creation of a network system for the production, collection, and sharing of varied data and knowledge about museums and their relations with society. It brings together different cultural institutions, promoting exchanges between art museums, science museums, and other museums of different thematic classifications in the cultural field. Website: http://www.fiocruz.br/omcc/cgi/cgilua.exe/sys/start.htm

6 Museum 1: Museum of Astronomy and Related Sciences; Museum 2: the cultural center, Oi Futuro.

7 Museum 1: Museum of the Universe; Museum 2: Museum of Arts and Crafts.

8 Museum 1: Museum of Life; Museum 2: Museum of Science and Life.

References

Achiam, M., & Marandino, M. (2014). A framework for understanding the conditions of science representation and dissemination in museums. *Museum Management and Curatorship, 29*, 66–82.

Allard, M., & Boucher, S. (1991). *Le musée et l'école*. Québec: Hurtubise HMH.

Almeida, M. A. (2008). Mediações da cultura e da informação: perspectivas sociais, políticas e epistemológicas. *Tendências da Pesquisa Brasileira em Ciência da Informação, 1*(1), 1–24.

Bakhtin, M. (1986). *Marxismo e filosofia da linguagem: Problemas fundamentais do método sociológico na ciência da linguagem*. São Paulo, Brazil: Editora HUCITEC.

Bakhtin, M. (2000). *Estética da criação verbal* (2nd ed.). São Paulo, Brazil: Editora Martins Fontes.

Barthes, R. (1990). *O óbvio e o obtuso: ensaios críticos III*. Rio de Janeiro, Brazil: Nova Fronteira.

Barthes, R., & Compagnon, A. (1987). Leitura. In *Enciclopédia Einaudi* (Vol. 11). Lisbon, Portugal: Imprensa Nacional.

Beetlestone, J. G., Johnson, C. H., Quin, M., & White, H. (1998). The science center movement: Contexts, practice, next challenges. *Public Understanding of Science, 7*, 5–26.

Bernstein, B. (1996). *A estruturação do discurso pedagógico: classe, códigos e controle*. Petrópolis, Brazil: Vozes.

Bizerra, A. (2009). *Atividade de aprendizagem em museus de ciências* (Doctoral dissertation). Universidade de São Paulo, São Paulo.

Caffagni. C. W. do A. (2010). *O estudo de analogias utilizadas como recurso didático por monitores em um centro de ciência e tecnologia de São Paulo/SP* (Master's thesis). Universidade de São Paulo, São Paulo.

Campos, N. F. (2013). *Percepção e aprendizagem no museu de zoologia: uma análise das conversas dos visitantes* (Master's thesis). Programa de Pós-Graduação Interunidades em Ensino de Ciências, Universidade de São Paulo, São Paulo.

Cazelli, S. (1992). *Alfabetização científica e os museus interativos de ciência* (Master's thesis). Pontifícia Universidade Católica do Rio de Janeiro, PUC-Rio, Brazil.

Cazelli, S. (2005). *Ciência, cultura, museus, jovens e escolas: Quais as relações* (Doctoral dissertation). Pontifícia Universidade Católica do Rio de Janeiro, PUC-Rio, Brazil.

Cazelli, S., Marandino, M., & Studart, D. (2003). Educação e comunicação em museus de ciências: Aspectos históricos, pesquisa e prática. In G. Gouvêa, M. Marandino, & M. C. Leal (Eds.), *Educação e museu: a construção social do caráter educativo dos museus de ciências* (p. 83). Rio de Janeiro, Brazil: Editora Access/FAPERJ.

Cazelli, S., Valente, M. E., Gouvêa, G., Marandino, M., & Franco, C. (1999). Museum-school relationship and broadening scientific culture. In C. Dufresne-Tassé (Ed.), *Families, schoolchildren and seniors at the museum: research and trends* (pp. 1–318). Québec, Canada: Éditions MultiMondes.

Cerati, T. M. (2014). *Educação em jardins botânicos na perspectiva da alfabetização científica: análise de uma exposição e público* (Doctoral dissertation). Universidade de São Paulo, São Paulo, Brazil.

Chevallard, Y. (1991). *La transposición didáctica: del saber sabio al saber enseñado*. Buenos Aires, Argentina: Editorial Aique.

Chevallard, Y. (2007). Readjusting didactics to a changing epistemology. *European Educational Research Journal, 6*(2), 131–134.

Coelho, J. T. (1989). *O que é ação cultural.* São Paulo, Brazil: Brasiliense.

Contier, D. (2009). *Relações entre ciência, tecnologia e sociedade em museus de ciências* (Master's thesis). Universidade de São Paulo, São Paulo.

Costa, A. F. (2009). *Museu de ciência: Instrumentos científicos do passado para a educação em ciência hoje* (Master's thesis). Universidade Federal do Estado do Rio de Janeiro, UNIRIO, Rio de Janeiro.

Davallon, J. (1999). *L'exposition à l'ouvre: stratégies de communication et médiation symbolique.* Paris, France: L'Harmattan.

Dierking, L. D. (2005). Lessons Without Limit: how free-choice learning is transforming science and technology education. *História, Ciências, Saúde: Manguinhos, 12*(Suppl.0), 145–160.

Eco, H. (1993). *Interpretação e superinterpretação.* São Paulo, Brazil: Martins Fontes.

Falcão, D. (1999). *Padrões de interação e aprendizagem em museus de ciência* (Master's thesis). Universidade Federal do Rio de Janeiro, Rio de Janeiro.

Falcão, D., Colinvaux, D., Krapas, S., Queiroz, G., Alves, F., Cazelli, S., Valente, M. E., & Gouvêa, G. (2004). A model-based approach to science exhibition evaluation: a case study in a Brazilian astronomy museum. *International Journal of Science Education, 26*, 951–978.

Fortin-Debart, C. (2003). Le musée de sciences naturelles, un partenaire de l'école pour une education relative a l'environnement: du message scientifique au débat de société. *VertigO – la revue électronique en sciences de l'environnement, 4*(2). Retrieved from vertigo.revues.org/4494

Fracalanza, H., & Megib Neto, J. (2006). *O livro didático de ciências no Brasil.* Campinas, Brazil: Editora Komedi/UNICAMP.

Garcia, V. A. R. (2006). *O processo de aprendizagem no zôo de sorocaba: Análise da atividade educativa visita orientada a partir de objetos biológicos* (Master's thesis). Universidade de São Paulo, São Paulo.

García Blanco, Á. (1999). *La exposición, um medio de comunicación.* Madrid, Spain: Akal.

Gaspar, A. (1993). *Museus e centros de ciências: conceituação e proposta de um referencial teórico* (Doctoral dissertation). Universidade de São Paulo, São Paulo.

Gouvêa, G. (2009). *Imagens e práticas educativas em espaços formais e não formais de educação. Inmimeo* (Not published). Rio de Janeiro, Brazil: UNIRIO.

Gouvêa, G., Alves, F., Falcão, D., Valente, M. E., & Marandino, M. (2002). A study of the process of museographic transformation in two exhibitions of the MAST. In C. Dufresne-Tassé (Ed.), *Evaluation: Multi-Purpose Applied Research* (Vol. 1, pp. 108–124). Montréal, Canada: Université de Montréal.

Gouvêa, G., Izquierdo, M., & Martins, I. (2006). Estudo das linguagens imagéticas em contextos formais e não formais de educação – o caso do livro didático de ciências. In *Atas do X Encontro de Pesquisa em Ensino de Física.* Londrina, Brazil.

Gouvêa, G., & Marandino, M. (2000). Contemporany paradigms in scientific exhibitions. In Y. Girault (Ed.), *Des expositions scientifiques à l' action culturelle, des collections pour quoi faire?* (Vol. 1, pp. 227–235). Paris, France: Éditions du Muséum National D' histoire Naturelle.

Jacobucci, D. F. C. (2006). *A formação continuada de professores em centros e museus de ciências no Brasil* (Doctoral dissertation). Faculdade de Educação da Universidade Estadual de Campinas, Campinas, Brazil.

Köptcke, L. S. (2001–2002). A parceria educativa: O exemplo francês. In *Cadernos do Museu da Vida: O formal e o não-formal na dimensão educativa do museu* (pp. 70–79). Rio de Janeiro: Museu da Vida/Museu de Astronomia e Ciências Afins.

Köptcke, L. S., Cazelli, S., & Lima, J. M. de (2005). *Museus e seus visitantes: Relatório de pesquisa perfil-opinião*. Brasília, Brazil: Gráfica e Editora Brasil.

Krasilchik, M. (1987). *O professor e o currículo das ciências*. São Paulo, Brazil: Editora da Universidade de São Paulo.

Lemke, J. (1998). Multiplying meaning: Visual and verbal semiotics in scientific texts. In J. R. Martin & R. Veel (Eds.), *Reading science*. London: Routledge.

Lopes, M. M. (1991). A favor da desescolarização dos museus. *Educação e Sociedade, 40*, 443–455.

Lopes, M. M. (1997). *O Brasil descobre a pesquisa científica: os museus e as ciências naturais no século XIX*. São Paulo, Brazil: HUCITEC.

Lopes, M. M. (2009). Por que história nos museus e centros de ciências? In M. Marandino, M. E. Valente, & A. M. Almeida (Eds.), *Museu: lugar do público* (pp. 199–210). Rio de Janeiro, Brazil: Fiocruz.

Marandino, M. (2001). *O conhecimento biológico nas exposições de museus de ciências: Análise do processo de construção do discurso expositivo* (Doctoral dissertation). Faculdade de Educação, Universidade de São Paulo (USP), São Paulo, Brazil.

Marandino, M. (2004). Transposição ou recontextualização? Sobre a produção de saberes na educação em museus de ciências. *Revista Brasileira de Educação, 26*, 95–108.

Marandino, M. (2005). A pesquisa educacional e a produção de saberes nos museus de ciências. *História, Ciências, Saúde: Manguinhos, 12*, 161–181.

Marandino, M. (2014).

Marandino, M., Achiam, m., & Oliveira, A. D. (2015). The diorama as a means for biodiversity education. In S. D. Tunnicliffe & A. Scheersoi (Eds.), *Natural history dioramas: History, construction and educational role*. Dordrecht, The Netherlands: Springer.

Marandino, M., BIzerra, A. F., Navas, A. M., Fares, D. C., Monaco, L. M., Martins, L. C., Garcia, V. A. R., & Souza, M. P. C. (2008). *Educação em museus: a mediação em foco*. São Paulo: Pró-Reitoria Cultura e Extensão, USP, and GEENF/FEUSP.

Marandino, M., Campos, N. F., Leporo, N., Caffagni, C. W. A., Maia, R., & Oliveira, A. D. (2012). A percepção de biodiversidade em visitantes de museus: um estudo no Brasil e na Dinamarca antes da visita. *Tempo Brasileiro, 188*, 97–112.

Marandino, M., & Diaz Rocha, P. E. (2011). La biodiversidad en exposiciones inmersivas de museos de ciencias: implicaciones para educación en museos. *Enseñanza de las Ciencias, 29*(2), 221–236.

Marandino, M., & Monaco, L. M. (2009). Biodiversidade nos museus: discussões sobre a (in)existência de um discurso sobre conservação em ações educativas dos museus de ciências. In S. E. Selles, M. S. Ferreira, M. A. L. Barzano, & E. P. de Q. Silva (Eds.), *Ensino de biologia: histórias, saberes e práticas formativas* (pp. 263–278). Uberlândia, Brazil: EDUFU.

Marandino, M., Valente, M. E., Alves, F., Cazelli, S., Gouvêa, G., & Falcão, D. (2003). Estudo do processo de transposição museográfica em exposições do MAST. In G. Gouvêa, M. Marandino, & M. C. Leal (Eds.), *Educação e museu: a construção social do caráter educativo dos museus de ciências* (pp. 161–184). Rio de Janeiro, Brazil: Editora Access/FAPERJ.

Martins, L. C. (2006). *A relação museu/escola: teoria e prática educacionais nas visitas escolares ao museu de zoologia da USP* (Master's thesis). Faculdade de Educação, Universidade de São Paulo, São Paulo.

Martins, L. C. (2011). *A constituição da educação em museus: o funcionamento do dispositivo pedagógico museal por meio de um estudo comparativo entre museus de artes plásticas, ciências humanas e ciência e tecnologia* (Doctoral dissertation). Faculdade de Educação, Universidade de São Paulo, São Paulo.

Massarani, L., & Almeida, C. (Eds.). (2008). *Workshop Sul-Americano & escola de mediação em museus e centros de ciências.* Rio de Janeiro, Brazil: Museu da Vida, Casa de Oswaldo Cruz, Fiocruz.

McManus, P. M. (1992). Topics in museums and science education. *Studies in Science Education, 20*, 157–182.

Megib Neto, J. (2007). Três décadas de pesquisa em educação em ciências: Tendências de teses e dissertações (1972–2003). In R. Nardi (Ed.), *A pesquisa em ensino de ciências no Brasil: Alguns recortes* (pp. 341–356). São Paulo, Brazil: Escrituras Editora.

Monteiro, R. S. da. (2011). *Entre monoculturas e ecologias: As percepções do público dos museus de ciência e técnica sobre as relações ciência, tecnologia, sociedade e ambiente (CTSA)* (Master's thesis). UNIRIO, Rio de Janeiro.

Moreira, L. M. (2013). *O teatro em museus e centros de ciências: Uma leitura na perspectiva da alfabetização científica* (Doctoral dissertation). Universidade de São Paulo, São Paulo.

Mortensen, M. F. (2010). *Exhibit engineering: A new research perspective* (Doctoral dissertation). Department of Science Education, University of Copenhagen.

Nascimento, S. S. do. (2008). O corpo humano em exposição: promover mediações socioculturais em um museu de ciências. In L. Massarini & C. Almeida (Eds.), *Workshop Sul-Americano & escola de mediação em museus e centros de ciências.* Rio de Janeiro, Brazil: Fiocruz.

Nascimento, S. S. do, & Almeida, M. J. P. M. (2009). Museu e escola: um duplo olhar sobre a ação educativa. In *Atas do XI Reunión de la RedPOP* (Vol. 1, pp. 1–5). Montevideo, Uruguay: Unesco.

Nascimento, S. S. do, Weil-Barais, A., & Davous, D. (2002). Diferentes fazeres, diferentes saberes: A ação de monitores em espaços não escolares. *Ensaio Pesquisa em Educação em Ciências, 5*(2), 28–38.

Navas, A. M. (2008). *Concepções de popularização da C&T no discurso político: Impactos nos museus de ciências* (Master's thesis). Universidade de São Paulo, São Paulo.

Oliveira, A. D. (2010). *Biodiversidade e museus de ciências: um estudo sobre transposição museográfica nos dioramas* (Master's thesis). Programa de Pós-Graduação Interunidades em Ensino de Ciências, Universidade de São Paulo, São Paulo.

Oliveira, M. C. (2008). *Visita monitorada a um museu de ciências: O que é possível aprender* (Master's thesis). Programa de Pós-Graduação Interunidades em Ensino de Ciências, Universidade de São Paulo, São Paulo.

Ovigli, D. F. B. (2010). *Os saberes da mediação humana em centros de ciências: contribuições à formação inicial de professores* (Master's thesis). Universidade Federal de São Carlos, São Carlos.

Pereira, J. S., Siman, L. M. C., Costa, C. M., & Nascimento, S. S. (Eds.). (2007). *Escola e museu: diálogos e práticas*. Belo Horizonte, Brazil: Secretaria de Estado de Cultura/Superitendência de Museus/Pontifícia Universidade Católica de Minas Gerais/Cefor.

Pinto, S. (2014). *A construção do discurso da mediação em atividades itinerantes de divulgação científica* (Doctoral dissertation). NUTES/UFRJ, Rio de Janeiro.

Queiróz, G., Krapas, S., Valente, M. E., David, E., Damas, E., & Freire, F. (2002). Construindo saberes da mediação na educação em museus de ciências: o caso dos mediadores do museu de astronomia e ciências afins/Brasil. *Revista Brasileira de Pesquisa em Educação em Ciências, 2*(2), 77–88.

Salgado, M. O. (2011). *A transposição museográfica da biodiversidade no aquário de Ubatuba: estudo através de mapas conceituais* (Master's thesis). Programa de Pós-Graduação Interunidades em Ensino de Ciências, Universidade de São Paulo, São Paulo.

Simonneaux, L., & Jacobi, D. (1997). Language constraints in producing prefiguration posters for scientific exhibition. *Public Understandings of Science, 6*, 383–408.

Tojal, A. P. da F. (2007). *Políticas públicas culturais de inclusão de públicos especiais em museus* (Doctoral dissertation). Universidade de São Paulo, São Paulo.

Valente, M. E. (2003). A conquista do caráter público do museu. In G. Gouvêa, M. Marandino, & M. C. Leal (Eds.), *Educação e museu: a construção social do caráter educativo dos museus de ciências* (pp. 21–46). Rio de Janeiro, Brazil: Editora Access/FAPERJ.

Valente, M. E. (2008). *Museus de ciências e tecnologia no brasil: Uma história da museologia entre as décadas de 1950–1970* (Doctoral dissertation). Universidade Estadual de Campinas, Campinas.

Van-Praët, M. (2003). A educação no museu, divulgar "saberes verdadeiros" com "coisas falsas". In G. Gouvêa, M. Marandino, & M. C. Leal (Eds.), *Educação e museu: A construção social do caráter educativo dos museus de ciência* (pp. 47–62). Rio de Janeiro, Brazil: Access Editora, FAPERJ.

Van-Praët, M., & Poucet, B. (1993). Les musées, lieux de contre-éducation et de partenariat avec l'école. *Éducation & Pédagogies, 16*, 22–29.

Zamboni, L. M. S. (1997). *Heterogeneidade e subjetividade no discurso de divulgação científica* (Doctoral dissertation). Instituto de Estudos da Linguagem, UNICAMP, Campinas.

CHAPTER 23

Reconstructing Our Images of the World: The Fundamental Task of Non-Formal Science Education

César Carrillo-Trueba

Abstract

In the 20th century, a separation arose between different aspects of social life. Science, the humanities, art and morality separated the so-called objective from the subjective, primitive societies immersed in darkness from civilization and its light. The result was a fragmented, compartmentalized world, one with a system of knowledge ruled by hyperspecialization, the separation of specialists from society, science from philosophy, thought from emotion. Given that science is foundational to our image of the world, it appears to us as fragmented, and we are unable to comprehensively understand the impacts of science and technology on society.

This chapter proposes a reintegration of the scientific image of the world through non-formal education strategies, based around following axes: popularizing specialist knowledge among the general public; the integration of other disciplines into science, that is, a truly comprehensive interdisciplinarity; a necessary dialogue between science and philosophy, history, ethics, politics and other social sciences and humanities; a vision of the knowledge possessed by each culture, in which science is only one part of the universal concert of knowledge in a relativistic, non-absolutist vision; and an essential reconciliation between the objective and the subjective, thought and emotion, the many ways that humans have of understanding the world and their fellows. Non-formal education has a great deal of work ahead of it.

Keywords

science – hyperspecialization – fragmentation of knowledge – image of the world – reintegration of disciplines – science and society – interculturality – recreation of the image of the world – non-formal education

 | DOI: 10.1163/9789004409088_023

The excesses of the system of competition
and premature specialization,
under the false pretense of efficiency,
kill the spirit, impede all cultural life,
and even suppress the progress of science.

ALBERT EINSTEIN

∴

When Brother Bartolome Arrazola felt lost, he accepted that nothing more could save him. The mighty jungle of Guatemala, relentless and definitive, had imprisoned him. Faced with his topographical ignorance he sat calmly to wait for death. He wished to die there, bereft of hope, in isolation, with his thoughts fixed on distant Spain, in particular the convent of Los Abrojos, where Carlos the Fifth had once deigned to address him and tell him that he trusted in the religious zeal of Arrazola's redemptive endeavors.

When he awoke he found himself surrounded by a group of natives of inscrutable visage who were preparing to sacrifice him on an altar, an altar which to Bartolome had the look of the bed on which he would rest, at last, from his fears, from his destiny, from himself.

In his three years in Guatemala he had acquired a middling command of the native languages. He mustered himself and said a few words, which his captives understood.

Then an idea was born in his mind that he deemed worthy of his talent, his universal culture, and his deep knowledge of Aristotle. He recalled that a total eclipse of the sun was to occur that day, and in the deepest recess of his mind he resolved to avail himself of that knowledge to deceive his oppressors and save his life.

"If you slay me", he informed them, "I can make the sun go dark in the heavens".

The natives gazed fixedly upon him, and he glimpsed the incredulity in their eyes. He saw them gather to debate their course of action, and waited confidently, not without a certain disdain.

Two hours later, Brother Bartolome Arrazola's heart violently spewed his blood over the sacrificial stone (gleaming in the opaque light of the eclipsed sun), as one of the natives, without the slightest inflection, unhurriedly recited, one by one, an endless list of dates on which solar

> and lunar eclipses would take place, which the Mayan astronomers had predicted and recorded in their codices without Aristotle's invaluable aid.

This story, written in 1958 by the Honduran writer Augusto Monterroso, portrays a drama that has yet to reach its dénouement in many countries of the developing world: the clash of two cultures, with their respective forms of knowledge. The knowledge of the Spaniards was as exact as that of the Mayans; both forms of knowledge were complete and correct enough to ensure the survival of both peoples in their respective natural environments and to serve as engines for the development of each culture. Nonetheless, one imposed itself on the other, irreversibly annihilating centuries of accumulated knowledge produced and systematized through generations, and modifying knowledge which has survived to this day, nourished by and resistant to widely diverse influences, maintaining its own rate of change.

Their conviction that their religion was the one good, correct religion and their certainty that their knowledge was the only true knowledge prevented the conquistadors from comprehending the world view of the indigenous peoples of the New World, finding points of communication, and together forging a more interesting knowledge that would have been better suited to the natural conditions of their new environment. And if, as in Monterroso's fiction, the conquistadors occasionally came out on the losing end – as indeed they sometimes did – they ended up victorious all the same and hence imposed, with unfettered violence, their religion, their knowledge, their modes of production, and their means of cultivating the earth, in sum, their culture, on the original inhabitants of the New World.

Monumental temples were destroyed and deities cast down; flaming pyres of texts were built, as described by Brother Diego de Landa, chronicler of the conquest of the Yucatan, in an image that remains painful to this day:

> We found great numbers of books filled with their writings, and as they contained naught but superstition and lies of the devil, we put them all to the torch, which made a great impression upon them and caused them great sorrow. (Landa, 1560/1982, p. 105)

The legacy of the Mesoamerican cultures was lost forever.

1 The Bias of History

What is truly odd is that, while history records many events like this, the value it ascribes to the body of knowledge non-Western cultures possessed and possess is never deemed equivalent to the knowledge of Europeans – indeed, it is not even considered comparable. Historians continue to affirm that the Spaniards brought science to the Americas, that they had superior technology, because they have very specialized use of metal, and so on and so forth. Some prominent students of Mesoamerican cultures even go so far as to share this view, as does the renowned expert in Mayan culture Eric Thompson, who affirms that the Mayans' knowledge of astronomy was not the product of "the pursuit of truth per se", because it stemmed from different purposes: "the Mayan astronomer strove to expand his knowledge, not for the sake of knowledge in its own right, but as a means of controlling fate, in other words a form of astrology" (Thompson, 1954, p. 225).

The idea of linear history still predominates in most perspectives and disciplines, and the positivist notion of western history – which holds that the western world had to pass through animism, metaphysics, and all of the other stages of what is referred to as pre-scientific knowledge, to finally arrive at objective science – continues to be used as a frame of reference. In this scheme, the developing world, in order to develop 'appropriately', must shed its superstitions and embrace contemporary scientific knowledge, which, in its own assessment, is manifestly superior. This way of seeing things attributes an internal logic to the development of science and technology, and does not allow for the possibility that each type of knowledge evolves in a specific natural, historical, and social context, responding to equally specific needs and interests, and with a unique and unrepeatable rhythm and dynamic of change (see Carrillo-Trueba, 2006).

This point is treated with outstanding clarity by Anthony Aveni, who analyzes the different observations of the same astronomical phenomena made by Mesoamerican astronomers, on the one hand, and the creators of contemporary astronomy, Copernicus, Galileo, and Kepler, on the other. Aveni's highly engaging analysis of these observations results in his concluding that it is impossible to affirm that one system is superior to the other or that one type of knowledge is better than the other; as well, he notes that it is extremely difficult to compare the development of the two or the results of their respective investigations:

> We can ask: Why didn't the Mayans produce a Copernicus who recognised that the sun was the center of the solar system? [...] But then, if they could speak to us, they might ask: Why didn't the Europeans [...] make more of the fact that bright Venus – unlike Mars – disappears and reappears, always remaining close to the sun? [...] Why did their naked-eye sky watchers pick up the quintessential movements of Venus [...] while Galileo and Horrocks, careful observers though they were, failed to do so? (Aveni, 1992, p. 200)

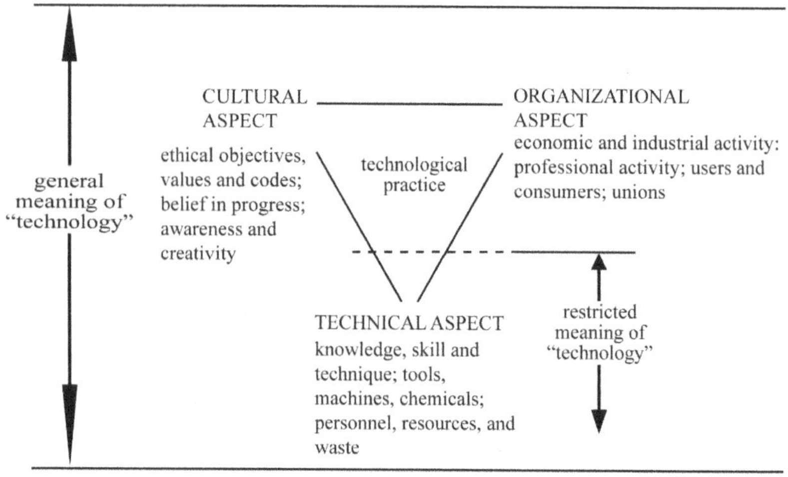

FIGURE 23.1 The V-Gowin diagram (reproduced from Pacey, 1983, p. 19)

The answer lies in cultural difference – an issue that is often overlooked in discussions concerning objective knowledge.

2 A Science above All Suspicion

Science is generally portrayed as separate from culture, as though it had an internal, neutral logic, a type of reasoning superior to and devoid of all social, political, ethical, and ideological intentionality, like its material arm, technology, which is seen as THE means of solving any practical problem, of dispensing prosperity, and as the driving force of progress. The introduction of a new technology is seen, by and large, as something inevitable, which, while it may not bring an immediate benefit, will unquestionably do so in the future. The use of technology to solve a problem conceals the social interests behind it; as Jürgen Habermas has observed, this is a means of purifying the intentions of the powerful classes (see Habermas, 1968).

We refer intentionally to technology, rather than to technique or mere apparatuses or utensils, because technology, as Arnold Pacey remarks, embraces a far broader meaning, comprising both its strictly material aspects and necessary scientific investigations, as well as the many economic, ethical, political, social, and ideological facets involved in its production (Pacey, 1983, pp. 18–22).

Also, it is through technology that most people relate to scientific knowledge. If we examine the way modern technology has spread over the world, we will find that it has invariably preceded scientific knowledge. In fact, until the early 20th century, most technological innovations arrived in the great metropolises well in advance of the explanatory knowledge connected with such inventions. And while it is true that in our century that trend has been reversed – because there is essentially no contemporary technology that has not required a strong injection of scientific research – even so, above all in the developing world, the sciences, "born of conceptual innovations", as Charles Morazé explains, "are received through their material manifestations" (Morazé, 1979, p. 142).

The problem, again, is the non-neutrality of technology, given that when a new technology is introduced a social relationship is also being introduced. We can appreciate this phenomenon by looking back into history.

3 The Earth Profaned

When the Spaniards came to Mesoamerica, one of the things that impressed them most was the irregular contour of the terrain. "It's like wrinkled paper", Hernán Cortés is said to have remarked. The great mountain chains that run through modern Mexico, reaching above thirteen thousand feet, and the astoundingly diverse landscapes found in the region – tropical dry forest, tropical rainforest, forests of conifers and oaks; arid zones, enormous valleys, river basins, mangroves, etc. – all served to instill wonder in the minds of those first encountering them. The conquistadors were also astonished, María de los Ángeles Romero-Frizzi notes, that so many people could survive in such uneven terrain, where – according to a 16th-century functionary – the land "cannot be planted on account of its gradient" (Romero-Frizzi, 1991, p. 161).

But they marveled even more, as many still do, at the Mesoamerican peoples' use of such rudimentary tools as the hoe or planting stick and their stone axes. "How do they manage to feed themselves?" the Spaniards apparently wondered. The management of water and plants, the rotation and association of crops, the construction of terraces, artificial islands, and ridges, among other things, are but a few of the technologies the Mesoamerican cultures had developed based on their vast stores of knowledge, which allowed them to manipulate their environments, of which they felt themselves to be an inseparable part. This latter is in stark contrast with the European view of nature as something distant from humanity, a lesser aspect of creation which must be dominated and possessed.

The Spaniards came with their seeds and their livestock, their knowledge and farming tools. They chose the places that best resembled their native lands – the temperate zones – and there they recreated their way of life. Wheat, grapes, barley, olives, citrus fruits, cows, swine, horses, mills, pickaxes, spades, and ploughs carved out a new landscape. Needless to say, they not only recreated their native conditions for themselves, but they imposed their crops and tools on the conquered peoples as well (see Crosby, 1986).

Changing millenarian habits is never easy. The Spaniards could oblige the natives to sow wheat – for Spanish consumption – but it was another thing entirely to force them to use the plough. A culture that offered apologies to the earth before felling a tree, or to the deities of the mountain when building terraces, could not accept an instrument, which, as some natives put it, "injures the land" (Romero-Frizzi, 1991, p. 164). It is known that some native leaders, clinging to power, sought to emulate the Spanish way of life by adopting their style of dress, riding on horseback, and introducing the use of the plough on their own croplands. However, what really caused the use of the plough to spread, as Romero-Frizzi accurately observes, was that, despite lessening the yield per unit of planted land, it raised the yield per person. In other words, planting an acre using a hoe required more person-hours, but the harvest was greater and the soil was preserved; whereas, with the plough, a person could plant an acre more quickly, but the yield and duration of soil fertility were diminished. This drawback was overcome centuries later with the introduction of chemical fertilizers, with the disastrous consequences we have seen. Nevertheless, in view of the scarcity of labor resulting from a series of wars and epidemics, and above all due to the development of the market economy, European technology prevailed (Romero-Frizzi, 1991, pp. 166–168).

The introduction of the plough was not possible in excessively steep and mountainous regions, and, as a result, colonial agriculture developed mainly in the temperate valleys. By the early 19th century, its use was widespread in such zones. The full consequences of such changes are impossible to determine with any reasonable degree of accuracy, but by that point the soil of the Valley of Mexico, which centuries before had been home to several thousand people, had eroded and deteriorated to such an extent that it could barely support a third of that population.

4 Purloined Seeds

"Hunger will be eradicated in the world!" brayed a slogan promoting the use of high-yield seed varieties, the pride of the Rockefeller Foundation, sponsored by the so-called Green Revolution. Launched in Mexico in the 1940s, it was a revolution based on growing varieties of wheat, corn, and rice that had been selected for maximum yield, whose purported attributes were the stuff of

fantasy: better assimilation of nutrients, greater numbers of grains per stalk, smaller and shorter stems, and a shorter or longer growth cycle – allowing growers to reap up to three harvests a year from the same land.

The minor drawbacks of these miracle seeds include the facts that they require larger amounts of fertilizers, herbicides, and water than normal seeds; machinery is needed to cultivate them; new seeds have to be purchased for each crop cycle; and they are highly susceptible to disease and pests, without overlooking the high costs entailed in their use, meaning that without bank financing it was impossible to aspire to the benefits of the so-called Green Revolution.

The implementation of the use of these seeds in Mexico resulted in the almost total reversal of the land redistribution implemented as part of the agrarian reforms enacted by president Lázaro Cárdenas in the 1930s, as farmland was once again concentrated in the hands of a small class of owners. This produced, on the one hand, a migration of dispossessed peasantry to cities, and, on the other hand, their transformation into itinerant or cross-border migrant day laborers. The environmental impacts of this shift – in the form of constant increases in the use of pesticides, impoverishment of soils, and loss of genetic diversity in the species cultivated, among others – have been extensively documented. Wondering whether it solved the problem of hunger is beyond pointless. What we can confirm is that wheat exports grew, as did production and the GDP, but that does not mean that the Mexican people were able to eat better. The only winners were landowners and multinational corporations (see Moore Lappé & Collins, 1977).

What is odd about all this is that, unlike the introduction of the plough, improved seeds were not imposed by a foreign invader; instead, heads of state and their cadres of experts and technocrats took it upon themselves to promote the Green Revolution. Why didn't anyone think about the environmental and social consequences of this misnamed revolution? Why were national, cultural, and social contexts never taken into consideration?

5 The Fragmented Mind

Answers to these queries can be arrived at from four different angles. First, because, as we have seen, science and technology are considered inherently neutral and positive. Second, because ethics, politics, philosophy, history, and sociology are thought to have no bearing whatsoever on science and technology. Third, because scientists and technocrats have an overly fragmented view of their disciplines, which renders them incapable of establishing non-superficial modes of communication with even closely related fields, not to mention

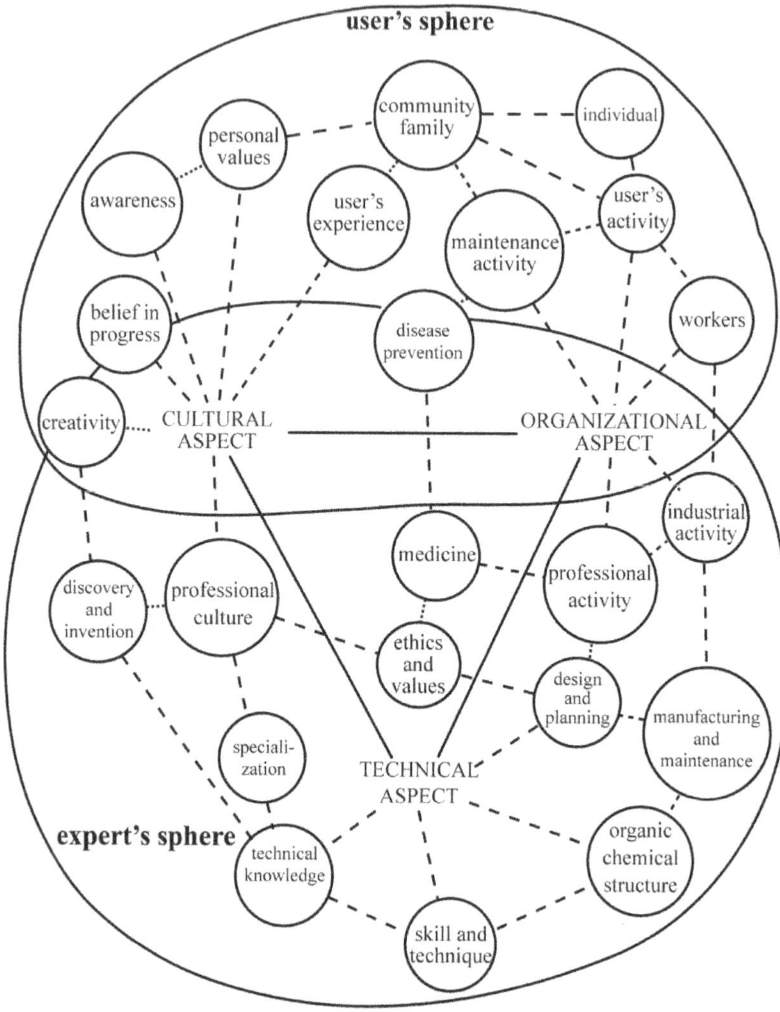

FIGURE 23.2 User's sphere diagram (reproduced from Pacey, 1983, p. 87)

more distant fields; without such means, solving problems is impossible. And, fourth, because experts by definition have extensive knowledge of their fields and do not need to consult with anyone else, much less a mere citizen living near the site where experts have decided to build a nuclear power plant or a toxic waste dump, to mention only two examples.

Arnold Pacey maintains that a technological innovation cannot be successful if it is conceived and designed without accounting for a series of factors, such as maintenance and equipment use; the knowledge and experience of users, workers, and/or patients; social and personal values, etc.; as shown in Figure 23.2, these constitute the user's sphere.

The normal approach to technology design does not necessarily consider the user's sphere, and this tends to produce technologies that are economically successful, but disastrous in terms of their social and environmental consequences. Such is the case with the Green Revolution and with biotechnology in general, which has been widely touted as the cure for all the world's ills, among which hunger seems to hold a permanent place, now to be overcome by means of transgenic crops (Lewontin & Levins, 2007, pp. 321–341). Yet again, technocrats and experts in the developing world emit an endless stream of grandiose promises, promises – as before – devoid of any reflection on the social implications of those technologies; while their counterparts in more-developed countries engage in at least some debate on these issues, as seen in the case of cloning.

What is the origin of this profound fragmentation in perceptions of the same phenomena?

6 A Fragmented World

Contemporary science emerged during the Renaissance as the product of a series of social changes in growing cities, where trade, manufacturing, money-lending, architecture, and other activities forged a new economy, along with a new society characterized by a culture that distanced itself from both feudal lords and the clerical elite. It was an endeavor driven by a new way of representing space – perspective – and a new idea of time, by an urge to measure and quantify, by attempts to transform humankind's relationship with the environment into an instrumental one, and to regulate social life. In this break with medieval thought, humankind is separated from nature –emancipated, one might say – the natural is separated from the supernatural, the objective from the subjective, the qualitative from the quantitative, and so on, successively, resulting in the creation of the fragmented world we inhabit. As Lewis Mumford remarks, "the instruments of science were useless in the realm of qualities. The qualitative was reduced to the subjective: the subjective was cast aside as unreal, and the unseen and immeasurable as nonexistent" (Mumford, 1934, pp. 64–65).

This process of separation culminated in the 18th century, the century of reason, mechanics, and the individual, when science attained a self-ascribed total degree of objectivity and was recognized as the source of liberation and prosperity, as Jürgen Habermas (1983) explains:

> The project of modernity formulated in the 18th century by the Enlightenment philosophers consisted of their efforts to develop an objective

> science, universal morality and laws, and an autonomous art in line with its internal logic [...] At the same time, this project sought to liberate the cognitive potential of each of those domains from its esoteric forms. The Enlightenment philosophers wanted to use that accumulation of specialized culture to enrich everyday life, in other words for the organization of everyday social life. (p. 28).

In other words, for progress.

As these areas were institutionalized, specialists were trained in each. Later, these experts took up the reins of society. In this process, the arts followed one path, morality and law another, and science and technology yet another. Specialists proliferated, creating ever-widening breaches between and within fields of expertise, and, above all, between them and the rest of the population.

The ensuing centralization of knowledge and power in cities produced a sharp devaluation of the traditional knowledge possessed by inhabitants of the rural world, leaving them on the margins and plunging them into an obscure dependency on the cities. Peoples all over the world shared the same fate as they were subjugated by European expansion and suffered the imposition of a new rationality, which, as we have seen, caused formerly shining constellations of forms of knowledge to be dimmed and even destroyed in a process that continues to this day. Today, local elites, brimming with modernity, are the perpetrators of persecution. The breach created between countryside and city would expand to encompass what later came to be known as the developing world, becoming a true abyss.

As a result of the processes described above – and, needless to say, others – we live in a fragmented world: fragmented nations, fragmented peoples, fragmented societies, fragmented production, fragmented families, fragmented selves, and ... fragmented knowledge.

In what concerns us here, to wit, non-formal science education, this fragmentation confronts us with five fundamental abysses:

1. The separation between experts or specialists and the rest of society, which is manifest in how decisions are made on matters such as the development of technological innovations. When a new technology is introduced, the opinions of its potential users or the persons indirectly affected thereby, along with its potential social impact or side effects, are never taken into account, as we have seen in the biotechnology and the Green revolution examples discussed above.
2. The excessive specialization of scientists and the hyper-specialization found in the production of knowledge, which produce an equally fragmented knowledge, to such an extent that even scientists in one area

or in closely related areas are incapable of understanding one another. Efforts have been made to overcome this problem by creating interdisciplinary groups which, as a general rule, have a merely formal, rather than true, conceptual integration. In addition, there are still those who, acting above all on the premises of accuracy and quantification, hold that there are superior disciplines, which further complicates interaction among scientists of varying disciplines (between exact sciences and social sciences, for example) and borders on a quasi-religious cult of the 'hard' sciences. In practical matters, an interdisciplinary perspective is indispensable for confronting the complexity of the problems that afflict us today (such as global warming).

3. The disassociation of scientific activity from the influence exercised on it by ethics, philosophy, ideology, and other facets of the social sphere, which is manifest in the widely shared idea that science and technology develop in a neutral way, isolated from ideas, values, concepts, and the social imagination. In short, this line of reasoning holds that that scientists are asocial beings, able to enter the laboratory and lay aside all their passions, ideas, and beliefs in order to create pure and objective knowledge, a notion that contrasts sharply with reality, as the sociology, history, and psychology of science have shown. Also, for many scientists, this neutrality and objectivity continue to confer a supremacy upon science that grants scientists the right to trumpet their opinions on ethical or political issues and to dispense judgment on those spheres of social life with scientistic pretensions.
4. The opposition that has been created between scientific knowledge and the forms of knowledge of other cultures, which are still very much alive in many parts of the world. Even now, modern-day awareness of the relativity of scientific knowledge notwithstanding, the latter are still not fully accepted as such by scientists despite their proven usefulness to western society in areas such as medicine (acupuncture and herbal medicine, for example) and ecology (sound management of ecosystems and conservation of biodiversity), among others.
5. The disaggregation of the subjective and the objective, which derives from the establishment of the quantitative as measurable and the qualitative as immeasurable, among other effects, has caused a total separation between the spheres of emotion and thought, between different forms of comprehending the world and reasoning, demoting human faculties such as the sense of smell, touch, corporality, feelings, and others that play key parts in our relationships with the environment and are intricately tied to the processes of knowledge.

7 Reconstructing Our Images of the World

In general terms, formal science education is wedded to this portrayal of science, reproducing and at times enlarging these fractures. Attempts to reexamine this state of affairs are few and far between – and are often confronted with requirements for establishing courses or academic programs that almost invariably end up themselves further compartmentalizing and fragmenting the problem due to the very structure of the system of teaching. As John Ziman notes, in describing science, technology, and society education:

> It is an integral part of the philosophy of STS [Science, Technology and Society] education that it should not be tamed or domesticated to the point where it becomes indistinguishable from the conventional academic curriculum. Much of its value to students and teachers lies in the sincerity, divergences, uncertainty, and spontaneity it can introduce to the classroom and the laboratory. (Ziman, 1980, p. 150)

It is here that the virtues of non-formal science education, its great freedom and creativity, can play a central role. Museums can combine emotion with rational thought, bring together the arts and the sciences, and provide a venue for contact between disciplines and cultures, bringing their publics into more direct contact with problems of direct interest, while providing good examples of possible ideas and solutions (see Carrillo-Trueba, 2003). Similarly, documentaries are capable of presenting discourses at different levels while simultaneously appealing to feelings and emotions, without being limited by the need to concern themselves with respecting disciplinary boundaries; they thereby act as integrating agents for knowledge and information that is generally learned in far more rigid ways in schools, where knowledge is often taken out of its proper context. The same can be said of radio, the various forms of printed media (books, magazines, newspapers), the internet, the theater, and other media that are used for non-formal science teaching today.

In practice, these media are not always used as means of greater integration. Often, they present decontextualized, fragmented information, which echoes the flaws of formal education: depicting science as something finished, conclusive, and nearly unquestionable; exalting the great advances of science and technology without mentioning their implications for our understandings of the world and society and without further consideration of the processes that produced such knowledge, the conditions in which it was created, or, in short, the complexities entailed throughout (see Latour, 1987, p. 33).

For these reasons, non-formal science education can serve as a tool of integration by mending these fissures; bringing disciplines closer together; combining ethics with science, the knowledge of indigenous peoples with scientific knowledge, and emotion with rational thought; and, above all, to narrow the ever-widening gap between the public and developments in science and technology. The latter point is fundamental, given that everyday life is increasingly immersed in an overwhelming technological context. For this reason, it is crucial to make a wealth of knowledge available to the broader public, thereby helping people to better understand the world we live in, the rapport humanity has established with it, and the vast potential of the many new and emerging conceptual tools created by scientific activity. The acquisition of that knowledge will lead its beneficiaries to take a more active part in the numerous social issues that concern or ought to concern society at large, and even in defining the orientations and priorities science should adopt in a democratic society (see Ziman, 1980, p. 155).

To accomplish this, we need to accept that science is neither perfect nor complete and that it is by no means the only way to understand the world. Furthermore, as we strive to re-connect a fragmented world by advancing non-formal science education and public understandings of science, we must always bear in mind a range of key points touched upon above, including the following: that scientific theories are not eternally valid; scientists do not work for the good of humanity, but fulfill a social role like many others; new technologies can have deleterious effects; scientific and technological development is not a linear process; we live in a multicultural world, in which each individual or social group possesses distinct forms of knowledge, values, and beliefs that must be respected; and, finally – to use the term coined by Eduardo Galeano – that we are *sentipensantes*, or sensing/thinking beings.

We cannot portray science as if we are selling a product. We are not advertising agents for science and technology. In fact, the more clearly the limitations and factors that shape scientific and technological development – with all their deficiencies and flawed aspects – are portrayed and understood in the public realm, the clearer will be the role those activities can and should play in society; with such knowledge, citizens will increasingly feel empowered to engage with science and technology on more critical levels. Far from devaluing science and technology, this approach can help underscore their real and potential contributions to the construction of a just and democratic society, a non-fragmented world in which all individuals and cultures have a place.

Reconstructing the images of the world that humanity has created – knowledge writ large and, in a narrower sense, science – by reintegrating the different spheres that make up our social and cultural lives would be a good first step in such a journey.

Acknowledgement

This chapter was translated from Spanish by Padraic Smithies.

References

Aveni, A. (1992). *Conversing with the Planets.* New York, NY: Times Books.

Carrillo-Trueba, C. (2003). Propuestas para un museo de historia natural del siglo XXI. *Elementos*, 9(48), 33–38.

Carrillo-Trueba, C. (2006). *Pluriverso, un ensayo sobre el conocimiento indígena contemporáneo.* México: UNAM.

Crosby, A. W. (1988). *Imperialismo ecológic: La expansión biológica de Europa, 900–1900.* Barcelona, Spain: Crítica.

Einstein, A. (1979). *Comment je vois le monde.* Paris, France: Flammarion.

Habermas, J. (1973). *La technique et la science comme "idéologie".* Paris, France: Gallimard.

Habermas, J. (1988). La modernidad, un proyecto incompleto. In *La posmodernidad.* México: Kairos/Colofón.

Landa, Brother D. de. (1982). *Relación de las cosas de Yucatán* (ca. 1560). México: Porrúa.

Latour, B. (1987). *La Science en action.* Paris, France: La Découverte.

Lewontin, R., & Levins, R. (2007). *Biology under the influence.* New York, NY: Monthly Review Press.

Monterroso, A. (1971). Obras completas (y otros cuentos). México: Joaquín Mortiz.

Moore Lappé, F., & Collins, J. (1977). *Food first.* New York, NY: Houghton Mifflin Co.

Morazé, C. (1979). *La science et les facteurs de l'inégalité.* Paris, France: UNESCO.

Mumford, L. (1982). *Técnica y civilización.* Madrid, Spain: Alianza Universitaria.

Pacey, A. (1991). *La cultura de la tecnología.* México: FCE.

Romero-Frizzi, M. (1991). La agricultura en la época colonial. In T. Rojas (Ed.), *La agricultura en tierras mexicanas desde sus orígenes hasta nuestros días* (pp. 139–215). México: CNCA/Grijalbo.

Thompson, E. S. (1988). *Grandeza y decadencia de los mayas.* México: FCE.

Ziman, J. (1985). *Enseñanza y aprendizaje sobre la ciencia y la sociedad.* México: FCE.

Index